McGraw-Hill Ryerson

Environmental Science

A Canadian Perspective

Educational Consultants

Marlene Faulkner
Peel District School Board

David Hall
Luther College High School, Saskatchewan

Melissa Hunter
Simcoe Muskoka Catholic District School Board

Ann Jackson
Catholic District School Board of Eastern Ontario

Ellen Murray
Toronto District School Board

Tiffany Schoonings
Niagara Catholic District School Board

Corey Ziegler
Regina Board of Education, Saskatchewan

Educational Advisors

Kathy Blanchard
Rainbow District School Board

Antonietta Cillo
Durham Catholic District School Board

Therese Forsythe
Annapolis Valley Regional School Board, Nova Scotia

Misty Gallant
Anglophone East School District, New Brunswick

Aaron Liscum
Toronto District School Board

Sarah Lovsin
Wellington Catholic District School Board

Lawrence McGillivary
Anglophone East School District, New Brunswick

Holly Newsome
Ottawa-Carleton District School Board

Mirka Orde
Hamilton-Wentworth Catholic District School Board

Rosalind Poon
Richmond District School Board #38, British Columbia

Lindsay Reynan
Lambton Kent District School Board

Robert Vesna
York Region District School Board

Nancy Wagenaar
Keewatin-Patricia District School Board

Senior Authors

Michelle Anderson
Science Writer and Educator The Ohio State University

Jonathan Bocknek
Science Writer and Senior Program Manager

Doug Fraser
Formerly of District School Board Ontario North East

Katherine Hamilton
Science Writer and Educator Formerly of University of Saskatchewan

Christine Weber
Science Writer

Assessment Consultant

Anu Arora
Peel District School Board

Technology and ICT Consultant

Catherine Fan
Educator Formerly of McMaster University

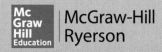

McGraw-Hill Ryerson

Environmental Science 11

The information and activities in this textbook have been carefully developed and reviewed by professionals to ensure safety and accuracy. However, the publisher shall not be liable for any damages resulting, in whole or in part, from the reader's use of the material. Although appropriate safety procedures are discussed and highlighted throughout the textbook, the safety of students remains the responsibility of the classroom teacher, the principal, and the school board district.

ISBN-13: 978-1-25-906777-8
ISBN-10: 1-25-906777-7

1 2 3 4 5 6 7 8 9 0 TCP 1 10 9 8 7 6 5 4 3

Printed and bound in Canada

Care has been taken to trace ownership of copyright material contained in this text. The publishers will gladly accept any information that will enable them to rectify any reference or credit in subsequent printings.

EXECUTIVE PUBLISHER: Lenore Brooks
PROJECT MANAGER: Katherine Hamilton
SENIOR PROGRAM CONSULTANT: Jonathan Bocknek
DEVELOPMENTAL EDITORS: Michelle Anderson, Nicole Fallon,
Katherine Hamilton, Christine Weber
MANAGING EDITOR: Crystal Shortt
SUPERVISING EDITOR: Janie Deneau
COPY EDITOR: Linda Jenkins
PHOTO RESEARCH/PERMISSIONS: Linda Tanaka
REVIEW COORDINATOR: Melanie Berthier
EDITORIAL ASSISTANT: Michelle Malda
MANAGER, PRODUCTION SERVICES: Yolanda Pigden
PRODUCTION COORDINATOR: Sheryl MacAdam
SET-UP PHOTOGRAPHY: David Tanaka
COVER DESIGN: Mark Cruxton, type+image.
INTERIOR DESIGN: Vince Satira
ELECTRONIC PAGE MAKE-UP: Brian Lehen Graphic Design Ltd.

COVER IMAGES: Main Image/Frog © Chrystal Bilodeau/shutterstock
Band Images: LED Streetlight © Charles Stirling/Alamy
Dock © The Globe and Mail Inc./The Canadian Press
Yard Waste/Compost © Radius Images/Corbis
Canola/Biofuel © Dave Reede/All Canada Photos/Corbis
Honeybee © arlindo71/iStock
Sunflare © Photo by NASA/GSFC/SDO/Rex Features/The Canadian Press
Back cover image / grass©Bruce Heinemann/Getty Images

Acknowledgements

Pedagogical Reviewers

Lynn Abrahams
Toronto District School Board

Gabriel Ayyavoo
Toronto Catholic District School Board

Janice Bradshaw
Peterborough Victoria Northumberland and Clarington Catholic District School Board

Virginia Dawe
Toronto District School Board

Michelle Driscoll
Peterborough Victoria Northumberland and Clarington Catholic District School Board

Erin Elliot
York Catholic District School Board

Kim Evans
Ottawa-Carleton District School Board

Patricia Gaspar
York Region District School Board

Dorthy Lai
Durham District School Board

Mike McArdle
Dufferin-Peel Catholic District School Board

Lucy Nguygen
Toronto District School Board

Jennifer Parrington
Durham District School Board

Kenneth Pham
Toronto District School Board

Krista Porter
Halton District School Board

Barbara Scott-Cole
Renfrew County District School Board

Ryan Smith
Trillium Lakelands District School Board

Steve Stephenson
Lambton Kent District School Board

Sandra Wells
Durham District School Board

Accuracy Reviewers

Caroline Barakat, PhD
Health Studies
University of Toronto, Scarborough

Gary Bull, PhD
Forest Resource Management
University of British Columbia

Maurice DiGiuseppe, PhD
Education
University of Ontario Institute of Technology

Brajesh Dubey, PhD
Environmental Engineering
University of Guelph

Doug Hayhoe, PhD
Education
Tyndale University College and Seminary

David Jackson, PhD
Engineering Physics
McMaster University

Stephen Jeans, PhD
Program Development Consultant
SEEDS Foundation, Canada
Also of University of Alberta

Chelsea Nilausen, BSc, BEd
Forest Resource Management
University of British Columbia

Safety Reviewer

Brian Heimbecker
Dufferin-Peel Catholic District School Board

Lab Testers

Alexandre Annab
Dufferin-Peel Catholic District School Board

Holly Newsome
Ottawa-Carleton District School Board

Bias Reviewer

Nancy Christoffer
Markham, Ontario

Catholicity Reviewer

Bernie Smith
York Catholic District School Board

Special Features Writers

Andrew Borkowski
Nicole Fallon

Contents

Mini-Activities and Labs

Mini-Activities

Inquiry/ThoughtLabs

Special Features

Safety in the Environmental Science Lab and Classroom

Become familiar with the following safety symbols and procedures. It is up to you to use them and your teacher's instructions to make your activities in *Environmental Science 11* safe and enjoyable. Your teacher will give you specific information about any other safety rules that need to be used in your school.

Safety Symbols

Be sure you understand each symbol used in an activity or Lab before you begin.

	Disposal Alert This symbol appears when care must be taken to dispose of materials properly.
	Biological Hazard This symbol appears when there is danger involving bacteria, fungi, or protists.
	Thermal Safety This symbol appears as a reminder to be careful when handling hot objects.
	Sharp Object Safety This symbol appears when there is danger of cuts or punctures caused by the use of sharp objects.
	Fume Safety This symbol appears when chemicals or chemical reactions could cause dangerous fumes.
	Electrical Safety This symbol appears as a reminder to be careful when using electrical equipment.
	Skin Protection Safety This symbol appears when the use of caustic chemicals might irritate the skin or when contact with micro-organisms might transmit infection.
	Clothing Protection Safety A lab apron should be worn when this symbol appears.
	Fire Safety This symbol appears as a reminder to be careful around open flames.
	Eye Safety This symbol appears when there is danger to the eyes and safety glasses should be worn.
	Poison Safety This symbol appears when poisonous substances are used.
	Chemical Safety This symbol appears when chemicals could cause burns or are poisonous if absorbed through the skin.
	Animal Safety This symbol appears when live animals are studied and the safety of the animals and students must be ensured.

WHMIS Symbols for Hazardous Materials

Look carefully at the WHMIS safety symbols shown here. WHMIS stands for "Workplace Hazardous Materials Information System." WHMIS symbols and the associated material safety data sheets (MSDSs) are used throughout Canada to identify dangerous materials. These symbols and the material safety data sheets show you all aspects of how to handle hazardous materials safely. Your school is required to have these sheets available for all chemicals. MSDSs are also available on the Internet. Make sure you read each MSDS carefully and understand what the WHMIS symbols mean. Use both of these resources to know how to take appropriate safety precautions.

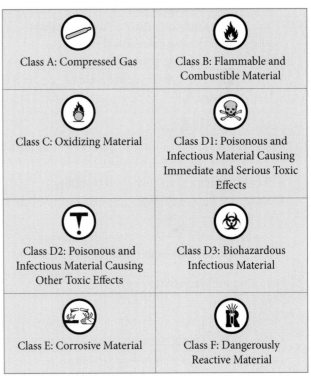

Class A: Compressed Gas	Class B: Flammable and Combustible Material
Class C: Oxidizing Material	Class D1: Poisonous and Infectious Material Causing Immediate and Serious Toxic Effects
Class D2: Poisonous and Infectious Material Causing Other Toxic Effects	Class D3: Biohazardous Infectious Material
Class E: Corrosive Material	Class F: Dangerously Reactive Material

1. **Working with your teacher …**
   - Listen carefully to any instructions your teacher gives you.
   - Inform your teacher if you have any allergies, medical conditions, or other physical problems that could affect your work in the environmental science classroom. Tell your teacher if you wear contact lenses or a hearing aid.
   - Obtain your teacher's approval before beginning any activity or labwork you have designed yourself.
   - Know the location and proper use of the nearest fire extinguisher, fire blanket, first-aid kit, fire alarm, spill kit, eyewash station, and drench hose/shower.

2. Starting an activity or labwork …

- Before starting an activity or labwork, read all of it. If you do not understand how to do a step, ask your teacher for help.
- Be sure you have checked the safety symbols and WHMIS information and have read and understood all Safety Precautions.
- Begin an activity or labwork only after your teacher tells you to start.

3. Wearing protective clothing …

- When your teacher tells you to do so, wear protective clothing, such as gloves, a lab apron, and safety goggles. Always wear protective clothing when you are using materials that could pose a safety problem—including unidentified substances—or when you are heating anything.
- Tie back long hair. Avoid wearing scarves, ties, or long necklaces.

4. Acting responsibly …

- Work carefully with a partner. Make sure your work area is clean, dry, and well organized.
- Handle equipment and materials carefully.
- Make sure stools and chairs are resting securely on the floor.
- If other students are doing something that you consider dangerous, report it to your teacher.

5. Handling edible substances …

- Do not chew gum, eat, or drink in your environmental science classroom or laboratory.
- Do not taste any substances or draw any material into a tube with your mouth.

6. Working in a environmental science classroom …

- Make sure you understand all safety labels on school materials or materials you bring from home. Also, familiarize yourself with the material safety data sheets (MSDSs), WHMIS symbols, and the safety symbols used in this textbook, found on page xii.
- When moving equipment for an activity or labwork, carry only one thing at a time, holding it carefully. Be aware of others and make room for students who may be carrying equipment to their work stations.

7. Working with sharp objects …

- Always cut away from yourself and others when using a knife or scalpel.

- If you have to walk carrying scissors or any other sharp object, always keep the pointed end of the object facing away from yourself and others.
- If you notice sharp or jagged edges on any equipment, take special care with it and report it to your teacher.
- Never use your hands to pick up broken glass. Use a broom and dustpan to dispose of broken glass in the proper containers, as your teacher directs.

8. Working with electrical equipment …

- Make sure your hands and work area are dry when touching electrical cords, plugs, or sockets.
- Turn OFF all electrical equipment before connecting it to or disconnecting it from a power supply. Pull the plug, not the cord, when unplugging electrical equipment.
- Report damaged equipment or frayed cords to your teacher.
- Place electrical cords where people will not trip over them.

9. Working with heat or an open flame …

- When heating an item, wear safety goggles and any other safety equipment that the textbook or your teacher advises.
- Always use heatproof containers. Apply heat gently and do not let containers boil dry.
- Point the open end of a container (such as a test tube) that is being heated away from yourself and others.
- Handle hot objects carefully. Be especially careful with a hot plate that looks as though it has cooled down—it may still be hot.
- If you use a Bunsen burner, make sure you follow all instructions for lighting it and using it safely.
- Use EXTREME CAUTION when you are near an open flame. Wear heat-resistant safety gloves and any other safety equipment that your teacher or the Safety Precautions in this textbook advise.
- If you do receive a burn, inform your teacher, and apply cold water to the burned area immediately.
- If your clothing catches fire, STOP, DROP, and ROLL. Other students may use the fire blanket to smother the flames. Do not wrap the fire blanket around yourself while in a standing position. This could cause a "chimney effect," bringing fire directly into your face.

- If someone else catches fire, do not use a fire extinguisher. Co-operate with others in using the fire blanket to extinguish the flames.

10. **Working with various chemicals …**
    - If any part of your body comes in contact with a substance, wash the area immediately and thoroughly with water. If you get anything in your eyes, do not touch them. Wash them immediately and continuously for 15 minutes at an eyewash station, and inform your teacher. If you wear contact lenses and you get material in your eyes, take your lenses out immediately to prevent the material from being trapped behind the contact lenses. Flush your eyes continuously for 15 minutes at an eyewash station.
    - Always handle substances carefully. If you are asked to smell a substance, never smell it directly. Hold the container slightly in front of and beneath your nose (at least 20 cm away), and waft (wave) the fumes toward your nostrils with your hand.
    - Hold containers away from your face when pouring liquids.
    - While outdoors, always follow your teacher's instructions in regards to where to walk and where not to walk. Be aware of your surroundings and make sure you are always within sight of your teacher.

11. **Working with living things …**
    On a field trip:
    - Do not disturb the work area any more than is absolutely necessary.
    - If you have to move something, do it carefully, and always replace it carefully.
    - If you are asked to remove plant material, remove it gently, and take as little as possible.
    - While working near or collecting from streams or other waterways, make sure you are wearing rubber boots and that you remain in shallow water (do not enter water that is higher than 2-3 inches above your ankles.

    In the classroom:
    - Make sure that living organisms receive humane treatment while they are in your care.
    - If possible, return living organisms to their natural environment when your work is complete.

12. **Designing and building …**
    - Use tools safely to cut, join, and shape objects.
    - Handle modelling clay correctly and wash your hands after use.
    - Follow proper procedures when using mechanical systems and studying their operations.
    - Use special care when observing and working with objects in motion.
    - Do not use power equipment such as drills, sanders, saws, and lathes unless you have specialized training in handling such tools.

13. **Cleaning up in the environmental science classroom or laboratory …**
    - Clean up any small spills according to your teacher's instructions. Large spills require clean-up by a trained individual with the aid of an appropriate spill control kit.
    - Clean equipment before you put it away, according to your teacher's instructions. Turn off the water and gas. Disconnect any electrical devices.
    - Wash your hands thoroughly after doing an activity or labwork.
    - Dispose of materials as directed by your teacher. Never discard materials in a sink or drain unless your teacher requests it.

14. **Ensuring safety in your online activities …**
    The Internet is like any other resource you use for research. You should confirm the source of the information and the credentials of those supplying it to make sure the information is reliable before you use it in your work. Unlike other resources, however, the Internet has some unique issues you should be aware of, and practices you should follow.
    - When you copy or save something from the Internet, you could be saving more than information. Be aware that information you pick up could also include hidden, malicious software code or other malware that could damage your system or destroy data.
    - Avoid sites that contain material that is disturbing, illegal, harmful, and/or was created by exploiting others.
    - Never give out personal information online. Protect your privacy, even if it means not registering to use a site that looks helpful. Discuss ways to use the site while protecting your privacy with your teacher.
    - Report any online content or activity that you suspect is inappropriate or illegal to your teacher.

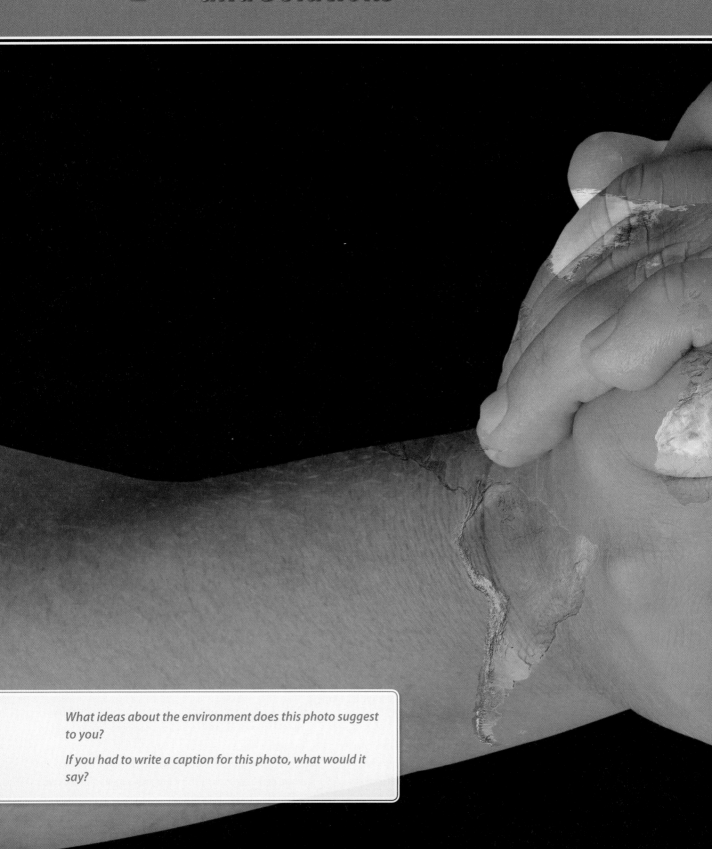

What ideas about the environment does this photo suggest to you?

If you had to write a caption for this photo, what would it say?

BIG IDEAS

- Environmental issues are complex and may involve conflicting interests or ideas.
- Scientific knowledge enables people to make informed decisions about effective ways to address environmental challenges.

Overall Expectations

- Analyze environmental issues, and how society influences scientific investigation of the environment.
- Investigate the role of science and other disciplines in addressing environmental issues.
- Demonstrate an understanding of environmental challenges and how we acquire knowledge about them.

Unit Contents

Chapter 1

Considering Environmental Issues

Chapter 2

Science, Society, and Solutions

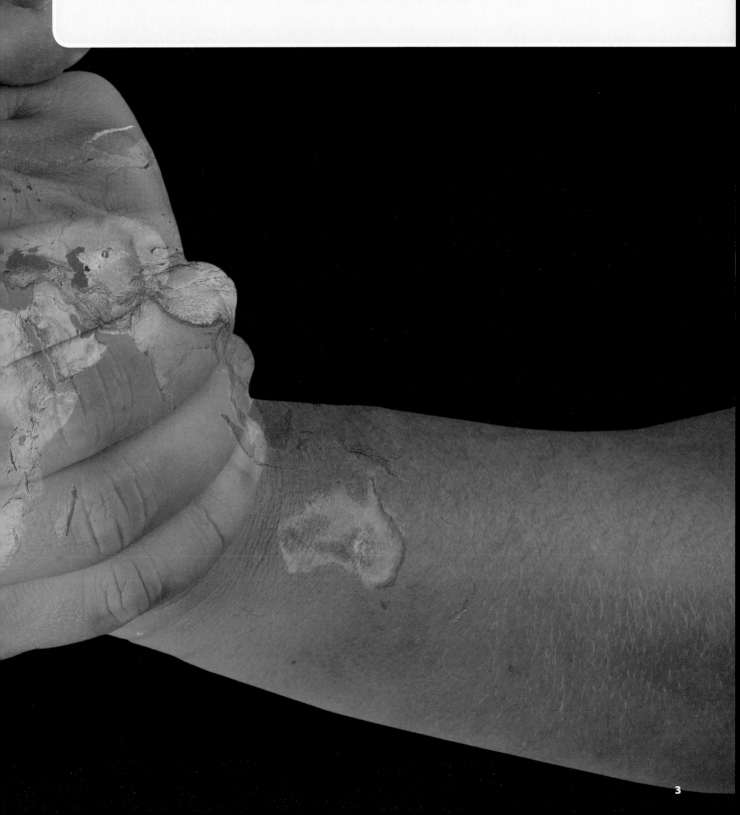

Topic 1: Biotic and Abiotic Components of the Environment

The environment of an individual organism includes *biotic* (living) and *abiotic* (non-living) components. Individual organisms affect and are affected by these components as they interact with one another and with the components. Individuals are part of a population—any group of individuals of the same species living in the same geographical area at the same time. Populations of different species are part of a community of organisms in a particular habitat (environmental setting). **Figure 1** summarizes all these conceptual relationships.

Figure 1 The interaction of biotic and abiotic components of the environment

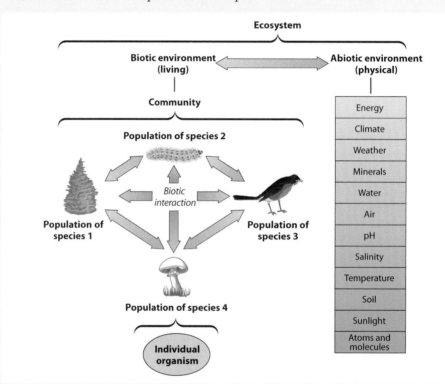

Topic 2: Earth's Spheres

Natural processes move matter in continuous cycles from the biotic and abiotic parts of the environment and back again. At any time, matter occupies one of four "spheres" that make up Earth.

- *atmosphere*: the gaseous part of Earth, which is concentrated mainly within 10 km of the surface but also extends hundreds of kilometres higher
- *geosphere* (or lithosphere): the solid, mainly rocky part of Earth
- *hydrosphere*: all of the water (liquid, as well as solid) that exists on and within the geosphere
- *biosphere*: all of the areas on and under the geosphere, in the atmosphere, and in the hydrosphere that are inhabited by and support life

Topic 3: Nutrient Cycles

Nutrients are elements, chemical compounds, and ions that all organisms need for growth, cellular maintenance, and all other life processes. Nutrients are accumulated for short or long periods of time in the atmosphere, geosphere, hydrosphere, and biosphere. These accumulations are referred to as stores. Biotic processes such as decomposition and abiotic processes such as river run-off can cause nutrients to flow in and out of stores. Taken together, the continuous flows of nutrients in and out of stores are called *nutrient cycles*. You also may see nutrient cycles referred to as exchanges. Refer to **Figure 2**.

Note that in some information resources, you will see nutrient cycles referred to as biogeochemical cycles. This term reflects the fact that these cycles involve the biological, geological, and chemical processes of the planet.

Over the long history of Earth, nutrient cycles have functioned in an overall state of balance. The concentrations of nutrients flowing into the stores have been essentially the same as the concentrations flowing out of the stores.

Human activities such as land clearing, agriculture, urban expansion, mining, and motorized transportation can affect a nutrient cycle by increasing the amounts or concentrations of nutrients in the cycle faster than natural biotic and abiotic processes can move them back to the stores. Over periods of time, increased concentrations of nutrients can accumulate in the atmosphere, in bodies of water, and on land as a result of these activities. This accumulation can have significant environmental effects.

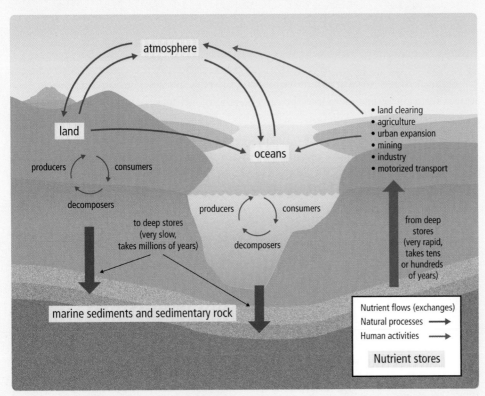

Figure 2 A general model of the flows (exchanges) involved in a nutrient cycle

Topic 4: The Cycling of Water, Carbon, Nitrogen, Phosphorus, and other Nutrients

There are six substances that limit the amount and types of life possible, because they are components of key organic molecules. These substances are water, carbon, hydrogen, oxygen, nitrogen, and phosphorus. **Figure 3** shows how water is exchanged among the hydrosphere, atmosphere, and geosphere during the water cycle..

Figure 3 The processes of evaporation, condensation, and precipitation cycle water continually through Earth's spheres.

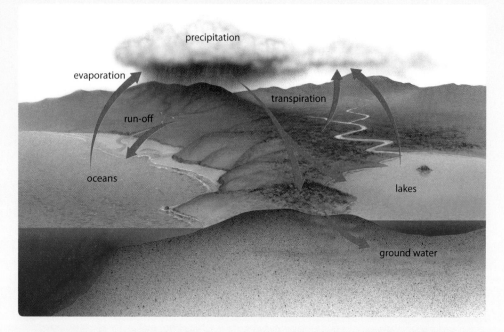

Figure 4 shows how carbon moves through Earth's spheres. Carbon dioxide gas moves from the atmosphere into the biosphere and back again. Carbon dioxide also moves back to the atmosphere when organisms die and their bodies decompose. Carbon enters the geosphere when the remains of organisms are trapped under layers of sediment. Over the course of millions of years, these remains are slowly transformed into coal, oil, and natural gas. Carbon dioxide is returned to the atmosphere when people burn these fuels, as well as wood, for energy.

Figure 4 The cycling of carbon

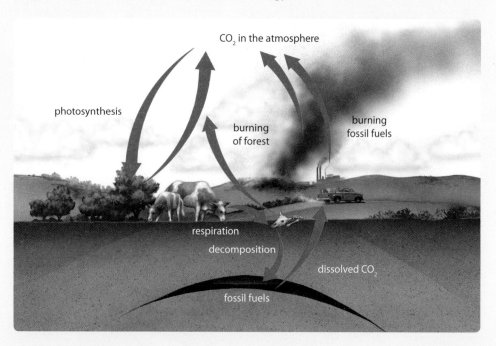

Figure 5 shows how nitrogen moves through Earth's spheres. The atmosphere is 78% nitrogen (N_2). Most organisms, however, cannot use nitrogen in this form. An important part of the nitrogen cycle involves processes that convert N_2 into forms that organisms can use. These forms include ammonium (NH_4^+), nitrite (NO_2^-), and nitrate (NO_3^-). Human activities dramatically increase the concentration of nitrogen in the atmosphere and the biosphere through the use of fertilizers (which contain nitrates), combustion of fossil fuels (which emits gaseous nitrogen oxides), and the clearing of forests and grasslands.

Figure 6 shows that certain elements—including phosphorus, calcium, iron, and magnesium—are stored in the geosphere. When rock material is broken down through natural weathering processes, these nutrients are released into soil and water. Plants and plant-like organisms in water absorb these nutrients, and animals obtain them when they eat plants, plant-like organisms, and other animals. Decomposers return these nutrients to soil and water as they break down dead organisms. Human activities also release these nutrients—especially phosphates—into soil and water.

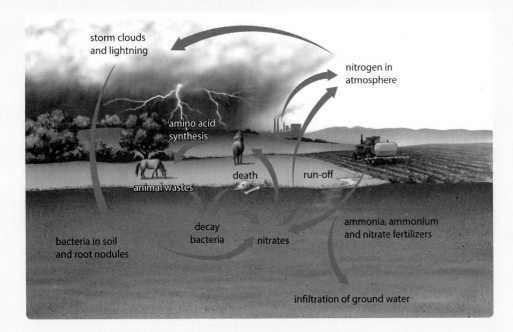

Figure 5 The cycling of nitrogen

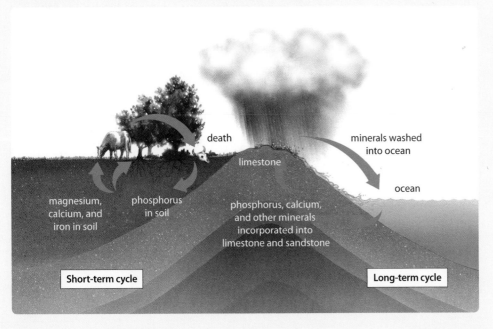

Figure 6 The cycling of phosphorus and other nutrients

Topic 5: Biodiversity

Biodiversity is a general term that encompasses all plants, animals, micro-organisms, and ecosystems. Biodiversity is often discussed at three levels. Genetic diversity refers to the variety of different versions of the same genes within individual species. Species diversity describes the number of different kinds of organisms within communities. Ecosystem (or ecological) diversity assesses the rich variety of ecosystems where organisms live.

Topic 6: Ecosystem Services

Ecosystem services are the benefits that organisms (including humans) receive from the environment and its resources. Ecosystem services are the natural result of all the activities that take place in the biosphere. Refer to Table 1. Keep in mind that these services are inseparably linked together on a global scale. Each is as important as any other, and none is expendable or replaceable.

Table 1 Examples of Ecosystem Services	
Ecosystem Service	**Example**
Atmospheric gas supply	Regulation of carbon dioxide, ozone, and oxygen
Climate regulation	Regulation of greenhouse gases
Cultural benefits	Aesthetic, spiritual, and educational value
Disturbance regulation	Storm protection, flood control, drought recovery, and other aspects of environmental response to disturbances
Food production	Crops, livestock, fish
Genetic resources	Medicines, genes for disease resistance
Habitat (living space)	Habitat for migratory species and for locally harvested species, overwintering grounds, nurseries
Nutrient recycling	Carbon, nitrogen, and other nutrient cycles
Pollination	Pollination of crops such as apples, blueberries, and clover
Raw materials (natural resources)	Fossil fuels, timber, minerals
Recreation	Ecotourism, sport fishing, hiking and other outdoor activities
Soil erosion control	Retention of topsoil
Water regulation	Supplying of water for agricultural (such as irrigation) and industrial (such as milling) processes
Water supply	Supplying of water by reservoirs, watersheds, and wells
Waste treatment	Sewage treatment

Topic 7: Carrying Capacity, Ecological Footprints, and Sustainability

No population can keep growing indefinitely. The largest size of a population that its environment can support is the *carrying capacity* of that environment for that species. Carrying capacity is determined by factors that include food and energy supply, oxygen supply, living space, disease, predators, and competitors.

One way to increase carrying capacity of a specific environment for a specific population is to alter that environment so that more energy and resources can be consumed. Another way involves changing the behaviour of the population. An ecological footprint is a measure of the impact of a population (or an individual) on its environment. Data used to measure an *ecological footprint* include energy consumption, land use, and waste generation. A population's footprint reflects the collective behaviour of its individual members. It describes the amount of productive land and water needed to support a population's (or individual's) standard of living.

The average person in developed countries such as Canada has one of the largest ecological footprints in the world, as shown in **Figure 7**. Footprints this large on a planet that has finite resources is likely to be unsustainable. A growing world population puts more stress on ecosystem services. As the footprints of people in developing countries also increase in size, these stresses will become greater. Thus, modern societies are now trying to establish footprints that reflect the concept of *sustainability*—using Earth's services in ways and at levels that can continue forever. Another aspect of sustainability is the idea of keeping natural resources and ecosystems healthy over time and maintaining human living standards in balance with economic growth. Thus, discussions of sustainability often take these three dimensions—the environment, society, and economics—into account.

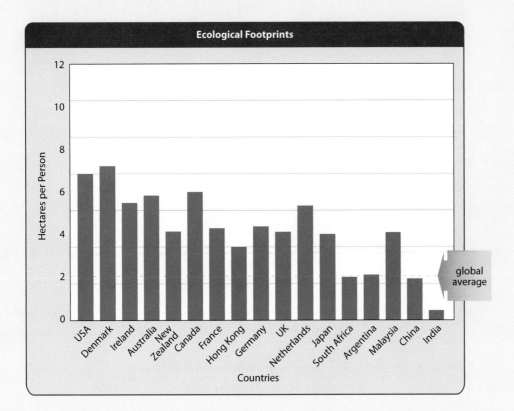

Figure 7 Environmental footprints of selected countries

CHAPTER 1

Considering Environmental Issues

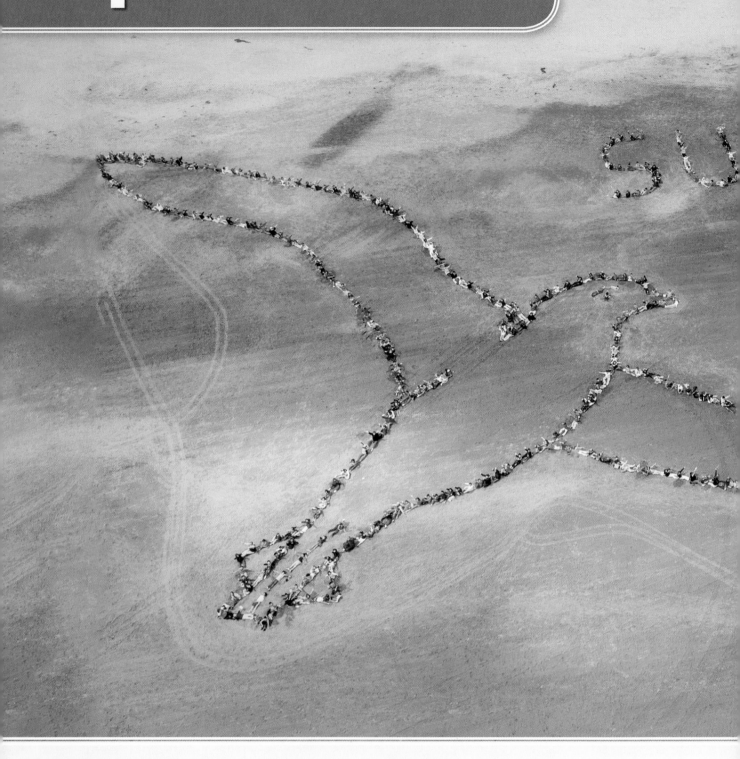

Many people used their bodies to form the words and the shape of the bird in this photo.

What might have motivated these people to do this? What does this image mean, and what does it mean to you?

You and Your Environment

1. Think about the choices you made today before coming to school. Here are a few examples you could consider:
   - Which lights did you turn off or leave on?
   - Did you leave water running while brushing your teeth or washing your hands, and for how long?
   - Did you walk to school, ride a bike, or come by bus or car?

2. List all the choices you made.

3. Which choices, if any, do you think had an impact on other people? Which, if any, do you think had an impact on the environment? In each case, explain your ideas to a partner, and listen back to your partner's explanations.

4. As a class, share your ideas about your possible impact on the world around you—your environment.

In this chapter, you will

- identify examples of environmental issues and their causes and effects
- describe a relationship between the Industrial Revolution and human population growth
- explain the paradigm shift that favours sustainability

Introducing Environmental Issues

Setting the Stage

During the time that this book was being written, the headlines shown in **Figure 1.1** were reported in the mainstream media, and they were widely picked up and rebroadcast in countless blogs and tweets on the Internet. Depending on where you live, none of these stories might have affected you. Consider for a moment, however: Is that really true? Could it be false? Take some time to think about **Figure 1.1** and the questions asked about each headline before you continue reading the rest of this page.

The Environment: It's All about You

In the movie *A Field of Dreams*, the main character, Ray Kinsella, is asked to do a series of tasks that seem to defy logic and understanding. Nevertheless, Ray does them, because he is convinced that in doing so he will come to understand their mysterious purpose. Near the end of the movie, Ray is upset that after all he's done, he is not being invited to take part in what he believes to be the event that will explain the great mystery.

Speaking with the baseball legend Shoeless Joe Jackson, who plays a key role in the film, Ray says, "I did it all. I listened to the voices, I did what they told me, and not once did I ask what's in it for me!" Shoeless Joe replies, "What are you saying, Ray?"

And Ray sputters, "I'm saying … *what's in it for me*?"

If you are like most students, you have wondered from time to time—perhaps out loud to your teachers—about how what you are studying is relevant to your life. Even in courses that you choose to take, you might have asked questions such as What are we learning about this for? How does it apply to me and my life?

These are fair questions to ask. Sometimes the answers might not seem that way. Maybe you have been told that you'll understand in a few years; or you'll need it for university (or college, or work); or that it'll build character; or even that you're learning it because you have to.

There can be some truth in these kinds of answers. But there is a better answer, and it may be more true of environmental science than of any other subject. The answer is that, where the environment is concerned, it is *always* about you. This idea is similar to a saying that is common to many of the Indigenous people of North America: *We are all connected.* In this chapter, you will explore what the idea of connectedness could mean for the you, the environment, and environmental issues.

Mini-Activity 1-1 **Preview the Unit Openers**

Look again at the unit-opening photo on pages 2 and 3. In what ways could it be related to the idea of connectedness? What else does it suggest to you?

Find the other four unit-opening photos in this book. What ideas about the environment and environmental issues do you think they tell on their own and together?

Figure 1.1 These three reports were making news in late 2012 and early 2013.

Applying *What other environment-related news have you heard about or been affected by recently?*

Superstorm Sandy slams Northeast, triggers massive blackouts and flooding

This superstorm was 1320 kilometres wide. It was linked to the deaths of at least 219 people, and it caused more than 50 billion dollars of damage in the Caribbean, the United States, and Canada. It also left more than eight million people without power for days (and in many cases weeks).

If you were living out of the path of Superstorm Sandy and were not directly affected by it, does that make it irrelevant to your life?

Canadians produce more garbage than anyone else

According to a study from the Conference Board of Canada, in 2009, you and each of your fellow Canadians generated more than 770 kg of solid garbage waste. You used a lot of water, too—more than 1100 cubic metres. (Imagine a cubic metre as a cube made with metre sticks. Then imagine how much water can fit in 1100 of those.)

If you were just a child in 2009, can you dismiss this news as not your problem?

Lake Winnipeg most threatened in world in 2013

Lake Winnipeg drains water from rivers in four provinces (Alberta, Saskatchewan, Manitoba, and Ontario) and four states (Minnesota, North Dakota, South Dakota, and Montana). The lake supports many multi-million dollar industries, including tourism, but it is heavily polluted by agricultural fertilizers and pesticides. As a result, the lake supports a thriving population of algae (the green in the photo), which starve the lake of oxygen and poison organisms with toxins they produce.

If you do not live in any of the regions that are linked with Lake Winnipeg, does this situation affect you?

A Planet and Its Human Population

resources everything in the environment that supports and sustains the life of organisms

environment the non-living and living (abiotic and biotic) surroundings that affect an organism's survival

Like all organisms, you need certain essential **resources** to sustain and maintain your life. These resources include air, water, food, and shelter. Most organisms are adapted to their immediate **environment**, which supplies the resources they need. For example, songbirds live in grassy meadows, search for the seeds of grasses and other plants to eat, weave nests out of dried grasses and twigs, and drink water from ponds or nearby streams. Even songbirds that live in towns and cities are able to find these same basic resources from the trees, grasses, and pools of water that are common in areas populated by humans.

There are some organisms that make more substantial changes to their environment to better meet their resource needs. For example, beavers build dams across streams to form ponds where none existed before. Changing the environment in this way kills some trees and displaces various organisms that live in the immediate area of the dam. At the same time, the beaver dam creates a new type of environment for other kinds of organisms. Some organisms benefit from these changes, and others may be harmed by them. The changes themselves are neither "good" nor "bad." They simply have consequences. Change is a natural process of the planet and the entire universe.

Of all organisms on Earth, however, humans have an unequalled ability to change and to cause change. This ability enables our species to live just about anywhere and everywhere in permanent settlements that range in size from hundreds to many millions of people. As shown in **Figure 1.2**, there are currently more than 7.1 billion of us on Earth, and about 80 million more are added each year. As a consequence of our great numbers and our widespread patterns of settlement, you and all others humans have the greatest impact on the resources that support life—*all* life—on the planet.

Did You Know?

"You cannot step into the same river twice, for fresh waters are ever flowing in upon you."

– *Herakleitos (535–475 B.C.E.)*

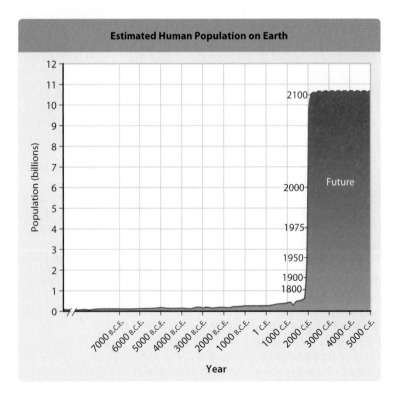

Figure 1.2 Graph showing how the human population has changed over the millennia

Interpreting *What patterns do you see in this graph, and how can you explain them?*

The Human Population and Environmental Issues

Many planets and moons of our solar system experience, or have experienced, environmental events like those that occur on Earth. These include violent quakes, windstorms, volcanic eruptions, flooding, collisions with meteorites and asteroids, planetary warming, and rain (although the rain is liquid methane, rather than water). Unlike on Earth, however, environmental events such as these pose no threat to life.

Natural environmental events have affected, and continue to affect, human populations and the populations of many other organisms. Several times in Earth's history—long before human feet walked the planet—most of the life on Earth has been erased through extinctions brought on by natural, planet-wide upheavals. Such occurrences are rare, however.

During the past 10 000 years or so, natural events such as earthquakes, eruptions, flooding, and drought periodically bring hardship, illness, and often death to human (and other) populations around the globe. In most cases, however, the kinds of environmental challenges outlined in **Figure 1.3** are the result of the ways that we humans use Earth's natural resources to support our own population.

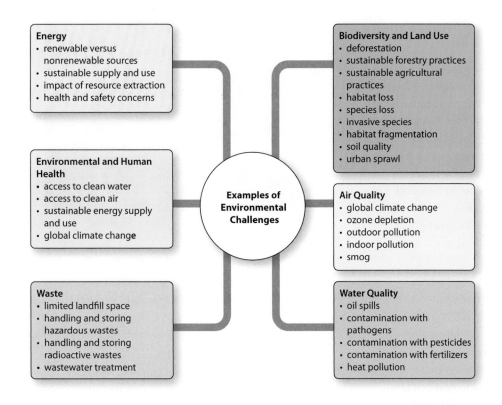

Figure 1.3 This concept map depicts many of the major environmental challenges that society must solve.

Applying What other environmental challenges do you think should be added?

Energy
- renewable versus nonrenewable sources
- sustainable supply and use
- impact of resource extraction
- health and safety concerns

Environmental and Human Health
- access to clean water
- access to clean air
- sustainable energy supply and use
- global climate change

Waste
- limited landfill space
- handling and storing hazardous wastes
- handling and storing radioactive wastes
- wastewater treatment

Examples of Environmental Challenges

Biodiversity and Land Use
- deforestation
- sustainable forestry practices
- sustainable agricultural practices
- habitat loss
- species loss
- invasive species
- habitat fragmentation
- soil quality
- urban sprawl

Air Quality
- global climate change
- ozone depletion
- outdoor pollution
- indoor pollution
- smog

Water Quality
- oil spills
- contamination with pathogens
- contamination with pesticides
- contamination with fertilizers
- heat pollution

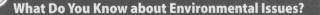

Mini-Activity 1-2   What Do You Know about Environmental Issues?

Sort the environmental issues shown in **Figure 1.3** into groups based on how much you know or think you know about them. For example, your groups could be "I Know a Lot," "I Have Some Understanding," and "I've Heard about It but Need More Information." Use these or any group titles that reflect the state of your understanding.

The point of this activity is to get an idea of the range of knowledge and understanding about environmental issues in your specific class. Your teacher may use this information to help you and your classmates tailor an exploration of environmental issues that suits the needs and interests of your class.

Case Study Resolving a Crisis: Calling All Canadians

Firefighters Battle Northern Ontario Forest Fires

In the summer of 2011, more than 100 forest fires raged across Northern Ontario. The fires consumed 300 000 hectares and required almost 3300 people to evacuate. New fires ignited on a daily basis as hot, windy weather and thick smoke hindered firefighting efforts. Eager to offer their help, 500 ground and aerial crews from British Columbia, Alberta, Saskatchewan, Québec, and Yukon came to help the 1500 Ontario personnel.

Toronto Hydro Workers Help Restore Power in the United States

Superstorm Sandy, a tropical cyclone, began its assault in October 2012. It devastated the Caribbean, large parts of the United States, and Canada from Southern Ontario to the Maritimes. Shortly after the storm dissipated, 64 Toronto Hydro workers with 30 vehicles headed south to help restore power to parts of the United States that had been without electricity for several days. One group went to New York City to work on the underground electricity network, which is similar to Toronto's. The other group went to Massachusetts to repair overhead power lines.

Engineers Without Borders Brings Clean Water to Malawi

Engineers Without Borders (EWB) Canada was formed in 2000 by two engineering graduates from the University of Waterloo, Parker Mitchell and George Roter. Their goal is to use Canadian engineering talent to bring essential technologies to developing countries. Today EWB Canada has over 45 000 members, with chapters in major cities and universities across Canada.

One example of EWB Canada's projects is its work in Malawi, Africa, which has more than 40 000 rural

wells that often break down and run dry, as well as insufficient support for sanitation. EWB Canada worked with the Malawi government to create a database that maps water coverage and shows where repairs or new installations are most needed. It also co-ordinates information sharing in the sanitation sector on a national level and brings together field staff, local and national governments, and non-governmental organizations (NGOs).

Research and Analyze

1. Do research about an individual or small group that has taken action to help with an environmental problem. Questions to ask: What is the problem, and what caused it? Who identified it? What have they done to help?

2. Why are national and international co-operation often helpful in dealing with environmental crises? (Consider money, resources, and expertise.)

Communicate

3. Research one of the "without borders" organizations that has an environmental focus, and write a journal entry or blog post about its goals and main projects.

4. What can you, as an individual or as part of a group, do to help during environmental crises? What can you do to prevent crises before they happen?

Summary

- Humans and all other organisms need certain essential resources from nature to sustain life.

- Most organisms are adapted to particular environments, and they use the resources provided by those environments alone.

- Unlike other organisms, humans change not only their immediate environment, but also any other environments they visit and choose to inhabit.

- The ways that our sizeable human population uses Earth's natural resources have caused many environmental challenges that require solutions.

Review Questions

1. A headline that appeared as this book was being written proclaimed "Public concern for environment lowest in 20 years." The report was based on a survey conducted in person or by phone with 6774 people in 12 countries between the years 1992 and 2012. The graph below shows the results. **T/I** **C** **A**

   a) The graph shows average findings across the 12 countries for which tracking data were available since 1992. Do you think this affects the results? Why or why not?

   b) The surveys were completed before Superstorm Sandy in October 2012. If the survey had been done after Sandy, do you think the results would have been different? Explain.

   c) Read the note that accompanies the single asterisk (*). Do you think these countries represent an appropriate sampling to represent a global opinion about the environment? Does it matter if not all questions were asked in all countries in all years? Explain your opinion.

   d) Many people in the world do not have phones or live in places that are not accessible for in-person interviews. In what ways, if any, could this affect the reliability or integrity of the survey results? Justify your answer.

2. The graph in **Figure 1.2** compresses thousands of years of history into a tiny space. One class made a large-scale model of the graph by using a stack of continuous paper. The students taped one end of the paper to their gym floor and marked it 7000 B.C.E., when the human population was about 5 million. They used the paper perforations as natural divisions for time on the x-axis. Each sheet represented 100 years. For the y-axis, 1 mm represented 25 million people. **K/U** **T** **C** **A**

   a) Use at least three reliable sources of information to find data for the human population from about 7000 B.C.E. to the present. Why are you asked for at least three sources? What would make them reliable? Why is reliability important?

   b) Use the students' approach, or design one of your own (perhaps with less paper waste), to make a large-scale model of human population growth.

   c) How can a large-scale graph help you better appreciate how the population of your species has changed over time? Explain your answer.

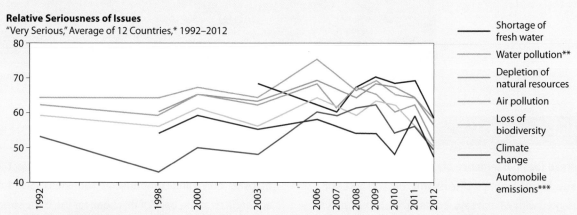

Relative Seriousness of Issues
"Very Serious," Average of 12 Countries,* 1992–2012

Legend: Shortage of fresh water; Water pollution**; Depletion of natural resources; Air pollution; Loss of biodiversity; Climate change; Automobile emissions***

*Average of Brazil, Canada, China, France, Germany, India, Indonesia, Mexico, Nigeria, Turkey, UK, and USA. Not all questions were asked in all countries in all years.
**Not asked in Brazil, Canada, and France
***Not asked in Brazil and Canada

A Shift in How We Think about Issues

Environmental issues often stem from a misunderstanding or a lack of understanding of key ideas of science. Two of the most fundamental of these ideas are the scientific laws that describe the behaviour of matter and energy. These are described in the information box on the next page. Take some time to read the information box. Then return to this page to carry on reading.

The invention of the steam engine was part of a transformative period in human history. Before its invention, most people lived by farming in small rural communities. After its introduction, however, many people began moving from farms to work in factories and related jobs in cities. As more factories were built, more people moved from farms to work in them. Thus, cities grew and new industries grew with them.

This period in human history, during the 1700s and 1800s, is called the Industrial Revolution. As it spread to more and more regions and countries around the globe, the industrialized societies that we know today became more commonplace. If you glance back at the human population graph in **Figure 1.2**, you can see that the Industrial Revolution marked the start of the dramatic explosion in our presence on Earth. It also marked the start of urban air and water pollution. (Refer to **Figure 1.4**.) Some people might even suggest that this period in history marked the start of most, if not all, of the environmental issues that you and the rest of society are confronting today.

At the same time that the factories of the Industrial Revolution were generating materials and products for cities and their infrastructures, another change was taking place. Philosophers in Europe were inventing methods to construct knowledge that emphasized reason and analysis. Modern science was one outcome of this emphasis.

Figure 1.4 The Industrial Revolution resulted in great volumes of pollution. It also created conditions that would eventually lead to concerns for worker rights and safety.

Mini-Activity 1-3   **The Origin of Environmental Issues?**

Reread the final sentence of the third paragraph on this page. In small groups, discuss whether you agree or disagree with what it says. Come up with examples to support your ideas.

You might try using one or more cause-and-effect maps to record and communicate your ideas. (See the appendices at the back of this book for the basic structure of cause-and-effect maps.)

Information Box: Environmental Issues and the Laws of Matter and Energy

The Law of Conservation of Matter and Its Implications

In the late 1700s and early 1800s, experiments conducted by a variety of chemists demonstrated that matter is conserved in chemical reactions. For example, in experiments involving the fermentation of grapes to make wine, scientists observed that although the matter of the reactants (water, sugar, yeast) changed their properties, the total amount of matter did not change. The mass of the wine was equal to the combined masses of all the ingredients that went into making it. Investigations of this kind led to the law of conservation of matter, which states that matter is not created or destroyed in a physical or chemical change. There is an equal quantity of matter before and after the reaction.

This law has important implications for society and the environment. New air, water, minerals, and other material natural resources cannot be produced or removed by any use of science or technology. All the matter that exists on Earth, now, is all the matter that ever existed and ever will exist. We can extract materials from the environment and change them into other useful forms, but we cannot make matter where it did not already exist. In the same way, we can bury our waste materials in the ground or send them into the air as incineration smoke, but the matter of which those materials were made is still here. In nature, matter can only be endlessly recycled.

The Energy Laws and Their Implications

During the 1860s, scientists were experimenting with steam engines. The steam engine was the first practical and affordable machine that could convert the chemical energy of fuel (wood and coal) into kinetic energy to run machines that could replace human and animal labour. As well as converting chemical to kinetic energy, the engine also generated waste heat that could not be put to any use.

Two ideas about energy resulted from work with steam engines. The first is the law of conservation of energy (also known as the first law of thermodynamics). It states that there is no increase or decrease in the quantity of energy when energy is converted from one form to another. The total energy going into an energy converter (such as an engine) and the total energy coming out of it are always equal. This law applies to organisms, too. For example, the amount of chemical energy in the food you eat is equal to the sum of the work done by your body (walking, thinking, and so on) plus the waste heat you generate.

The generation of waste heat led to a second idea about energy called the second law of thermodynamics. It states that when energy is converted from one form to another, some of the useful energy is always lost as unusable heat. During any energy conversion, then, the first law says that there is no loss to total energy, but the second law says that there is always a loss of useful energy. In essence, you cannot get more energy from a system than you put into it, and in fact you will always get less.

All energy conversions, therefore, are inefficient. No organism, machine, or device can run indefinitely or without degrading its supply of energy. You and all other organisms must always eat, because living systems require an ongoing supply of energy. In the same way, non-living systems as small as nano-transistors and as large as the Internet require energy on an ongoing basis to function. Even the process of environmental clean-up produces waste. The clean-up of an ocean oil spill requires the burning of diesel fuel to power the clean-up equipment.

The Need for New Mindsets

ThoughtLab 1A, Do You Tread Lightly on Earth?, on page 24

Think again about the matter and energy laws outlined on the previous page. On a pre-industrial planet Earth, a relatively small human population could probably behave as if Earth's resources are inexhaustible. However, the Earth you share with other people and millions of other species stopped being that planet about 400 years ago. Nevertheless, many human societies around the globe—and especially in developed countries such as our own—continue to use resources as if the laws of matter and energy do not apply to us, or as if they can be rewritten somehow to "get around them."

The laws of science *do* apply to us—and there is no way to renegotiate or change them.

Considering Connections

"We must now speak environment, economy, foreign policy, health, and human rights in the same breath. Everything is connected." These words come from Canadian Inuit leader Sheila Watt-Cloutier (**Figure 1.5**). In a petition she made before the United States Senate in 2004, she delivered a message that remains as true today as it did then.

"Use what is happening in the Arctic—the Inuit story—as a vehicle to reconnect us all, so that we may understand that the planet and its people are one. The Inuit hunter who falls through the depleting and unpredictable sea ice is connected to the cars we drive, the industries we rely upon, and the disposable world we have become." (Testimony of Sheila Watt-Cloutier, Chair, Inuit Circumpolar Conference to the Senate Committee on Commerce, Science and Transportation, Washington DC, September 15, 2004)

Figure 1.5 Sheila-Watt Cloutier

paradigm a world view that is accepted by most people in a society and that influences their thoughts, behaviour, and actions

This idea of connections and connectedness is at the heart of a way of life and a mindset that has always been a part of Aboriginal cultures in North America and around the world. A similar mindset, or **paradigm**, has been developing, slowly, within and among many other cultures and societies since the first Earth Day in 1970 signalled the start of the modern environmental movement. At that time, the prevailing paradigm of infinite development and growth was challenged. This challenge presented the people of Earth with a new paradigm that sees our planet and all its human and non-human inhabitants, as well as its resources, as a single interconnected whole.

Mini-Activity 1-4 Making Connections

Reread the quotation from Sheila Watt-Cloutier.

1. In what ways could an Inuit hunter falling through the sea ice be connected to the activities of people—like you—who live south of the Arctic?

2. Many Aboriginal people talk about how all living (and non-living) things are connected to and depend on one another. What does "being connected" mean to you?

3. Do you think you are connected to everyone and everything in the world? Can you prove that you are not? Share your ideas with a partner or in small groups. Can your class reach a full or partial agreement on these questions? Why or why not?

Sustainability: The New Paradigm

Natural ecosystems are *sustainable*—they are able to continue to exist indefinitely, recycling their materials, as long as they have a continued and constant source of energy. Natural ecosystems are always changing, so what is sustainable for some organisms at some times might not be sustainable for them at other times. However, the system itself is always sustainable.

The idea of sustainable ecosystems is similar to another Aboriginal idea. In the words of Oren Lyons, Faithkeeper of the Onondaga Nation (**Figure 1.6**), "The Peacemaker taught us about the Seven Generations. He said, when you sit in council for the welfare of the people, you must not think of yourself or of your family, not even of your generation. He said, make your decisions on behalf of the seven generations coming, so that they may enjoy what you have today."

Most human ecosystems—the cities and towns we live in and the industrial complexes that support and drive them—have not been sustainable. They have functioned from a mindset that assumes endless growth and unfettered consumption are possible. Most human societies have rarely thought about or concerned themselves with one generation into the future, let alone seven.

You are fortunate to be living during a momentous, precious time in Earth's history. You are part of a rare event: a paradigm shift. This shift involves the way people think about the sustainability of ecosystems, about how we use Earth's resources, and about the consequences of our actions. The shift involves the choices we make, the actions we take, and the responsibility we accept for making and taking them. The shift involves the development of a mindset that sees all of Earth's billions of organisms and intricate systems as inextricably connected—that sees Earth itself as one great, life-support system.

In the next chapter, you will consider the roles of science, society, and the individual in contributing to the environmental solutions that can help to carry the shift forward.

Ray Kinsella in *A Field of Dreams* asks "What's in it for me?"

As you investigate environmental science and environmental issues in this book, this course, and throughout your life, you might periodically ask yourself "What's in it for *me*?"

Figure 1.6 Oren Lyons

Mini-Activity 1-5   **Paradigms and People**

There have been only a few paradigm shifts in Earth's history. Which ones do you know of or can you think of? Use a reference book or the Internet to find some examples. (Here are a couple of search terms to start you off: *heliocentric* and *plate tectonics*.)

Also research people who have played a role in helping to establish or advance the shift toward a paradigm that emphasizes sustainability. People you might consider include Rachel Carson, Aldo Leopold, James Lovelock, Amory Lovins, and David Suzuki.

Case Study Rapa Nui: A Cautionary Tale

Rapa Nui is a small island in the southeastern Pacific Ocean. Known by its more famous name, Easter Island, it lies more than 2000 km from the nearest inhabited island and more than 3500 km from the mainland of South America. It is one of the most isolated, uninhabited islands on Earth.

The island was first settled by Polynesian people who sailed there from other island groups thousands of kilometres away. Its English name comes from the first Europeans, who arrived at Easter time in 1722. Today, Rapa Nui is known for two main things: the immense stone statues known as *moai* that stand watch over the land, and a tale of its destruction by its inhabitants.

Archeological work from 2011

Version 1: A Cautionary Tale

For years the case of Easter Island has been held up as an example of how humans can ruin their environment. In this view, the island was first settled around 800 C.E., but the population stayed small in size. About the year 1200, the population began increasing rapidly, and the culture focussed on building *moai*. Most of the palm forests were cleared for farming. The wood was used for fuel, boat building, and fashioning large frames to transport the heavy statues. These practices were not sustainable. Deforestation led to soil erosion and overexposure to salt winds from the ocean. Without wood to build fishing boats, people could not fish for food. The environmental collapse brought on by human actions caused the society to decline into famine, warfare, and even cannibalism.

Version 2: A More Complex Picture

Recent research suggests that the original settlers did not arrive until about 1200 C.E., and they started cutting down trees and building *moai* soon after. They brought with them rats, as either stowaways or a food source. With no natural predators on the island, the rat population increased dramatically. The rats fed on the seeds of the palm trees, thus worsening the island's deforestation. The humans and the rats both contributed to the decline of the island's ecosystem. The Europeans, who brought conflict and new diseases, worsened the society's collapse.

Research and Analyze

1. What factors made Rapa Nui vulnerable to the kinds of events that led to its ruin?

2. Use a flowchart to summarize the chain of events that led to ecological degradation on Easter Island.

Communicate

3. Easter Island has been used as a cautionary tale about how humans can cause ecological disaster. New evidence shows a more complicated picture. Write an opinion piece (about 250 words) on how the example of Easter Island can still teach us lessons about sustainability.

Summary

- The Industrial Revolution marked the start of dramatic increases in Earth's human population and played a major role in the development of environmental issues confronting modern society.

- The law of conservation of matter indicates that Earth's material resources are finite. The energy laws indicate that all energy use is limited by inefficiencies that can be reduced but never overcome.

- A paradigm is a concept or world view that is accepted by most people in a society and that influences their thoughts, decision-making, and actions.

- A paradigm shift is a period in history that marks a transition from one way of thinking and acting about the world and how it works to a different way of thinking and acting about the world and how it works.

Review Questions

1. Read the following statement about the Industrial Revolution:

   "The Industrial Revolution was the source of many positive changes in human society, but it also had unintended consequences that have had negative effects in many environmental contexts."

   Agree or disagree with this statement, and give reasons to support your ideas. C A

2. The previous environmental paradigm was based on ways of thinking about and using the environment that are unsustainable. Describe three examples of this, and explain your choices. T/I C A

3. The new environmental paradigm emphasizes sustainability. Explain how you think the new environmental paradigm views each of the following. T/I C

   a) human population growth
   b) biodiversity
   c) waste disposal

4. The paradigm shift toward sustainability has the potential to be as significant as the shift that took place during the Industrial Revolution. T/I C A

   a) In what ways do you think people during the Industrial Revolution had to change their thinking as the world around them became more urban-centred and more mechanized?
   b) Do you think we fully understand all the ways that industrialization affected the world? Explain.
   c) Identify and describe at least three effects or outcomes of a paradigm shift toward sustainability. Consider effects that might be negative as well as positive, and explain your reasoning.

5. In 2008, the Royal Astronomical Society of Canada adopted a program that recognizes two types of dark-sky designations: Dark Sky Preserves and Urban Star Parks. Being a Dark Sky Preserve means that an area is dedicated to reducing the effects of artificial lighting on the nighttime environment. A C

   a) Is light a form of pollution? Give reasons for your opinion.
   b) Do you think the establishment of a program like this is an outcome of a paradigm shift in our thinking about the environment? Why or why not?
   c) What benefits might a Dark Sky Preserve have for people and for other animals? Explain your answer in both cases.
   d) In what ways do you think an Urban Star Park differs from a Dark Sky Preserve? What benefits might an Urban Star Park have for people and for other animals? Explain.

Materials

- computer with Internet access

Do You Tread Lightly on Earth?

An ecological footprint is a measure of your impact on the natural environment. Your footprint represents how much land is required to produce the natural resources that you use for food, clothes, housing, heat, transportation, education, and recreation. Your teacher will direct you to a number of websites that calculate ecological footprints. They tend to give answers in terms of the area needed to sustain an individual, and the number of Earth-like planets needed to support humanity if everyone had the same lifestyle as the person using the footprint calculator.

Pre-Lab Questions

1. What is the current population of Earth?
2. Survey five people who are not in your class to find out the three most important issues affecting them. Pool your data with the rest of the class. They will be used in a later question.

Question

What might a sustainable lifestyle for humans look like?

Procedure

1. Your teacher will direct you to a few ecological footprint calculator sites on the Internet. Copy the data table shown on the next page into your notebook, and fill in the answers to the questions as you go to each site.
2. What is the size of the average Canadian's footprint?
3. What percentage of the average Canadian ecological footprint do *you* leave?

Analyze and Interpret

1. How many Earths would we need if everyone used the same amount of resources that an average Canadian uses?
2. How many Earths would we need if everyone used the same amount of resources that you use?

Conclude and Communicate

3. Think about the results you have found here compared to the issues revealed in the class survey from earlier in this Lab. Do you think the survey identified the most important issues? Why or why not?
4. Think of an approach that could be used to raise awareness about how humanity is affecting Earth's resources. Then suggest your approach to at least one person who might be able to help implement your approach, such as a politician, a community leader, or an influential business leader. Inform your class about any feedback or follow-up you receive from this person.
5. **Inquire Further** Find out about "Earth Overshoot Day" and make a chart of what day of the year it falls on. Draw a conclusion based on this chart.

Sample Data Table

Questions	Footprint Calculator 1	Footprint Calculator 2	Footprint Calculator 3
URL of ecological calculator site			
What organization sponsors the site?			
What is the size of your footprint (in hectares) using this calculator?			
To which questions did you not have the answers? Find those answers and redo the survey. How accurate were your initial guesses, and how did the correct data affect the final answer?			
Try the survey again, adjusting your answers to make it so everyone on Earth can have a footprint like yours. Were you able to do this? What did you have to change to do so?			
Redo the survey as someone with your lifestyle in a tropical, developing country. (Don't change your answers to the questions, but pretend that you live in a tropical, developing country.) What happens to your footprint? Why does this happen?			
What was the ecological footprint for a person using the same resources that you use while living in the developing country that you chose?			
Change some of the parameters by changing the answers to some of the questions. What seems to make the greatest difference in the final ecological footprint calculation?			

Chapter 1   SUMMARY

Section 1.1   Introducing Environmental Issues

All organisms require material and energy resources to survive. However, Earth's human population uses material and energy resources in ways that privilege people at the expense of other organisms.

Key Terms
resources
environment

Key Concepts
• Humans and all other organisms need certain essential resources from nature to sustain life.

• Most organisms are adapted to living in particular environments, and they use the resources provided by those environments alone.

• Unlike other organisms, humans change not only their immediate environment, but also any other environments they visit and choose to inhabit.

• The ways that our sizeable human population uses Earth's natural resources have caused many environmental challenges that require solutions.

Section 1.2   A Shift in How We Think about Issues

Environmental issues are the result of a human mindset that has believed in infinite growth (growth that can continue forever) on a planet where resources are finite (limited). The new, emerging paradigm is shifting from unsustainable activity to an emphasis on sustainability.

Key Terms
paradigm

Key Concepts
• The Industrial Revolution marked the start of dramatic increases in Earth's human population and played a major role in the development of environmental issues confronting modern society.

• The law of conservation of matter indicates that Earth's material resources are finite. The energy laws indicate that all energy use is limited by inefficiencies that can be reduced but never overcome.

• A paradigm is a concept or world view that is accepted by most people in a society and that influences their thoughts, decision-making, and actions.

• A paradigm shift is a period in history that marks a transition from one way of thinking and acting about the world and how it works to a different way of thinking and acting about the world and how it works.

Knowledge and Understanding

1. In what ways are humans similar to and different from other organisms in terms of the need for and use of resources?

2. The concept of a paradigm is a challenging concept for most people. In your own words, and to the best of your ability, explain your understanding of what a paradigm is. Use an example to support your explanation.

Thinking and Investigation

3. The map shows the land area of the Netherlands compared to the theoretical area that the Netherlands would need to have in order to support its ecological footprint.

The land area of the Netherlands is about 34 000 km². However, the area of its ecological footprint is about 15 times larger.

a) Choose three countries from Figure 7 of the Essential Science Background on page 9, Use an encyclopedia or other information resource to find the land area and current population for those countries.

b) For each country, multiply the population information by the hectare per person value from Figure 7. Record the actual land area, population, and theoretical area the country would need to support its ecological footprint.

c) If the theoretical land area for any country was larger than the actual land area, what does this imply about the sustainability-related behaviour of its population?

d) If the theoretical area for any country was smaller than the actual land area, what does that imply about the sustainability-related behaviour of its population? (Note: Can you assume that, because the ecological footprint is smaller, everyone in that country has equal access to the resources they need to survive? Why or why not?)

Communication

4. Draw a footprint in your notebook.

a) At each toe, write one thing you do that reduces your own ecological footprint.

b) At the heel of your footprint, write the one thing you do (or don't do) that is responsible for your largest environmental impact.

c) Across the middle of your footprint, write the one thing you would find the most challenging to change in order to shrink your ecological footprint.

d) In your opinion, how important is it to shrink your ecological footprint? What might inspire you to work to shrink it? What might make you not work to shrink it?

5. Write a slogan and a brief public service announcement for a public awareness campaign that is designed to motivate people to contribute to a paradigm shift that emphasizes sustainability.

6. What other cultures and traditions do you know of (or are you a part of) that express ideas that are similar to those of the new environmental paradigm? Describe your awareness and understanding of these ideas. Do research if necessary to enhance your answer.

7. Did You Know? Reread the quotation on page 14. This same philosopher is famous for another quotation: "Only change is unchanging."

a) What do both quotations say about the idea of change?

b) Sometimes, environmental challenges can result from people resisting the idea that there are ways of doing things that are different from the way we are used to doing them. What role do you think change plays in the causes of and solutions for environmental challenges?

8. Did You Know? Reread the quotation on page 21. How is what Einstein says related to the idea of paradigms and paradigm shifts?

Pause and Reflect

What have you learned in this chapter that you could incorporate into your daily actions or choices?

CHAPTER 2

Science, Society, and Solutions

This photo shows a tiny portion of Canada's vast boreal forest, which extends from northern British Columbia all the way to the island of Newfoundland.

What might a hotel developer see in looking at this scene? What about a camper, an environmental activist, and a mining company? What do *you* see?

Different Approaches

Whether you live within minutes, hours, or days of forests, your life is closely connected to them—perhaps in ways you might not even realize. Use the questions below to help you consider how.

The three most common approaches to thinking about the environment are briefly described below.

- *Development approach*: as much nature as necessary should be converted for human use. Human creativity and ingenuity will solve environmental challenges.

- *Preservationist approach*: large portions of nature should be preserved intact as habitat for non-human organisms. Nature has value apart from human uses, but it also has value for recreational and aesthetic (restorative) purposes.

- *Conservationist approach*: promotes human well-being, but tends to consider a wider range of long-term human goods in making decisions about managing the environment.

  1. Which, if any, of these involve the idea of sustainability?

  2. What other approaches to environmental challenges might there be? *Should* there be more than these three? Discuss your ideas.

In this chapter, you will

- appreciate why environmental science encompasses so many fields of study
- describe scientific approaches to collecting and analyzing data
- describe the importance of critical thinking in solving environmental challenges

The Methods of Environmental Science

The Nature of Environmental Science

In previous grades, you have studied ecology. This is the field of science that investigates the way organisms interact with each other and their surroundings. Ecology has many things in common with environmental science, but they are not the same. As you can see in **Table 2.1**, ecology is just one of the many fields of study, or *disciplines*, that contributes to environmental science.

You might hear environmental science referred to as being *interdisciplinary*. **Table 2.1** shows why. Notice in the table that there are areas of study *within* science, as well as *outside* of science. This broad range of inquiry makes sense for a field of study such as environmental science. **Environmental science** is the study of how we humans interact with all the living and non-living parts of the environment. Environmental scientists collect valid data and communicate informed ideas to help them understand the natural world, our place in it, and our impact on it. This understanding also inspires them to search for solutions to environmental challenges. See **Figure 2.1**.

environmental science
the study of how humans interact with the environment

Figure 2.1 Different fields of study contribute to solutions for environmental challenges.

Inferring What other questions could be asked in each segment of this diagram?

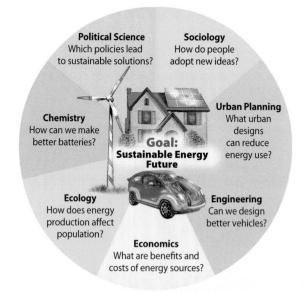

Table 2.1 Some Key Fields of Study That Are Part of Environmental Science			
Main Discipline	**Examples of Sub-disciplines**		
Biology	• Botany • Ecology	• Limnology (freshwater ecology) • Marine biology	• Microbiology • Zoology
Chemistry	• Biochemistry	• Geochemistry	• Nuclear chemistry
Earth sciences	• Climatology • Paleontology	• Meteorology • Hydrology (water sciences)	• Geology
Physics	• Atmospheric physics • Engineering	• Nuclear physics	
Mathematical and related sciences	• Computer science	• Statistics	
Social sciences	• Anthropology • Geography • Law	• Economics • Political science	• Ethics • Sociology

How Environmental Scientists Collect and Generate Data

Some environmental scientists may work mostly outdoors in natural settings. Others might do controlled experiments in laboratories, or use computers to design and analyze models. No matter where they work, they use the methods of science inquiry that all scientists do. **Table 2.2** describes some data collection methods.

 scientific inquiry—
see Appendix A

Table 2.2 Examples of Methods of Data Collection	
Method	**How It Works**
Air, water, and soil samples 	Taking samples of air, water, and soil provides key data about how their quality might be changing. Air samples help to track the presence of polluting gases and solids. Water samples may focus on the amount of oxygen, sediment, phosphates, or toxic chemicals. Soil samples might track the quality of topsoil, the presence and type of soil organisms, and the presence of toxic chemicals.
Ice core samples 	Permanent ice fields in the Arctic and Antarctic have existed for hundreds of thousands of years. Each year, a new layer of snow and ice is deposited. Each layer holds a record of what the atmosphere was like when the ice formed. This record of the past can be collected with special drills that take core samples from the ice. Dust and ash trapped in the ice indicate events such as volcanic eruptions and forest fires. Plant pollen tells the species of plants alive at the time. Temperature and humidity can be inferred from the size and shape of the ice crystals. Air bubbles trapped in the ice can be analyzed to determine how much oxygen, carbon dioxide, and other gases were in the atmosphere at the time the ice formed.
Quadrat sampling 	One way of measuring biodiversity is to use a quadrat—a known area that is marked with a pre-made square of plastic, or with stakes and string. Quadrats can range in size from 1 m² to 20 m², depending on the type of habitat surveyed. Different species and their numbers within the quadrat are counted. Counting is repeated many times in different places throughout the habitat to get an accurate representation of biodiversity.
Statistical data 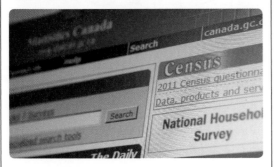	Statistical data such as birth and death rates and immigration and emigration rates are collected to learn more about how the human population is growing. Scientists also analyze data on where people are living to track changes in population in urban and rural areas. As well, medical records can help indicate the effects of air and water quality on human health.

Using Environmental Impact Assessments

An environmental impact assessment (EIA) is done before a project that may affect the environment. Under the Canadian Environmental Assessment Act of 2012, an EIA may be required when a project could affect aquatic species and their habitats, migratory birds, and/or Aboriginal peoples, especially for land and resources used for traditional purposes. An EIA may be conducted by federal or provincial governments. If a project involves nuclear materials, the Canadian Nuclear Safety Commission may conduct the assessment. If the project involves pipelines that cross provincial or international borders, the National Energy Board may conduct the assessment. For other projects, the Canadian Environmental Assessment Agency conducts them.

ThoughtLab 2A, Model an Environmental Impact Assessment, on page 41

Using Models and Statistics

Models may be used to analyze data and make predictions. For example, a global climate model (GCM) is used to make long-term predictions. It combines mathematics and physics to predict temperature, precipitation, wind speeds, and similar properties of weather and climate. The results are displayed as maps like the one shown in **Figure 2.2**. GCMs include the effects of greenhouse gases and oceans in their calculations. In order to test climate models, past records of climate change can and have been used.

Geographic Information Systems (GIS) are another kind of modelling. GIS uses computers and satellite technology to capture, store, query, analyze, and display the locations and features of natural and constructed landscapes. Common applications of GIS include resource management, population studies, and urban planning.

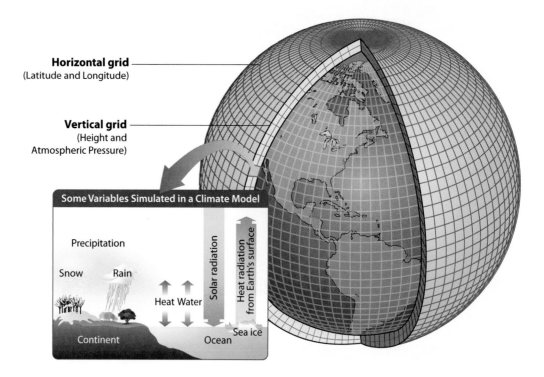

Figure 2.2 A global climate model divides the planet into a three-dimensional grid system that extends from the ocean floor to the outer edge of the atmosphere.

Pause and Reflect

1. What is environmental science?

2. Critical Thinking Do you think the interdisciplinary nature of environmental science makes it easier to find solutions to environmental challenges? Explain.

Summary

- Environmental science is the interdisciplinary study of how humans interact with all the biotic and abiotic parts of the environment.

- Environmental scientists collect data and communicate ideas to help them understand the natural world, the place of humans in it, and the human impact on it.

- An environmental impact assessment considers and reviews the factors of a project that has the potential to affect the health of the environment.

- Models such as the global climate model and geographic information systems enable environmental scientists to analyze data and make predictions.

Review Questions

1. In what ways are the fields of ecology and environmental science similar and different? **K/U** **A**

2. For some people, including many scientists, science is *the* way of knowing about nature and the world. For other people, including some scientists as well as philosophers and historians of science, science is a way of knowing about nature and the world.

   a) What difference does the use of "the" versus "a" make in the meaning of the two descriptions of science? **A** **C**

   b) Explain how environmental science especially draws on different kinds of knowledge. **K/U** **C** **A**

3. Consider an environmental issue such as global climate change through the perspective of the following disciplines: economics, politics, journalism, psychology, agricultural science, and ecology. **C** **A**

   a) What are some questions that each discipline could contribute to the scientific understanding of global climate change?

   b) What are some questions that each discipline could contribute to the societal understanding of global climate change?

   c) What differences are there between a scientific understanding and a societal understanding of climate change? Should there be differences? Explain your viewpoint.

4. All scientific models carry uncertainties that result from several sources. One is the precision of the measurements taken of the variables that pertain to the model. Others include the quality and quantity of the data and the complexity of the variables. The line graph shows the responses of 19 climate models to the same test of climate change. Each acronym in the legend represents a different computer model. Each model predicts how temperature will increase after 80 years of increasing levels of carbon dioxide. **T/I** **A**

   a) Do any of these models agree precisely with any other model(s)? How do you know?

   b) Which prediction(s) do you think you could rule out as being unlikely? Explain your thinking.

   c) Based on a wide variety of data and models, most climate scientists have concluded that global temperatures will continue to increase as time passes, if our current actions and behaviour stay the same as they are currently. What pattern(s) in the predictions of these 19 climate models do you see that could provide confidence in this conclusion?

   d) To what extent do you think uncertainties should be allowed or used to undermine that conclusion? Explain your reasoning.

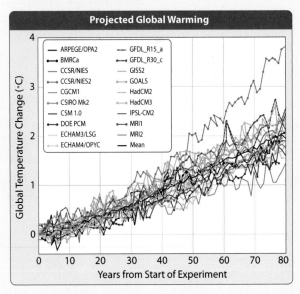

The Mindsets of Environmental Science

Scientific Habits of Mind

Just as there are certain methods that scientists use to collect and analyze data, there are certain habits of mind that are characteristic of science. These include:

- Skepticism: not believing everything you read, hear, or see

- Open-mindedness: being willing to consider different points of view

- Persistence: being willing to "stick with" a task, even if the results may take days, weeks, months, or even years to collect and analyze

- Humility and honesty: being aware that conclusions might not be supported by new evidence or information, and being willing to change your mind

Critical Thinking Skills

You are exposed to information from many different sources. Competing claims and conflicting ideas may battle for your attention. Our rapidly changing world can make it hard to know what to think and how to act. Scientists themselves may disagree about what certain data mean or how to interpret and draw conclusions from them. How can you distinguish between information, which is accurate, complete, and factual, and misinformation, which is incorrect, incomplete, based on unsupported opinion, or misleading?

Being able to seek new facts, learn new skills, generate data, evaluate information, and form your own ideas and hypotheses (and test or challenge them) are vital skills for everyone—students, as well as citizens. A number of skills, attitudes, and methods can help you evaluate information, form your own ideas, and make decisions. For example:

- *Analytical thinking* asks, "How can I break this problem down into smaller and more manageable parts?"

- *Creative thinking* asks, "How can I approach this problem in ways that are 'outside the box'?"

- *Logical thinking* asks, "How can reasoning help me think clearly?"

- *Reflective thinking* asks, "How can I use my experiences and learning to help me make judgments about what has happened and what should happen?"

critical thinking skills associated with thinking independently, systematically, and analytically

Figure 2.3 shows that these approaches to thinking are part of the dynamic process of **critical thinking**. Critical thinking asks questions such as "What do I want to accomplish, and how can I use my skills, experiences, and attitudes to accomplish it?" **Table 2.3** offers ideas for how you can develop and refine your critical thinking skills.

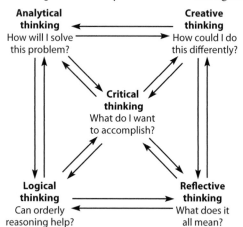

Figure 2.3 Different approaches to thinking are used to solve different kinds of problems or to study aspects of an issue.

Table 2.3 Using Critical Thinking Skills	
Critical Thinking Skill	**Examples of Questions To Ask Yourself**
Identify and evaluate statements and conclusions in an argument.	• What is the basis for the claims made? • What evidence is presented to support the claims, and what conclusions are drawn from this evidence? • If the statements and evidence are correct, does it necessarily follow that the conclusions are valid?
Acknowledge and clarify uncertainties, vagueness, and contradictions.	• Do the terms used have more than one meaning? • If so, are all participants in the arguments using the same meanings? • Are the uncertainties deliberate? • Can all the claims be true at the same time?
Distinguish among facts, opinions, and values.	• Are claims made that can be tested? (If so, they should be verified, when and where possible.) • Are claims made about the worth or lack of worth of something? (If so, these are value statements or opinions and probably cannot be verified objectively.)
Recognize and assess assumptions.	• Given the backgrounds and views of the people in the argument, what underlying reasons might there be for the statements, evidence, or conclusions they present? • Does anyone have a personal agenda? What do they think I know, need, want, or believe? • Are there underlying suggestions based on gender, ethnicity, economics, a belief system, or other factors that can affect the discussion?
Determine the reliability or unreliability of a source.	• What makes the experts qualified in this issue? • What special knowledge or information do they have? • What evidence do they present? • How can I determine whether their information is unbiased, accurate, and complete?
Recognize and understand the context in which things are presented.	• What are the basic beliefs, attitudes, and values that this person, group, or society holds? • How do these beliefs and values affect the way people view themselves and the world around them? • If there are conflicting beliefs or values, how can these differences be resolved?

Pause and Reflect

3. In your own words, define the term *critical thinking*.

4. Describe two situations in which you would use critical thinking skills.

5. Critical Thinking Does it surprise you that being able to think creatively is associated with critical thinking? Explain why or why not, and give at least one example to showcase an environment-related situation in which creative thinking would be an asset.

Did You Know?

"I long to accomplish a great and noble task, but it is my chief duty to accomplish humble tasks as though they were great and noble. The world is moved along, not only by the mighty shoves of its heroes, but also by the aggregate of the tiny pushes of each honest worker."

—*Helen Keller (1880–1968), author, activist, advocate for people who have disabilities*

With Power Comes Responsibility

This book has been designed to help you develop your skills of critical thinking about environmental challenges and to help you appreciate the crucial roles that science and opinions informed by science play in solving these challenges. **Table 2.4** shows how the shift that is transforming society and the world starts with individuals like you. It is your ideas and opinions, your skills, your decisions, and your actions that help to shape and propel the shift. You owe it to yourself, your family and friends, your fellow Canadians, and your planet to understand and use your power responsibly.

InquiryLab 2B, Effects of Arctic Ice Cap Melting, on page 42

Table 2.4 Examples of How Individuals Can Influence Society	
Role or Action	**Empowering Effect**
Consumer 	**Consumers have power.** • What choices do you make about the products you will and will not buy? • What reasons motivate your choices? • How can you find out more about the methods and materials used to make a product?
Volunteer	**Volunteers inspire through their commitment and example.** • Where do you, or can you, volunteer your time? • Who benefits from your willingness to share a part of yourself? • How could volunteering locally have global effects?
Advocacy group	**Advocacy groups can affect change, increase sustainability, and increase stewardship.** • How can you find out more about advocacy groups and the causes they represent? • How can the work you do as a member of an advocacy group lead to changes in legislation that help protect the environment, people, and other organisms?
Individual and voter	**Members of society have responsibility.** • In what ways are you a member of your community? your province? your country? your planet? • What responsibilities do you have as a member of society? • How can you educate yourself about candidates before you vote? • How could you find out more about local projects you could participate in, such as Meals on Wheels, bird counts, and litter clean-ups? What local projects interest you?

Case Study   How Do You Tell the News from the Noise?

In 2011, there were at least 170 million blogs on the Internet with 15 000 being added each day. More than a billion people are linked in social networks. Each day, several billion emails, tweets, text messages, videos, and social media postings link people in innumerable ways. Hand-held devices make it even easier to surf the Web, watch videos, or link to friends.

Ironically, many people use their digital connectedness to reinforce existing beliefs, rather than to educate themselves. A study on the state of media by the Center for Journalistic Excellence at Columbia University concluded that the news is growing increasingly biased. Rumours and falsehoods fly across the Internet at light speed. More than two thirds of all TV news segments consist of reports or interviews in which only a single point of view is presented without any background or perspective.

Choose a media report about the environment to read (or watch) and analyze. How can you detect bias in it? Ask yourself (or your friends) questions such as these as you practise critical thinking.

- What political positions are represented? Are they obvious or hidden?
- What special interests might be involved in the issue? Who stands to gain from presenting a particular viewpoint? Who is paying for the message?

- What sources are used as evidence in the communication? How reliable are they?
- Are facts and statistics cited in the presentation? Are they reliable? Are citations provided so you can check the sources?
- Is the story one-sided, or are alternate views presented?
- Are arguments based on facts and logic, or are they emotional appeals? Does it matter?

Analyze

1. What do you think of these bias-detecting questions? Which ones do you use? Which ones do you not use, and why? What others could be asked?

2. How many sources do you rely on for information? What could this mean about your understanding of a fact, situation, or issue?

Communicate

3. What factors influence the ways you form your opinions from the news and other sources of information?

Case Study No Challenge Too Big

Dunbarton High School, in Pickering, Ontario, is a long-standing EcoSchool whose students engage in a wide range of environmental activities. It didn't start out so big, though. In the beginning a handful of passionate students and teachers banded together to form the Enviro Club to do activities such as tree planting. The club grew as enthusiastic students spread the word and convinced their friends to join too. Today the Enviro Club has 40–50 core members, though other students join in for bigger projects.

Batteries and Cellphones

Teacher Karen Larter was teaching classes in electronics, and the question naturally arose: what happens to all these used batteries? A simple box was placed in a convenient spot outside the school library, and students and teachers were encouraged to bring in used batteries from home. Karen figures that as much as 25 kg of batteries are collected each year to be taken to the local hazardous waste transfer station for recycling.

Next up: used cellphones. Dunbarton chose to partner with the Toronto Zoo, which has a successful recycling program for unwanted devices. With the help of schools like Dunbarton, the Zoo has safely recycled more than 13 000 cellphones. All profits go toward preserving gorillas in their native habitat.

"Recycling Central": Expanding Recycling Efforts

As interest from the student body has grown, and with strong support from parents, students began to look at more and more items that might be suitable for recycling—common things that might appear to have no use and are commonly tossed in the trash with no thought. The school now has a central recycling station set up where students and teachers can bring in a variety of items to re-use or recycle. These include soda can tabs, old plastic pens and markers, milk bags, and bread bag tags. For certain items, such as printer cartridges, the school earns some money that can be put back into environmental efforts.

Making the School Greener

Students have also taken steps to make their school grounds greener—literally. Over the years they have planted more than 50 shade trees on the property. In 2010 an outdoor classroom complete with shade trees and rocks for seating was installed, followed in 2011 by a pollinator garden that uses native species to attract pollinating insects such as bees and butterflies. Dave Gordon, supervising teacher for the Enviro Club, says that they are currently looking into setting up a "bee condominium" in the garden.

Keeping Hydrated the Environmental Way

Students have chosen tap water over bottled water. Tap water is safe to drink and locally sourced, and it doesn't waste resources in packaging and transportation the way bottled water does. Beginning in fall 2013, students entering Dunbarton High School for the first time will receive a stainless steel, reusable water bottle that they can fill up at one of five tap water filling stations around the school, paid for in part by Enviro Club fundraising.

Partnering with the Community

The Enviro Club at Dunbarton participates in many community environmental events. They partner with a local Scout troop in an e-waste recycling program. Unwanted electronic items such as computers, printers, and TVs are collected from the community to be delivered to specialized companies that safely recycle them. In turn, the Scouts share the financial return they receive from the companies with the school. To raise further funds, Dunbarton students even got into the direct marketing business, but not with chocolate bars. Instead they sold rain barrels. The money raised from such efforts is used for further eco-activities and student scholarships.

Currently, the school is involved with the Lake Ontario Atlantic Salmon Restoration Program run by the OFAH (Ontario Federation of Anglers and Hunters). The aim is to re-establish a wild population of Atlantic salmon in Lake Ontario. Salmon had been eliminated from the lake over 100 years ago due mainly to development, deforestation, and dam construction. The OFAH program gets students involved in a "hands on" way. Stocks of very young salmon, called fry, are raised in classroom "hatcheries" over the winter, and students later help release the fry into local streams to increase the growing salmon numbers. About 2000 students in more than 90 schools participate in this conservation program.

Analyze

1. Take an inventory of environmental efforts going on in your school. Suggest ways these efforts could be expanded.

Communicate

2. Write a proposal for a new environmental initiative to start at your school. Think about the following: What is the problem or need? How could the problem be solved or the need be met? What changes in procedure at the school or what new equipment or supplies would be required?

Reviewing Section 2.2

Summary

- Scientific habits of mind include skepticism, open-mindedness, persistence, humility, and honesty.
- Critical thinking skills enable people to seek and learn new facts, learn new skills, evaluate information, and form valid ideas and opinions about facts and information.

- Individuals have the responsibility to use scientific habits of mind and critical thinking skills to make informed decisions and to take appropriate actions that can help to shape and propel the paradigm shift toward sustainability.

Review Questions

1. Many companies market their products with words like *green, natural,* or *non-toxic.* They use logos and packaging that promote the idea that the products are safe for people and the environment. In many cases, these claims may be "greenwashing." This term refers to products that use deceptive marketing techniques to create the impression that the production and/or use of a product reflects sustainable practices on the part of the manufacturer. Often, greenwashing is accomplished through the use of ambiguous terminology. For example, what do words such as natural and or non-toxic really mean? In some cases, promotion may take the form of misdirection. For example, a product might be described as being free from artificial colours and artificial flavours—but does that mean the product is free from other kinds of artificial ingredients? **T/I C A**

   a) What examples of greenwashing have you seen or heard, either in the media or on product labels in stores?

   b) How is the greenwashing accomplished? (In other words, how is the consumer being manipulated by the technique?)

   c) What power do you, as a consumer and as a potential voter, have to influence manufacturers, governments, or both in matters such as this?

2. Re-examine the Launch activity at the start of this chapter. Which approach to the environment—development, preservation, or conservation—do you think that you adopt in your own life? Do you think everyone in the world should share the same attitude you hold? Explain why or why not. **C A**

3. Environmental challenges are often presented in highly polarized ways, where each side is claimed by the other to hold an extreme position. When this happens, people are reduced to their positions or their occupations. For example, "loggers" are pitted against "environmentalists", and "developers" are pitted against "activists." **A C**

   a) What happens to meaningful dialogue and communication when people who are engaged on both sides of complex, real issues are reduced to stereotypes?

   b) Is it always fair to assume that an "environmentalist" has the best interest of the environment in mind? Is it always fair to assume that a "developer" only cares about profits? Is it always fair to assume that a "logger" doesn't care about organisms that live in forests? Use clear, reasoned thinking to state and justify your opinions.

4. The photo below appeared in many different newspapers. **T/I C A**

   a) What do you think and feel when you look at the photo?

   b) What "story" does the photo tell?

   c) What "story" or part of a story does the photo not tell?

   d) Is a single picture or a single source of information (such as a news article that goes with the photo) enough for you to form a meaningful opinion about an environmental issue? Explain your thinking.

Safety Precautions

- Always wash your hands with warm water and soap before and after handling plant materials.

Materials

- computer with Internet access
- library resources

Model an Environmental Impact Assessment

The ways that people use land and water can have far-reaching effects. How many resources should be used for human needs and how many for wildlife? Decisions should take into account environmental, economic, and social considerations.

Pre-Lab Questions

1. What is an environmental impact assessment?
2. What projects near where you live involve a conflict over land use?

Question

What can you discover by modelling an environmental impact assessment?

Procedure

1. Choose a local project that involves a conflict over land use. Agree on four to six roles your class can use to discuss the project and its effects. Half the roles will support the project; half will oppose it. For example, if the project is a mine near a fishing lake, supporters might include local businesses, unemployed miners, and companies that use minerals from the mine. Opponents could include local fishers, residents concerned about water quality, and a motel owner with cabins for rent on the lake.

2. Form groups so that each group represents one role.

3. Do research on the project. Focus on questions that your group may be asked. For example: What economic benefits will the project bring? What effects will it have on soil, vegetation, and water quality?

4. Prepare a presentation for the Environmental Impact Assessment Board. Your teacher will play this role. Use charts and graphs to display data when possible.

5. Present your case to the class. Allow time for questions.

Analyze and Interpret

1. Describe some advantages and disadvantages of the process you modelled in this Lab.

2. What factors do you think are most effective in determining the outcome of decisions over land use? Explain your answer.

Conclude and Communicate

3. Based on the information presented, as a class, draft a report that recommends whether the project should be approved, abandoned, or modified.

Suggested Materials

- access to the Internet and other research materials

Effects of Arctic Ice Cap Melting

A polar ice cap is a high-latitude region of land or water that is covered in ice. Ice caps form as snow accumulates year after year. As part of a cyclical process, each year some of the ice melts in the summer, and then reforms in the winter. In the last several decades, scientists have tracked significant changes in the size of the Arctic ice cap. Data analyzed from satellite images show that the area of permanent ice cover is decreasing at a rate of 9% each decade. At this rate, scientists predict that summers in the Arctic will be ice-free by 2030. In this Lab, you will investigate the environmental, social, and economic effects of the melting Arctic ice cap.

Pre-Lab Questions

1. What changes are occurring in the Arctic ice cap?
2. What are some possible effects of the Arctic ice cap melting?

Question

How does the melting of the Arctic ice cap affect sustainability?

Procedure

Part A

1. Research why the Arctic ice cap is melting.
2. Summarize the results of your research from step 1. Use diagrams, illustrations, animations, or other formats as appropriate in your summary.

Part B

1. Research the environmental effects of the melting of the Arctic ice cap. Use the following questions to guide your research. Use your critical thinking skills to evaluate sources of materials for bias or inaccuracies.

   - How have animals, such as polar bears, whales, walruses, and seals, been affected?
   - How will Greenland's ice sheet be affected by the Arctic ice cap melt? What effects will the changes to Greenland's ice sheet have?
   - How could the melting of the Arctic ice cap affect climate in other parts of the world?

2. Summarize the results of your research in a format approved by your teacher.

Arctic Ice-Cap Coverage, 1980

Part C

   1. Research more about the social and cultural effects the ice cap melt could have. Use the following questions to guide your research. Use your critical thinking skills to evaluate sources of materials for bias or inaccuracies.

   • How have communities along Arctic coastlines been affected by the melting of the ice cap, and what might future effects be?

   • How does the melting ice cap affect hunting and fishing by Aboriginal people?

   • How does the melting ice cap endanger cultural traditions of Aboriginal people?

   2. Summarize the results of your research in a format approved by your teacher.

Part D

   1. Research more about the economic effects the ice melt could have. Use the following questions to guide your research. Use your critical thinking skills to evaluate sources of materials for bias or inaccuracies.

   • How have communities along Arctic coastlines been economically affected by the melting of the ice cap? How will they be affected in the future?

   • How does the melting ice cap affect the oil, natural gas, and mineral industries?

   • How does the melting ice cap affect the shipping industry?

   2. Summarize the results of your research in a format approved by your teacher.

Part E

   1. Research what actions can be taken to slow, stop, or reverse the current trend of the melting.

   2. Summarize the results of your research in a format approved by your teacher.

Analyze and Interpret

   1. Why do you think the melting of the Arctic ice cap could be considered a global problem?

   2. What challenges are involved in efforts to reduce the rate of melting?

Conclude and Communicate

   3. Create an educational report to increase public awareness of this issue. You could write an editorial piece, make a short video, write a blog entry, create a public service announcement, or use another format approved by your teacher.

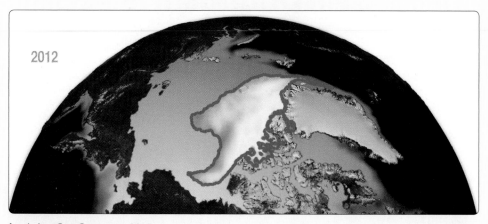

Arctic Ice-Cap Coverage, 2012

Section 2.1   The Methods of Environmental Science

Environmental science uses the methods and tools of disciplines from within and outside of science to investigate and propose solutions to environmental challenges. No single discipline can solve these challenges in isolation, but science is uniquely positioned to offer data and knowledge upon which any solution must be based.

Key Terms
environmental science

Key Concepts
- Environmental science is the interdisciplinary study of how humans interact with all the biotic and abiotic parts of the environment.
- Environmental scientists collect data and communicate ideas to help them understand the natural world, the place of humans in it, and the human impact on it.
- An environmental impact assessment considers and reviews the factors of a project that has the potential to affect the health of the environment.
- Models such as the global climate model and geographic information systems enable environmental scientists to analyze data and make predictions.

Section 2.2   The Mindsets of Environmental Science

Solutions to environmental challenges require habits of mind and critical thinking skills that are characteristic of scientific inquiry and investigation. Solutions also require the motivation and dedication of individuals who recognize the power they have to influence society, from the products they buy, to the decisions they make during elections, to the actions they take as volunteers and in their lines of work.

Key Terms
critical thinking

Key Concepts
- Scientific habits of mind include skepticism, open-mindedness, persistence, humility, and honesty.
- Critical thinking skills enable people to seek and learn new facts, learn new skills, evaluate information, and form valid ideas and opinions about facts and information.
- Individuals have the responsibility to use scientific habits of mind and critical thinking skills to make informed decisions and to take appropriate actions that can help to shape and propel the paradigm shift toward sustainability.

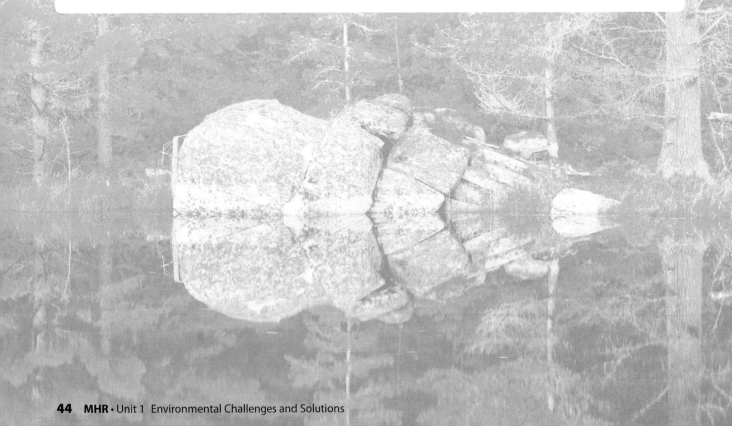

Knowledge and Understanding

1. Identify at least three disciplines from outside of science that are part of environmental science.

2. How are scientific habits of mind similar to and different from critical thinking skills?

Thinking and Investigation

3. According to a leading information technology research and advisory company, about 2.2 billion computers, tablets, and cellphones were shipped to worldwide markets in 2012. That number was expected to increase about 9% in 2013 and about 40% by 2017. Consumption—of electronic gadgets or anything else—has been evolving as people have found new ways to make their lives simpler and/or to use their resources more efficiently. We consume a variety of resources and products today as we move beyond meeting basic needs.

a) Do you think consumption beyond minimal or basic needs is necessarily a negative thing? Why or why not?

b) Issues around consumerism require us to ask questions about the behaviour of society (both as consumers and suppliers). One example of a question is, What are the effects of certain forms of consumerism on the environment? Come up with at least three other questions of your own.

c) Is consumerism a part of all human societies, or just societies in developed countries such as Canada? Give reasons to justify your answer.

Communication

4. Should environmental scientists be advocates for environmental policy? One argument in favour of this idea suggests that scientists are citizens first and scientists second. Therefore, as part of society, they have a responsibility to advocate to the best of their abilities and in a transparent manner. Those who oppose the idea of scientists acting as advocates say that advocacy undermines a scientist's credibility as being impartial and objective.

a) Offer and explain your own opinion about this.

b) Even if scientists decide to refrain from being advocates on an issue, does this mean that they should not use their scientific data and insights to help guide policy decisions? In other words, can someone offer opinions about an issue without advocating (either for or against) on that issue? Explain your reasoning.

5. **Did You Know?** Re-read the quotation from Helen Keller on page 36. Consider that it is common to admire great and noble qualities in certain individuals—people such as sports figures, activists, world leaders, and other inspirational and charismatic people. Sometimes it is easy to believe that there is something almost superhuman in these individuals, which makes them more special, more capable, and more powerful than "ordinary" people. Thus, we elevate these individuals to the status of heroes.

a) In what ways can recognizing the heroes around us motivate and inspire us in our thoughts and actions?

b) In what ways can recognizing the heroes around us have the opposite effect?

c) What is your opinion of the idea expressed in the quotation?

Application

6. Factors that are part of an environmental impact assessment for a project include the following:

- environmental effects that could be caused by the project
- the importance of these effects—how mildly or severely they may cause changes in the environment
- clean-up efforts to address any major damage that could be done, including the effects of accidents or mechanical failures
- the results of any other studies that might help scientists understand more about possible effects of the project comments from the public

Environmental impact assessments are often described as representing a proactive and preventative approach to the management and protect of the environment.

a) What does it mean for something to be proactive and preventative?

b) Environmental impact assessments can be costly in terms of expense and time. Do you think they should be conducted for all projects that could affect the environment? Explain your opinion.

Pause and Reflect

What have you learned in this chapter that you could incorporate into your daily actions or choices?

Canadians in Environmental Science

David Schindler: Ecologist in Action

When Dr. David Schindler talks, politicians, businessmen, and scientists listen. As a professor of Ecology at the University of Alberta in Edmonton, David has been a leader in Canada's most important environmental battles for almost fifty years. He was born in Fargo, North Dakota in 1940 and was studying engineering when a summer job with a biology professor convinced him to switch into science. In 1968, after finishing his PhD at Oxford University, he became head of the Experimental Lakes Area project in northwestern Ontario. By dividing a lake in half with plastic barriers and adding chemicals to one half, David and his team proved that phosphorous used in sewage treatment and phosphates from laundry detergents were causing the algal blooms that had damaged the ecosystems of Lake Ontario and Lake Erie. These chemicals were banned by the federal government in 1973.

He then lead a panel for the U.S. National Research Council that showed emissions from power plants burning fossil fuels were the cause of acid rain that was killing fish in Canadian and American lakes. Since 1989, David has conducted research on the effect of dioxins from pulp mills, the importance of Canada's boreal forest in fighting climate change, and the environmental impact of Alberta oil sands development.

David has been a vocal defender of the environment. His views have often led him into conflict with business and government when they try to deny or ignore scientific research. He says that his public battles have been a bit like "playing chess with a gorilla. The game is boring and you know you're going to win, but you have got to duck once in a while when they get angry and take a swing at you."

David Schindler's team divided this northwestern Ontario lake in half with a plastic barrier to demonstrate that phosphorous and phosphates were causing algal blooms in Lake Ontario and Lake Erie.

Environmental Science at Work

Focus on Environmental Challenges and Solutions

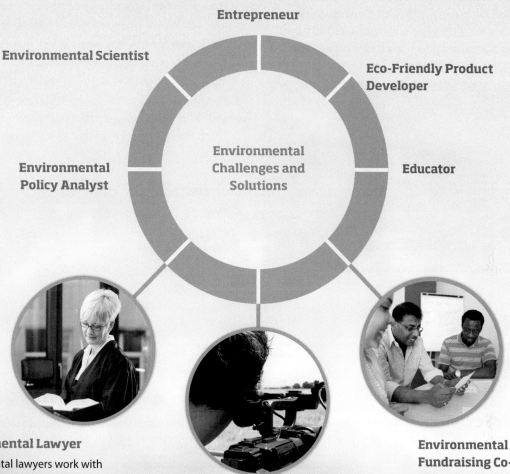

Entrepreneur

Environmental Scientist

Eco-Friendly Product Developer

Environmental Challenges and Solutions

Environmental Policy Analyst

Educator

Environmental Lawyer

Environmental lawyers work with clients on a variety of matters, including pollution, land use concerns, waste disposal, and resource extraction. Many lawyers in this field have a background in environmental studies or science. Environmental lawyers may work for law firms specializing in environmental law, or for government departments, environmental activist groups, or industry.

Environmental Documentary Filmmaker

Environmental documentary filmmakers play an important role in informing the public about environmental issues. They may act as both the director and producer for documentaries that focus on environmental issues. Producers oversee the entire film. Directors develop the overall vision for the documentary and make sure the filming team records all essential material.

Environmental Fundraising Co-ordinator

Environmental fundraising co-ordinators often work in the not-for-profit sector and must act as the voice and face of their organizations. The fundraising co-ordinators apply for grants, solicit funds from potential donors, and co-ordinate fundraising events. They must have skills in event planning, communications, budgeting, and marketing.

For Your Consideration

1. What other jobs and careers do you know or can you think of that involve dealing with environmental challenges and solutions?

2. Research a job or career in this field that interests you. What essential knowledge, skills, and aptitudes are needed? What are the working conditions like? What attracts you to this job or career?

Create a Social Media Web Page

Is there a particular environmental issue or program that interests you and that you would like to learn more about? In this project, you will develop a social media web page about this issue or program and update the page as you progress through the course.

Question

How can you create a social media web page for an environmental issue?

Initiate and Plan

1. Skim through the unit and choose one environmental issue that you would like to feature on your social media web page. You may want to also look through the chapters in the other units, to see if there is an issue described there that interests you.

2. Visit one or more social media websites and observe the different features. Brainstorm how you can apply these features to your chosen issue.

Perform and Record

3. Create your social media web page. As a minimum, it should include at least the following:
   - a description of the issue
   - a map showing the locations in Canada that are affected
   - 3 posts describing the ecosystems it affects and how it affects these systems
   - 3 images that can be photos or art related to the environmental issue
   - 3 likes: this could represent the different actions being taken to address or solve the issue

Analyze and Interpret

1. Exchange social media web pages with another student. Write comments in response to each post, like, and image. Each comment should show your knowledge of environmental science.

2. Exchange comments with your classmates, and add the comments you receive to your social media web page.

3. People use social media to communicate with other people. Explain how a social media page would be helpful to inform many people about an environmental issue.

Communicate Your Findings

4. With your teacher's permission, post your social media web page online.

Assessment Checklist

Review your project when you complete it. Did you ...

- ☑ **T/I** choose an environmental issue that interests you?
- ☑ **A** observe the features on social media websites and brainstorm how these features can be applied to an environmental issue
- ☑ **C** select appropriate images that reflect the significance of the environmental issue?
- ☑ **T/I** create your social media page?
- ☑ **T/I** look at another student's social media page and comment on the posts, likes, and photos?
- ☑ **K/U** show your knowledge of environmental science in your social media web page and comments?
- ☑ **A** explain how a social media web page can be used to inform many people about an issue?

An Issue to Analyze

Assessing Environmental Reports

Environmental issues often generate heated debates. Different stakeholder groups—industry, citizens, environmentalists, government—may have significantly different opinions about environmental matters. In order to get a balanced picture of an issue, you need to become a critical reader. You must ask questions like "Who is writing the article?" Is the writer an editorial writer (someone who gives an opinion), a news reporter, or a science writer? Is the writer an expert or non-expert in this subject area? You must also distinguish between facts and opinions. In this project, you will examine an article written by a specific kind of writer as assigned by your teacher. You will identify the type of article, the kinds of information presented (facts, opinions, or both), and bias. You will then analyze the writer's main arguments and predict the effect the article would have on a reader.

Issue

How does the source of information influence what is reported in an article?

Initiate and Plan

1. In a group, select an article about an environmental issue written by the type of writer assigned by your teacher.
2. Before you begin your analysis,
   - determine where the article is published. If the source is a newspaper or magazine, is it local or national?
   - determine the scope of the article. Does it report a local, national, or international event?

Perform and Record

3. Summarize the article. What is the article about? Does it report an event? Is it a commentary on an issue? Does the article report new information? Are data presented and can they be verified? Are the sources of data described?
4. Determine if different views on the topic are presented. Is a range of perspectives presented or is the article one-sided in the view it presents?

Analyze and Interpret

1. Analyze any bias you find in the article. Bias can occur in several ways:
   - introducing inaccuracies or presenting facts that are not correct
   - interpreting the data to favour one perspective over another
   - claiming information as true without having data or facts to support it omitting facts
2. What are the arguments presented in the article? Does the writer make assumptions? How are arguments manipulated?
3. Identify keywords or other clues that will help you decide the author's credentials. Identify the type of writer and his or her credentials. Were the credentials stated clearly in the article or would a reader have to do further research to find out?
4. What would you predict would be the effect of this article on a reader? Why?
5. What additional information do you think a reader should be aware of when reading this article?

Communicate Your Findings

6. Share your findings and analysis with the class, using a format of your choice. Possibilities include a video documentary, social media website, or blog. Make sure to keep your audience in mind when designing your presentation.

Assessment Checklist

Review your project when you complete it. Did you …

- ✓ **K/U** identify where the article was published and whether it was about a local, national, or international event?
- ✓ **T/I** summarize the article and identify the different viewpoints in the article?
- ✓ **T/I** analyze the article for bias?
- ✓ **A** predict what the effect of article may be on the reader?
- ✓ **T/I** identify the type of writer and what credentials she or he has?
- ✓ **C** present your findings using an appropriate format, keeping the audience in mind?

Knowledge and Understanding

Choose the letter of the best answer below.

1. Which of the following statements about humans and the environment is correct?
   a) Humans have only been a part of the natural environment since the Industrial Revolution.
   b) Humans have always been a part of the natural environmental.
   c) Humans are limited to living in only a few environments on Earth.
   d) Humans have always had a significant impact on Earth's natural resources.
   e) Humans do not depend on the same resources that other organisms do.

2. Which of the following consequences can be attributed to the Industrial Revolution?
   a) Human populations began to move from living mostly in rural areas to living in cities.
   b) Air and water pollution began to increase significantly.
   c) Conditions were created that would eventually lead to concerns for worker rights and safety.
   d) Larger, more industrialized cities developed.
   e) All these consequences can be attributed to the Industrial Revolution.

3. Which of the following is the best synonym for the term paradigm?
   a) viewpoint
   b) mindset
   c) behaviour
   d) action
   e) theory

4. Environmental science is an interdisciplinary field of inquiry. This means
   a) that it is limited to scientists in different fields
   b) that it involves the world of the public
   c) that it requires strict attention to the methods used by experimental scientists
   d) that it involves a variety of fields of study including, but not limited to, science
   e) that it is essentially the same scientific field of study as ecology

5. Which of the following is an example of a scientific habit of mind?
   a) logical thinking
   b) analytical thinking
   c) skepticism
   d) calculated thinking
   e) environmental thinking

Answer the questions below.

6. What distinguishes a sustainable environment from an unsustainable environment?

7. Briefly explain, in two or three sentences, how the law of conservation of matter and the energy laws apply to Earth and the activity of organisms on Earth, including and especially humans.

8. Use the concept of sustainability to explain the terms *paradigm* and *paradigm shift*.

9. Identify three methods that scientists use to collect data, and in one or two sentences describe each of these methods.

10. Choose an organism, and use it to explain how different parts of the environment are connected.

Thinking and Investigation

11. Plan (but do not actually conduct) an investigation to find answers to one of the questions below. Be sure to identify all variables, the data you will need to collect, and what method you will use to collect your data. Decide which format you will use to record your findings—for example, a chart.
    a) What are the abiotic and biotic parts of your schoolyard environment?
    b) How do the abiotic and biotic parts interact with each other?
    c) What human activities have had an impact on the environment?

12. a) Imagine you are an environmental scientist asked to investigate why the population of fish in a local stream has been declining. The stream is located in a rural farming area. What questions could you ask that would help you design experiments and collect data to determine the cause of the decline in the fish population?
    b) Now imagine that the stream is located within a ravine that is part of a large city. In what ways does this fact change the types of questions that you would ask to determine the cause of the fish decline? Explain your answer.

Communication

13. a) Give an example from your own life experiences that makes you feel that you are in the midst of a paradigm shift regarding the environment and sustainability.

b) Paradigm shifts involve science and technology. What examples of technology do you think may influence the paradigm shift that is taking place now?

c) What other factors do you think play important roles in paradigm shifts? Support your opinions with facts, either from your own knowledge or based on research.

14. Suppose that your community is looking for ideas to promote sustainable development in the region where you live.

a) Describe an activity that is sustainable in your community.

b) Describe an activity that is unsustainable in your community.

c) Explain how the sustainable activity is more efficient and responsible socially, economically, and environmentally.

15. The graph below shows the growth of a population.

a) Write a title and caption for the graph.

b) Infer the conditions in which the population is changing, and explain your reasoning.

c) Can the pattern of change represented by this graph continue forever? Explain why or why not.

16. Explain why each of the following is or is not an appropriate source for conducting research. Under what conditions could any or all be appropriate sources?

a) interviewing several friends for their opinions

b) using Wikipedia

c) using blogs

Application

17. Sometimes people change an area to enhance its ecosystem services. For example, wildlife officials may stock a lake with fish to provide recreation for fishing enthusiasts. In California, environmental scientists conducted a four-year study to learn more about the effects of the introduction of non-native trout to mountain lakes in the western United States. The results of the study showed that, after the introduction of the trout, there were reduced population numbers of several amphibian species and changes in the number and variety of aquatic insect species. In particular, the trout consumed aquatic insects in their very early, larval stages. Other organisms, including amphibians and other fish, also rely on insect larvae as a food source. As well, birds and bats that live near the lakes eat adult insects. The scientists concluded that all of these species must now compete with the non-native trout for food. The presence of trout was linked with a decrease in the number of birds and in the activity of some types of bats.

a) Infer how the introduction of the non-native trout affected the sustainability of the lake ecosystem.

b) Should one or two ecosystem services of an ecosystem be enhanced at the cost of some of its other services? Explain why or why not.

c) As an empowered individual, how would you respond if you knew this was happening in a lake near your home?

18. The philosopher, poet, and cultural critic, George Santayana (1862–1952), is famous for (among other things) the following quotation: "Those who cannot remember the past are condemned to repeat it."

a) In what ways does this quotation apply to your current understanding of and opinions about environmental science?

b) Use this quotation as the title for a supported opinion piece that you either write as an essay, compose as a speech, or develop as the script for a video documentary or as the actual production of a video documentary.

UNIT 2 Energy Use and Conservation

Based on this astounding satellite image from 2012, in which parts of the world are lights at night concentrated?

Do these regions of light concentration surprise you? Why or why not?

BIG IDEAS

- The impact of energy production and consumption on environmental sustainability depends on which resources and energy production methods are used.

Overall Expectations

- Assess the impact on society and the environment of the use of renewable and nonrenewable energy resources, and propose a plan to reduce energy consumption.

- Investigate methods of conserving energy and improving energy efficiency.

- Demonstrate an understanding of energy production, consumption, and conservation with respect to renewable and nonrenewable sources.

Unit Contents

Chapter 3

Energy Resources

Chapter 4

Energy Use for a Sustainable Future

Topic 1: Forms of Energy

Different terms are used to classify, describe, and distinguish a variety of forms of energy. At the simplest level, energy may be classified as either kinetic or potential. *Kinetic energy* is energy of motion. Any moving object has kinetic energy. *Potential energy* is stored energy or energy reserved for future use. For example, a motionless object (such as a book on a table) has the ability (the potential) to move—for example, as a result of being pushed off the table or being lifted above the table.

Potential energy can be stored in different forms. For example, the potential energy of a book held above a table is called gravitational potential energy. The potential energy of a book falling off a table is also gravitational potential energy. Other examples of potential energy include the elastic potential energy of a coiled spring, the chemical potential energy of a battery, the electrical potential energy of two charges separated by a distance, and the nuclear potential energy of a fuel rod used in nuclear power stations.

Different forms of potential energy, as well as kinetic energy, are summarized in **Table 1**. Forms that appear in red type may be classified as potential or kinetic, depending on a given situation. Note also the term mechanical energy. Mechanical energy is the sum of the total kinetic and potential energies of a system. For example, as a book falls off a table, its potential energy continuously changes to kinetic energy. For purposes of simplicity in describing systems with moving components, this textbook uses the term kinetic energy as a synonym for mechanical energy.

Table 1 Forms of Energy	
Kinetic Energy	**Potential Energy**
Electrical: the energy stored by a separation of positive and negative charges	Chemical: the energy contained within the bonds between atoms
Mechanical: the sum of the kinetic and potential energy of a system	Elastic: the energy stored within a stretched or compressed object such as an elastic band or a car bumper
Solar (radiant): the energy of electromagnetic waves travelling through space	Electrical: the energy stored by a separation of positive and negative charges
Sound: the energy of vibrations or disturbances of the atoms and other particles that make up matter	Gravitational : the energy due to the position of an object
Thermal (Heat): the energy of the random motion of atoms and other particles that make up an object or system	Mechanical : the sum of the kinetic and potential energy of a system
	Nuclear: the energy stored within the nucleus of an atom
	Thermal (Heat): the energy of the random motion of atoms or particles that make up an object or system

Topic 2: Energy Transformations and Converters

The change of one form of energy to another form of energy is called an energy transformation. An *energy transformation* involves some amount of energy—input energy—going into a system and some amount of energy—output energy—coming out of that same system.

An *energy converter* is a device that converts (transforms) energy from one form to another for a specific purpose. Notice in **Figure 1** that energy converters involve several energy transformations, and losses of usable energy occur with each transformation. These losses, expressed as percentages, describe the efficiency of an energy transformation. It is calculated by using amounts of energy input and output, according to the following equation: efficiency = energy output/energy input × 100%.

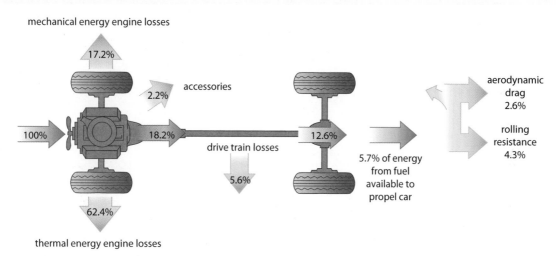

(a) Energy path diagram of typical combustion engine to transmission

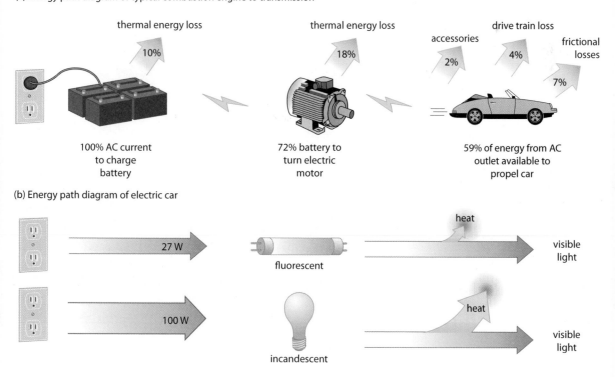

(b) Energy path diagram of electric car

(c) Energy path diagrams for incandescent and fluorescent lights

Figure 1 Energy converters transform energy into both useful and wasted (unusable) forms.

CHAPTER

3 Energy Resources

Our modern society is very dependent on a large and steady supply of energy.

What are some possible negative effects of this dependency?

Using Energy

List five items you use in your everyday life that use energy. Include items from different areas of your life, such as transportation, school, completing household chores, and/or communicating with friends. For each item, try to identify the source of energy that powers it. For example, if the item uses electricity, from which energy source is the electricity generated? Is it a fossil fuel, nuclear power, hydropower, solar power, or some other source? If you do not know the energy source, how could you find out what it is? Why might it be important for you to know the sources of energy you use every day?

In this chapter, you will

- examine the role and value of energy in society
- investigate important sources and forms of energy
- consider the advantages and disadvantages of a variety of nonrenewable and renewable energy supplies

Understanding Energy Resources

energy resource a resource that provides energy to bring about movement or change

An **energy resource** provides energy to bring about movement or change. When people talk about energy resources, in most cases they are referring to something that provides energy for humans. This includes everything from the sandwich shown in **Figure 3.1A** to the gasoline that fuels the car in **Figure 3.1B**. How much an energy resource can move or change something depends on how much energy it contains. For instance, both food and gasoline contain chemical energy. Food provides enough energy to enable you to carry out your daily activities. On the other hand, gasoline is an oil product that contains enough energy-rich molecules to move a vehicle at significant speed.

Figure 3.1 A sandwich (**A**) and gasoline (**B**) are both energy resources. However, because energy-rich molecules are concentrated in gasoline, this resource is able to bring about more movement or change.

Classifying Energy Resources

nonrenewable energy resource an energy resource that is non-replaceable once consumed

Energy resources are classified based on whether they can be renewed. The two categories are nonrenewable and renewable. **Nonrenewable energy resources** are those that require thousands or millions of years to reform, if they can at all. Since a human lifetime is so short in comparison, once these resources are used up, they are gone for good. As shown in **Figure 3.2**, fossils fuels are one type of nonrenewable energy resource. These fuels—coal, oil, and natural gas—supply about 80% of the world's energy needs. The other type of nonrenewable energy resource shown in **Figure 3.2** is uranium, which is used to generate electricity in nuclear power plants.

Figure 3.2 The two types of nonrenewable energy resources—fossil fuels and uranium.

Applying Develop a definition for the term *nonrenewable energy resource* that refers to sustainability.

Fossil Fuels
Most of Earth's coal, oil, and natural gas formed between 280 and 360 million years ago. Since these fuels formed from the remains of ancient plants, animals, and micro-organisms, they are called fossil fuels. They contain energy-rich compounds that were originally derived using energy from the Sun through photosynthesis.

Uranium
Uranium is a heavy element that is mined around the world to provide fuel for nuclear reactors. Large amounts of energy are stored in the nucleus of certain uranium atoms. This energy is released when the atoms are split through a process called *nuclear fission*.

Renewable energy resources are produced on a continual basis or can be replenished fairly quickly. They are not at risk of being used up over the course of a human lifetime. As shown in **Figure 3.3**, renewable energy resources include solar, wind, hydro, tidal, geothermal, and biomass.

renewable energy resource an energy resource that is available on a continuous basis

Figure 3.3 The potential use of renewable energy resources is virtually unlimited.

Solar
Solar energy is energy produced by the Sun. The Sun is expected to produce solar energy for about another five billion years.

Wind
Wind energy is the energy of moving air. In the form of tornadoes and hurricanes, wind has enough energy to destroy buildings and lift vehicles.

Hydro
Hydro energy refers to the energy of falling or moving water. A river or waterfall is a source of hydro energy.

Geothermal
Geothermal energy is thermal energy from Earth's interior. Evidence of this energy includes hot springs, geysers, and volcanoes.

Biomass
Biomass is the chemical energy stored in organic materials of living or dead organisms or their byproducts. These include plants and plant materials, as well as manure. Like the energy in fossil fuels, biomass energy originally came from the Sun.

Wave and Tidal Energy
This is the energy of the regular movement of incoming and outgoing waves and tides.

Mini-Activity 3-1 **The Perfect Energy Resource**

In your opinion, what are the properties of the "perfect" or ideal energy resource, and why? In small groups, make a list of at least eight characteristics of an energy resource that you think makes it the best for all or most possible uses. Note: You are *not* being asked to name the renewable or nonrenewable resource that you think is best. Your task is to create a list of the most desirable properties that an energy resource should have.

Share your group's list with other groups in the class. See if your class can reach a consensus on all the properties that make the ideal energy resource. Keep your list (or lists) handy during this unit, so you can compare the advantages and disadvantages of each energy resource you investigate with the ideal.

An Overview of Energy Consumption by Humans

Every form of life and all societies require a constant input of energy. If the flow of energy through organisms or societies ceases, they stop functioning and begin to break down. Some organisms and societies are more energy efficient than others. In general, complex industrial civilizations, like those throughout North America, South America, Europe, and Asia, use more energy than hunter-gatherer or subsistence agricultural societies. If modern societies are to survive, they must continue to use energy. However, they may need to change their patterns of energy consumption as current sources become limited.

The use of energy resources by humans and human societies has changed in significant ways over time. Many energy-related discoveries and innovations have benefited society. All have taken their toll on human health and the environment. **Figure 3.4** shows an overview of the production and use of energy resources by humans on a global scale.

Figure 3.4 The time line shows energy production and use over the course of human history.

1 000 000–500 000 B.C.E. Fire-starting technology allowed our ancestors to use wood to generate both heat and light. Food could now be cooked and our ancestors could survive in colder environments.

4000–2500 B.C.E. The energy of the wind was first harnessed in ancient Egypt. Masts and sails on boats made of reeds enabled people to travel up and down the Nile River quickly. By 2500 B.C.E., sailboats made of wood enabled the Egyptians to sail into the Mediterranean Sea and establish new trade routes.

1 500 000 B.C.E. Our ancestors used only food and sunlight as energy resources.

5000 B.C.E. The energy of animals was being used for agriculture purposes. More food could now be produced with less human effort. This allowed the human population to expand, and more people began to live together in small settlements.

200 B.C.E.–200 C.E. Wood was now burned on a large scale for heat. This resulted in the earliest documented environmental health problems. During the ancient Roman Empire and the Chinese Han dynasty, wood was burned in large amounts to work metal and to heat homes and baths.

Answer the questions below based on the information given in **Figure 3.4** and the data shown in **Table 3.1**.

Table 3.1 Daily Consumption of Energy (per person)($\times$ 1000kJ)						
	1 000 000 Years Ago	100 000 Years Ago	5000 Years Ago	1000 Years Ago	100 Years Ago	40 Years Ago
Transportation				4.2	58.8	264.6
Machinery*			16.8	29.4	100.8	382.2
Heating		4.2	16.8	50.4	134.4	277.2
Food**	8.4	8.4	16.8	25.2	29.4	42
Total	8.4	12.6	50.4	109.2	323.4	966

* Agricultural and industrial ** Including animal feeds

1. 100 000 years ago, humans were hunting and using fire as a source of heat for warmth and to cook food. How is this change in the use of energy reflected in the data?

2. 5000 years ago, humans began growing crops and using animals as a source of energy to help plough fields. How did this affect the amount of energy used by humans?

3. What events occurred in the 1700s and 1800s that would account for the increase in energy use in transportation, machinery, and heating by humans as shown in the "100 Years Ago" column of the table?

4. The last column of the table shows data from 40 years ago. How do you think these numbers would compare to numbers from today? Explain your reasoning.

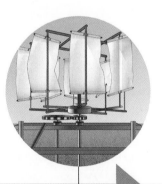

1698–1900 C.E. Steam engines spearheaded the Industrial Revolution and commercial-scale coal mining and powered locomotives, machinery, and ships. Air pollution also increased dramatically. Coal smoke mingled with fog and blanketed the city of London. In 1873, nearly 300 lives were lost to bronchitis.

500–900 C.E. The earliest known windmills were developed in Persia (now called Iran) to grind grain and pump water. More food and water became available with less effort.

1859 C.E. The first continuously operating gasoline-powered engine was built by French engineer J. J. Étienne Lenoir. This paved the way for the development of early automobiles fueled by gasoline. Soon society would be transformed by fast, affordable personal transportation. By the 1950s, these machines became linked to smog and other air pollutants.

1945 C.E. The ZEEP nuclear reactor was built in Canada. It was one of the first nuclear reactors in the world. It was an early phase in the dream of a world with no shortage of clean, affordable energy. Since then the problems of harmful nuclear wastes and reactor accidents have tempered this dream significantly.

Case Study Whale Oil: Historical and Traditional Uses

Whale oil comes from the blubber of baleen whales such as bowhead and right whales and from the head cavities of sperm whales. Whale oil was used as fuel in oil lamps, and to make products like candle wax, industrial soap, lubricants, and margarine. Use of whale oil for lighting declined in the second half of the nineteenth century as cheaper and longer-lasting fuels like coal gas and kerosene were developed.

Regulations on Whaling

Commercial whaling, which harvested whales for meat and oil, began in the seventeenth century. As whaling technology became more effective, huge numbers of whales were caught each year and many species were nearly hunted to extinction. With whale stocks dangerously low, the international community began putting rules in place to protect whales and impose quotas. The International Whaling Commission (IWC), established in 1946, finally banned commercial whaling in 1986.

Canada was initially a member of the IWC, but when the federal government banned traditional subsistence whaling in 1980, the Inuit protested strongly. Canada left the IWC in 1982 and signed treaties with the Inuit to grant them limited whaling rights. The Inuit in Canada follow strict limits on the number of whales they can hunt and must have a federal fisheries license. The IWC now allows some subsistence whale hunting as well. For example, it grants quotas for Inuit whaling in Alaska and Greenland.

Whaling in Inuit Culture

The ancestors of the modern Inuit, the Thule culture, used harpoons and lances to hunt bowhead whales from a boat called an *umiak* (similar to a kayak). The Thule culture spread out from Alaska, across the Canadian North and all the way to Greenland. Whaling was a dangerous undertaking, so a successful whale hunt was cause for community celebration. No part of the whale was wasted—meat and blubber were used for food, oil for light and heat, and bone for tools and the construction of boats and shelters.

Whaling still has social, ceremonial, and spiritual importance for the Inuit. The traditional whale oil lamp, the *qulliq*, shown below, appears on the coat of arms for Nunavut. Today, modern technologies are used in indigenous whale hunting to minimize distress to the animals. However, traditional Inuit attitudes of sustainability still remain, including hunting only for need, avoiding waste, and always showing respect for the animals taken.

The traditional lamp of the Inuit, the *qulliq*, burns whale or seal oil.

Research and Analyze

1. Research the current state of whaling in Canada. What is the government's latest position in regards to Inuit whaling and the IWC?

2. Beginning in the eighteenth century, European whalers made contact with, and later employed, Inuit in the whale hunt. Research how this interaction affected Inuit culture.

Communicate

3. Some conservationists believe that traditional practices do not justify whale hunting. Decide if you are for or against whaling by the Inuit in Canada. In a small group, prepare for a debate. Write your opening remarks and outline 3 to 4 points that support your position.

Summary

- An energy resource provides energy to bring about movement or change.

- Energy resources are classified based on their ability to be renewed. The two categories are nonrenewable and renewable.

- Nonrenewable energy resources are those that require thousands or millions of years to reform, if at all. Fossils fuels—coal, oil, and natural gas—are one type of nonrenewable energy resource. The other type of nonrenewable energy resource is uranium, which is used to generate electricity in nuclear power plants.

- Renewable energy resources are produced on a continual basis. Renewable energy resources include solar, wind, hydro, tidal, geothermal, and biomass.

- The use of energy resources by humans and human societies has changed in significant ways over time. Many energy-related discoveries and innovations have benefited society. All have taken their toll on human health and the environment.

Review Questions

1. What is an energy resource? K/U

2. Using examples, distinguish between renewable and nonrenewable energy sources. K/U

3. Use a Venn diagram to compare and contrast biomass energy with fossil fuel energy. K/U C

4. Among the earliest energy sources used by humans were wood, wind, and coal. Compare how humans used them and suggest at least one advantage and one disadvantage of each. K/U

5. Electricity generated in Ontario comes mostly from nuclear power, hydropower, and the burning of natural gas. Which of these is/are renewable sources of energy? K/U

6. Earth's interior will cool over the next few billion years. In light of this, do you think that geothermal energy is correctly classified as renewable? Explain your reasoning. T/I A

7. Using data from Table 3.1, plot a graph illustrating the change in *total* daily human energy consumption over time. Plot energy consumption along the *y*-axis and time along the *x*-axis. What factors account for the rise in the use of energy from 100 years ago to 40 years ago? T/I C

8. Imagine there was a large-scale power failure that left your province without electricity for two weeks during the summer months. A C

   a) In what ways would your daily life be affected?

   b) What would be the most serious consequences for you and for your community?

   c) What alternative energy supplies, if any, could be used?

   d) How might the problems be different if the event took place in January?

9. Human energy consumption has increased dramatically over time. As shown in the graph below, so has the size of the human population. The major factors resulting in this recent explosive growth are modern medicine, access to clean drinking water, and modern agriculture. A

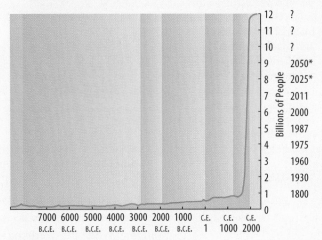

*predicted according to one projection scenario

   a) Which of these three factors was most influenced by the availability of new and abundant energy supplies? Explain you reasoning.

   b) What impacts do you think this trend has had and will have on total human energy consumption?

10. What would be some of the most significant consequences for society if fuels were unavailable for transportation? How might your community change? A C

Nonrenewable Energy Resources

Fossil Fuels

Fossil fuels, including coal, oil, and natural gas, have properties that make them a convenient energy choice. For example, they are very concentrated sources of chemical energy. This makes gasoline a great source of energy for motor vehicles and many other modes of transportation. Fossil fuels are also relatively inexpensive to produce. As a result, they provide an affordable and reliable source of electrical and thermal energy for homes, businesses, and industry. In view of these benefits, it is clear why this nonrenewable energy resource supplies more than 80% of the world's energy. However, the use of fossil fuels also comes with significant disadvantages, as discussed throughout this section.

Coal

Coal forms over millions of years from ancient deposits of terrestrial plants buried under sediments. Its formation is described in **Figure 3.5**. Canada is particularly rich in coal, with deposits found in Nova Scotia, New Brunswick, Ontario, Saskatchewan, Alberta, and British Columbia. In Canada, coal is used mainly to generate electrical power and for industrial uses, such as steel production.

Figure 3.5 Millions of years ago, dense swamps covered Earth. Masses of plant debris collected under the water. Over time, the plant material was submerged and covered by seas and sediment. The mass of the sediment and water compressed the organic matter, and the water evaporated. The end result is the coal we use today.

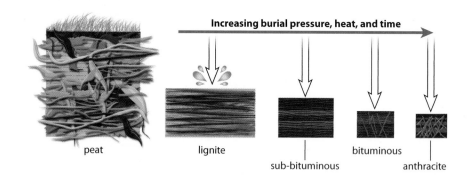

Increasing burial pressure, heat, and time

peat   lignite   sub-bituminous   bituminous   anthracite

Since coal formation involves burying plant material under layers of sediment, coal must be mined from below Earth's surface. The two methods of mining coal are *surface mining* and *underground mining*. In surface mining, the soil and rock over a vein of coal are removed in order to access the coal below. Surface mining is sometimes called *strip mining*. When the overlying soil and rock layers are too thick, underground mining is the method of choice. In this process, coal is extracted from a mine that is dug deep below the surface.

Coal is abundant and relatively inexpensive to mine, although underground mining is a dangerous occupation. On a local level, coal mining often causes water pollution and land degradation. Mining coal also results in a dangerous condition known as black lung disease, shown in **Figure 3.6**. This respiratory condition occurs when fine coal dust accumulates in miners' lungs over many years. Coughing, shortness of breath, and respiratory failure may result.

Regionally, burning coal is also a major source of air pollutants. It is linked to serious health risks, including premature deaths. On a global scale, burning coal is the world's largest source of greenhouse gases. Due to these disadvantages, use of coal is declining in parts of Canada. For instance, as of 2013, Ontario was in the process of phasing out the use of coal-fueled power plants. Canada still mines coal for export to other countries, however.

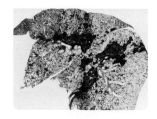

Figure 3.6 The large dark spots in this lung resulted from breathing in coal dust while mining.

Oil

In 2011, the world consumed more than 610 million barrels of oil a week. Stacked on top of each other from Earth's surface, this number of barrels would reach past the Moon. Petroleum, also known as crude oil, or simply oil, has its origins with microscopic organisms that thrived in ancient seas. The remains of these organisms were buried in sediment on the sea floor. In a process similar to coal formation, heat and pressure over a period of millions of years transformed this material into oil.

Today, oil is found mainly in underground deposits of sedimentary rock called *reservoirs*. Reservoirs are accessed through land or offshore drilling rigs such as the one shown in **Figure 3.7**. After it is excavated, oil is pumped to the surface and transported by pipeline or oil tanker to a refinery. Refining involves the physical separation and processing of oil to produce substances such as gasoline, plastics, asphalt, and motor oil. Oil is also found on Earth's surface in oil sands. Most of Canada's oil is in oil sands in Alberta, where heavy oil, or bitumen, is tightly bound to fine particles of sand or clay.

 Thought Lab 3A, Extraction of Nonrenewable Energy Resources, on page 83

Figure 3.7 The *Hibernia Platform* off the coast of Newfoundland is the world's largest oil rig. In Canada, offshore oil is found beneath the sea floor of the Pacific, Atlantic, and Arctic Oceans.

 Thought Lab 3B, Considering a Controversy: The Alberta Oil Sands, on page 84

Oil extraction and use has many effects on human and ecosystem health, some of which are listed below.

- Extracting and processing oil from oil sands degrades large areas of land. These processes are energy intensive and require large quantities of water and chemicals. This causes significant water and air pollution. Scientists have found increases in cancer-causing pollutants in lakes within 100 km of the Alberta oil sands.

- Leaks from pipelines and accidents on oil rigs and tankers also release oil into the environment, often with deadly consequences to aquatic organisms and sea birds.

- When burned, oil produces a variety of pollutants. These contribute to the greenhouse effect and the formation of smog.

- Pipelines used to transport oil on land, such as the one shown in **Figure 3.8**, can also interfere with normal ecosystem activities, such as migration or hunting prey. The pipelines block routes that animals normally take to pursue these activities.

Figure 3.8 An oil pipeline

Analyzing What are some advantages and disadvantages of oil pipelines?

Natural Gas

methane hydrates natural gas that is also trapped within crystals of frozen water

Natural gas forms, and is associated, with both oil and coal deposits. **Figure 3.9A** shows how this gas is trapped above crude oil in rock underground. Natural gas is also trapped within crystals of frozen water as **methane hydrates**. Methane hydrates, like those in **Figure 3.9B**, have been found in very large deposits beneath Arctic permafrost and deep in the ocean. Once natural gas is brought to the surface, it is processed to separate it from contaminants such as hydrogen sulfide and water vapour.

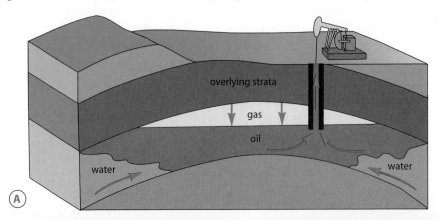

Figure 3.9 Diagram (A) shows natural gas in reservoirs deep underground. Methane hydrates (B) are a promising natural gas resource, but they are also fragile. Scientists suggest that improper extraction could disrupt the sea floor and release methane from other hydrates nearby. Such a large release could have serious consequences for climate change.

Canada is a significant producer of natural gas. **Figure 3.10** shows the location of Canada's natural gas reserves, as well as the locations of other fossil fuel resources. Natural gas is used for gas furnaces, hot water heaters, stoves, and dryers in homes and businesses. It is also used to generate electricity, produce fertilizer, and it can fuel vehicles that are modified for this purpose.

hydraulic fracturing involves pumping fluid into the ground under high pressure to cause layers of shale deep underground to fracture

Like other fossil fuels, natural gas is a nonrenewable resource. It is also a source of greenhouse gases. Methane, the main component of natural gas, is a potent greenhouse gas. Escape of methane gas during the drilling, extraction, and transportation of natural gas is a significant concern. A relatively new extraction process called **hydraulic fracturing**, also known as fracking, has raised concerns among scientists and citizens about its effects on both the environment and human health. Fracking is an attractive extraction option, however, because it permits access to large new reserves and lower prices for natural gas.

Burning natural gas also releases carbon dioxide. However, burning natural gas releases much less of this greenhouse gas than oil and almost half as much as coal.

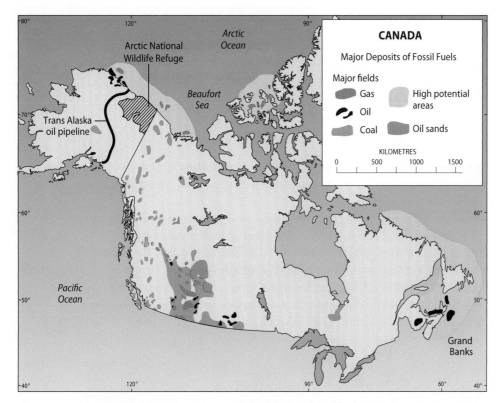

Figure 3.10 This map shows the locations of Canada's main fossil fuel reserves.

Advantages

- Coal and natural gas are relatively inexpensive to access and process.
- Coal and oil are highly concentrated energy sources. Gasoline is an ideal transportation fuel.
- Canada has large reserves of coal, oil, and natural gas.
- Natural gas causes less environmental damage when extracted and used. It releases less carbon dioxide than coal and oil when burned and contributes less to the greenhouse effect.

Disadvantages

- Supplies of fossil fuels are limited and nonrenewable. Coal reserves are estimated to last 150 years at current consumption rates.
- Burning coal and oil products releases harmful pollutants into the air.
- Fossil fuels release large amounts of greenhouse gases, which contribute to climate change.
- Fossil fuel extraction pollutes waterways and results in land degradation. Oil spills can cause environmental damage on an especially large scale.
- Like oil pipelines, cross-country natural gas pipelines disrupt ecosystems.

Pause and Reflect

1. Explain how coal extraction and combustion affect human health.

2. How do oil and natural gas pipelines affect ecosystems?

3. Critical Thinking Which fossil fuel is the least harmful to the environment? Explain your reasoning.

Case Study Hydraulic Fracturing

What Is Fracking?

Hydraulic fracturing, also known as fracking, is a relatively new method of extracting oil and gas resources. The first fracking experiments were conducted in the mid-20th century, but the process did not become economically viable until the late 1990s.

As shown in the illustration below, fracking involves pumping fluid into the ground under high pressure to cause layers of shale deep underground to fracture. This gives access to hydrocarbon sources such as natural gas that would be inaccessible by conventional drilling techniques. The injected fluid is composed of water (90%), sand (9.5%), and chemicals (0.5%). The sand grains move into the cracks and keep them propped open while still allowing the hydrocarbons to flow through. The chemical additives—some of which are known toxins or carcinogens—have a variety of roles, including controlling fluid viscosity, reducing friction, and acting as disinfectants.

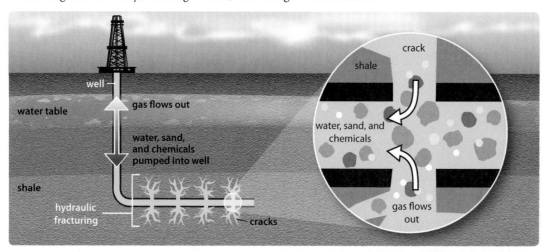

Economic Benefits

Fracking technology, along with new directional drilling techniques, allows access to fuel sources that were previously inaccessible or not economically feasible to extract. Canada is currently the world's third-largest producer of natural gas. However, as conventional sources deplete, the industry has been shifting to "unconventional" sources of gas—those that come from geological formations that are more difficult to produce. The use of fracking to harvest these resources means not only economic benefits across the country, but also that Canada is not dependent on importing fuels. Fracking can provide jobs in the gas industry and help ensure security for Canada's future energy needs.

Environmental Concerns

Fracked wells may use flaring (a controlled burn of gases) and venting, both of which release pollutants and greenhouse gases into the air. Vast amounts of water are required for fracking—a single well may use up to 30 million litres of water over its lifetime. Once the rock layer has cracked, fluids known as "flowback" return to the surface and can contaminate ground water.

Flowback is composed of used fracking fluid and water naturally present in the rock formations that have been freed by the fracking process. This liberated water may contain dissolved radioactive elements. Flowback must be treated either on site or by local water treatment plants. Sometimes flowback is pumped back deep into the ground. This process has been known to cause minor earthquakes.

Fracking is still a relatively new and controversial process. As with many environmental issues, opinions vary widely. Read the points of view presented below, and then answer the questions.

Tap water contaminated with methane

Farming Family:
"We had excellent water for eight years, and then in 2005 the production from our house well dropped so much that it could not replenish itself.... Since we had another high-volume well that was plentiful for the livestock, we decided to pipe that into our house. Within a couple days, it became apparent that there was something in the water. A lab test revealed methane, ethane, propane, butane, and isobutane."

Canadian Association of Petroleum Producers: Guiding Principles:
"We will support the development of fracturing fluid additives with the least environmental risks. We will continue to advance, collaborate on, and communicate technologies and best practices that reduce the potential environmental risks of hydraulic fracturing."

Canadian Activist Group:
"The recent announcement of voluntary 'guiding principles' confirms the fracking industry is worried about the growing opposition. But these voluntary guidelines set by industry are classic greenwashing."

Local Citizens' Group:
"Our group would just like to put things on hold and make sure that it is safe.... The oil is not going anywhere—it has been there millions of years. Why not wait?"

Energy Analyst:
"Shale gas provides a new opportunity to meet rising global energy requirements. Before the emergence of shale gas as a major new source of energy supply in the mid- to late-2000s, energy prices were rising sharply worldwide and analysts were anticipating such severe shortages.... Virtually overnight, the shale gas revolution has reversed these global energy scarcity woes."

Research and Analyze

1. Choose one of the points of view and express it in your own words. What bias might this point of view have?

2. Research and take notes on one aspect of fracking that interests you, such as fracking techniques, economic benefits, or environmental concerns.

Communicate

3. Take on the role of one of the points of view. Gather together with others in your class with the same point of view. Hold a town-hall debate on the issue of fracking in your region.

Uranium and Nuclear Energy

fission the action of splitting an atom into two or more parts

Uranium consists mostly of the isotope uranium-238, with about 0.7% uranium-235—the isotope that fuels nuclear reactors. Inside a nuclear reactor, uranium-235 atoms are split. This process, **fission**, releases an enormous amount of thermal energy. Splitting one atom sets off a chain reaction that causes more atoms to split. A nuclear reactor contains and controls fission reactions and the energy they release, as shown in **Figure 3.11**. Canada developed its own reactor type, CANDU, which uses natural uranium, whereas most of the world's other reactors require fuel enriched in uranium-235. At this time, there are 18 CANDU reactors operating in Canada and about 20 in other countries.

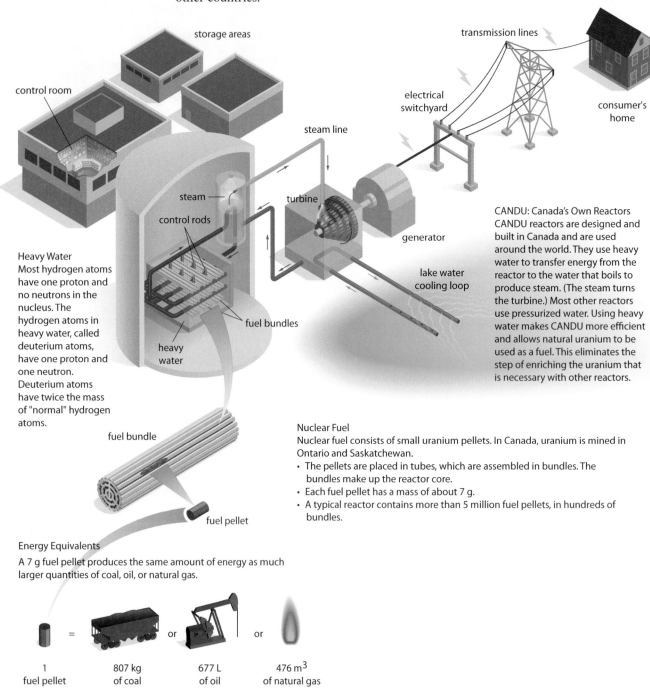

storage areas

control room

transmission lines

electrical switchyard

consumer's home

steam line

steam

control rods

turbine

generator

lake water cooling loop

Heavy Water
Most hydrogen atoms have one proton and no neutrons in the nucleus. The hydrogen atoms in heavy water, called deuterium atoms, have one proton and one neutron. Deuterium atoms have twice the mass of "normal" hydrogen atoms.

heavy water

fuel bundles

CANDU: Canada's Own Reactors
CANDU reactors are designed and built in Canada and are used around the world. They use heavy water to transfer energy from the reactor to the water that boils to produce steam. (The steam turns the turbine.) Most other reactors use pressurized water. Using heavy water makes CANDU more efficient and allows natural uranium to be used as a fuel. This eliminates the step of enriching the uranium that is necessary with other reactors.

fuel bundle

fuel pellet

Nuclear Fuel
Nuclear fuel consists of small uranium pellets. In Canada, uranium is mined in Ontario and Saskatchewan.
- The pellets are placed in tubes, which are assembled in bundles. The bundles make up the reactor core.
- Each fuel pellet has a mass of about 7 g.
- A typical reactor contains more than 5 million fuel pellets, in hundreds of bundles.

Energy Equivalents
A 7 g fuel pellet produces the same amount of energy as much larger quantities of coal, oil, or natural gas.

| 1 fuel pellet | = | 807 kg of coal | or | 677 L of oil | or | 476 m^3 of natural gas |

Figure 3.11 In a nuclear power plant, thermal energy released from nuclear fission boils water to steam. The steam turns a turbine connected to a generator to produce electrical energy.

Generating electricity in a uranium-powered nuclear plant releases no carbon dioxide, unlike a coal-burning power plant. Similarly, the nuclear plant produces very little air and water pollution when compared to burning fossil fuels. Despite these advantages, the production of nuclear energy remains socially and politically controversial due to its potential effects on the environment. These effects occur during the entire life cycle of nuclear fuel, from mining and processing uranium to reactor operations to waste disposal.

Mining and Processing

In uranium mining, a great deal of potentially harmful waste material is left behind after the uranium is removed. Enormous piles of low-level radioactive mine wastes and abandoned mill tailings are two of the disposal problems created by the nuclear fuel cycle. **Radioactive material** is any material that exhibits radioactivity. These materials occur naturally and can be produced technologically. Governments write legislation, in consultation with science experts, that details levels that are considered safe or unsafe. **Tailings** are mining waste left after the mechanical or chemical separation of minerals from crushed ore. The production of 1000 tonnes of uranium fuel typical generates 100 000 tonnes of tailings and 3.5 million litres of liquid waste. Some of this material may escape and be carried by wind or washed into streams, contaminating areas far from its original source. In addition to these wastes, tonnes of other low-level wastes, such as contaminated tools, clothes, and building materials, also need to be dealt with.

radioactive material any material that exhibits radioactivity

tailings mining waste left after the mechanical or chemical separation of minerals from crushed ore

 Case Study: Nuclear Accidents and Public Safety Concerns, on page 72

Waste Disposal After Fission

After uranium has undergone fission, the materials that remain are highly radioactive for thousands of years. These materials must be stored safely and kept isolated from the environment. When a nuclear reactor is shut down, a large amount of radioactive material must also be dealt with. The advantages and disadvantages of uranium and nuclear energy are summarized below.

Advantages
- Uranium fission produces low levels of pollution and greenhouse gases.
- A small amount of uranium produces a large amount of energy.
- Canada has expertise in nuclear reactor technology.

 Case Study: The Public and Nuclear Waste, on page 260

Disadvantages
- Disposal of radioactive nuclear waste is an unresolved problem.
- Nuclear plant accidents can release radioactive materials.
- The public has many concerns about the safety of nuclear reactors.
- Nuclear reactors are costly to build and maintain.

Mini-Activity 3-3 — Thorium Reactors

Use the following questions to guide your research about thorium reactors.

- How are they different from nuclear reactors that use U-235?

- What are some advantages and disadvantages?

- What controversies surround thorium reactors?

- Are thorium reactors being tested and/or used as a source of energy in the world today? Where?

Write a summary report on the results of your research.

Case Study   Nuclear Accidents and Public Safety Concerns

Nuclear isotopes produce radiation that can damage DNA and other biological molecules, leading to mutations and cancer. The potential biological effect of a radiation dose is measured in sieverts (Sv). There are many naturally occurring radioactive isotopes in sources such as minerals, food, soil, water, and cosmic rays from beyond the atmosphere. Medical applications such as X-rays also add to the radiation to which people are exposed. Radiation from these sources is 3 mSv to 6 mSv per year. Radioactive isotopes can contaminate soil and water, where they may persist for many years. People can experience an accumulated dose if exposed to radiation over a long period of time, so health problems may appear later in life.

Of the approximately 430 nuclear plants currently operating, most have done so safely. There have been three major accidents that have generated public concern over their safety.

Factsheet—Three Mile Island
March 28, 1979
TMI Unit 2, Pennsylvania, USA

What Happened

When a malfunction caused the main cooling water pump to stop working, heat could not be removed from the system and the reactor automatically shut down. As pressure began to increase, a relief valve was opened, but it failed to close. Cooling water poured out and the reactor overheated. The operators had no instrument to show the core's coolant level, so their actions made the problem worse. A severe core meltdown occurred and a small amount of radioactive gas was released.

Short-term Impacts

Misinformation led to confusion and panic among the public. Pregnant women and young children were ordered to evacuate. However, the average dose of radiation to the 2 million people in the area was very small (0.01 mSv).

Long-term Impacts and Lessons Learned

There were no long-term effects on human health or the environment because most of the radiation was contained. The accident inspired major improvements in the nuclear industry: advancements in plant design, a greater emphasis on safety and regular inspections, and the establishment of organizations dedicated to sharing expertise internationally.

Factsheet—Chernobyl
April 26, 1986
Unit 4, Chernobyl, Ukraine

What Happened

Operators disabled the reactor's automatic shutdown and cooling mechanisms before performing a test. During the test a dramatic power surge occurred. Operators tried to insert control rods to stop the reactor, but it was too late. Explosions threw out debris from the core. This material caught fire and burned for 10 days, leading to the main release of radioactive material.

Short-term Impacts

Chernobyl was the largest nuclear accident ever, releasing 400 times more radioactive material than the bombing of Hiroshima. Thirty-one workers were killed. Airborne material was carried by wind over a large area of Ukraine, Belarus, Russia, and parts of Europe. Within a radius of 30 km, 116 000 people were evacuated from their homes.

Long-term Impacts and Lessons Learned

The main long-term effect was an increase in thyroid cancer, especially in exposed children. Anxiety over the radiation effects and relocation caused major stress for residents of the region. Analysis of the Chernobyl accident has led to improvements in reactor safety, particularly in the Soviet-designed plants of Eastern Europe.

Factsheet—Fukushima
March 11, 2011
Daiichi Units 1–3, Fukushima, Japan

What Happened

On March 11, 2011, a major earthquake occurred off Japan's northeast coast. At the nearby Daiichi plant only three of six reactors were operating. The reactors shut down as designed and sustained no significant earthquake damage. However, a 15 m tsunami flooded the site, causing a loss of power. The flooding also disabled the backup generators and the mechanisms for removing excess heat. Most of the three cores melted in a few days and several hydrogen explosions occurred. Water in the site's waste storage pools started to boil away, causing fuel rods to melt and burn. Official "cold shutdown" was not achieved for nine months.

Short-term Impacts

There were no deaths due to the nuclear accident, but 100 000 people had to be evacuated from their homes. Although the fuel was essentially contained, radioactive substances leaked out into the air and water. As well, elevated radiation levels were found in some local foods and water sources. Nearly 20 000 people were exposed to radiation while conducting clean-up efforts on the site. A total of 167 workers received radiation doses over 100 mSv, of whom 6 received more than 250 mSv.

Long-term Impacts and Lessons Learned

Because the accident occurred recently, the long-term effects are not yet known. However, Japan has halted any expansion of nuclear power and is conducting extensive safety assessments of all existing reactors. A number of countries have since announced plans to reduce or abandon nuclear power and turn toward renewable energy sources.

The Fukushima reactors are an older design that needs electrical power for cooling during an emergency shutdown. Even though the reactors withstood the earthquake itself, the tsunami's disabling of the power grid and backup generators led to a serious nuclear accident.

Analyze and Conclude

1. Analyze the circumstances that led to each of the three nuclear accidents. Consider the roles of equipment malfunction, design flaws, human error, and natural disasters. What can you conclude about the safety of nuclear energy production and nuclear reactors by analyzing these factors?

2. Use point form to summarize how the local people and environment were affected by one of the accidents described. How might the circumstances have affected the mental health and social conditions of the people in the region?

Communicate

3. Imagine that your province has announced plans to expand nuclear power generation. Write a short letter to your member of provincial parliament or member of the legislative assembly to express your support for or concerns about public safety in regards to nuclear power.

Summary

- Fossil fuels, including coal, oil, and natural gas, have properties that make them a convenient energy choice.

- Although coal is abundant and relatively inexpensive to mine, extracting it and using it have disadvantages. On a local level, coal mining often causes water pollution and land degradation. Regionally, burning coal is also a major source of air pollutants. On a global scale, burning coal is the world's largest source of greenhouse gases.

- Oil extraction and use has many effects on human and ecosystem health. Extracting and processing oil from oil sands degrades large areas of land. When burned, oil produces a variety of pollutants. These contribute to the greenhouse effect and the formation of smog.

- The production of nuclear energy remains socially and politically controversial due to its potential effects on the environment. These effects occur during the entire life cycle of nuclear fuel, from mining and processing uranium to reactor operations to waste disposal.

Review Questions

1. Explain how the energy stored in fossil fuels came from the Sun. **K/U**

2. If the same geological processes that form fossil fuels are still taking place today, why are fossil fuels considered a nonrenewable resource? **K/U**

3. What are methane hydrates? How do they differ from other fossil fuel deposits? **K/U**

4. Why is coal used primarily to generate electricity while oil is used primarily to produce transportation fuels? **K/U**

5. Make a table to compare and contrast the environmental hazards of the use of coal, oil, and natural gas as primary energy sources. Consider the following for each energy source. **K/U** **C**

   a) the extraction methods
   b) processing and transportation
   c) effects of use on sustainability

6. If we had enough fossil fuels to last thousands of years many scientists would still recommend cutting back on their use. Why? **T/I** **C**

7. When operated safely, nuclear power plants release only very small quantities of pollutants. Why, then, are some people opposed to the construction of more nuclear reactors? **K/U**

8. The energy content in litre of gasoline is equivalent to the energy content of more than 30 000 AA batteries. **A**

   a) If one litre of gasoline can power a car over a distance of about 14 000 metres, how far could 34 AA batteries power the same car?
   b) Compare the approximate costs of these two options.
   c) Based on your answer to part (b) would you consider gasoline to be expensive?

9. Today, the high-level radioactive waste from Ontario's nuclear generators remains stored on site in canisters placed in large swimming pool–sized tanks of water. One hope is that this waste will eventually be suitable to be stored safely and permanently deep underground. Brainstorm possible challenges inherent in this approach. Consider the following. **A** **C**

   a) For how many years would the waste have to remain "safe"?
   b) Where do you think such a facility could or should be built?
   c) Would you agree to have the waste buried in your region if experts said it was safe? Why or why not?

10. The Deepwater Horizon accident, shown below, resulted in deaths of 11 people and is one of the largest oil spills in history. Research this accident and answer the following questions. **A** **C**

    a) What was the primary cause of the accident?
    b) Why was it so difficult to stop the leak?
    c) How long did it take for engineers to finally "cap" the well?
    d) In terms of lawsuits and fines, what happened to the companies involved?

Renewable Energy Resources

Solar Energy

Solar energy is produced and given off by the Sun. There are three main methods used to capture solar energy: passive solar collection, active solar collection, and photovoltaic cells. The first two gather solar energy and convert it into thermal energy. The thermal energy is then used directly for heating, or it can be indirectly converted into electricity. The third method uses a technology called *photovoltaic (PV) cells*. This method converts solar energy into electrical energy directly.

solar energy energy produced by the Sun

Passive Solar Collection

Passive solar collection transforms solar energy into thermal energy without any special devices. A greenhouse or car with closed windows are common examples of passive solar collectors. The glass windows let solar energy pass through. The inside air and surfaces absorb the energy and reradiate it as thermal energy. Since this energy cannot pass back out through the glass, the interior air heats up.

This same process is used to heat homes and other buildings, as shown in **Figure 3.13**. Sunlight entering windows during the day is absorbed and transformed into thermal energy. Thick, well-insulated walls prevent the energy from escaping. In some cases, these buildings also have a large water tank or concrete interior walls and floors to absorb excess thermal energy during the day and release it slowly at night.

> **Did You Know?**
> "I'd put my money on the Sun and solar energy. What a source of power! I hope we don't have to wait [until] oil and coal run out before we tackle that."
>
> —*Thomas Edison, American inventor*

Figure 3.13 Homes that collect solar energy passively have large south-facing windows and well-insulated walls and attics.

Applying What other features of the home itself and the land surrounding it can help keep homes warm in winter and cool in summer?

Active Solar Collection

Active solar collection concentrates solar energy. The simplest collectors use ground or rooftop systems such as the one shown in **Figure 3.14** for heating. Larger systems use mirrors to concentrate solar energy to generate electricity. Active solar-collector systems are most commonly used to provide heat energy for water heaters, pools, and homes.

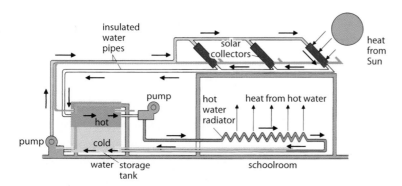

Figure 3.14 A simple active solar collector consists of a solar collector, a pump, a heat storage system, and a system of pipes to move the heat from one place to another. Energy from the Sun heats water that circulates through pipes. The hot water is pumped to the location that is to be heated, such as the schoolroom shown here. The cold water is pumped back through the pipes so it passes through the solar collectors, once again being heated by energy from the Sun.

Photovoltaic Cells

Photovoltaic cells use solar energy to generate a current through a circuit, much like a battery does. These cells are made of thin layers of silicon crystals. Only about 10% to 20% of absorbed solar energy is transformed into electrical energy. One application of PV cells is in consumer products that need small amounts of direct current. More than one billion hand-held calculators and several million watches, portable lights, and battery chargers use PV cells for power. PV technology is also widely used in situations or regions where connection to an existing electrical grid is impossible or too expensive. Examples include pumping water for small-scale remote irrigation, residential uses in remote villages, emergency radios, and orbiting satellites.

Until recently the high cost of manufacturing solar cells and their low efficiency limited their potential large-scale applications. However, the past decade has seen remarkable advances in manufacturing technologies fueled by a dramatic increase in demand. As a result prices continue to decline as the production of solar cells grows rapidly. In recent years, the amount of PV power installed worldwide has increased. In 2010, about 10 000 megawatts of PV cells were installed, bringing the total to 33 000 megawatts. As shown in **Figure 3.15**, large numbers of PV cells can be combined to serve as power plants that supply the electrical grid.

Figure 3.15 Completed in 2010, one of the largest PV solar power plants is just outside of Sarnia, Ontario. It currently generates 80 megawatts of power, enough to provide electricity to over 12 000 homes in the area. The plant has 450 hectares of solar panels and cost 400 million dollars to build.

Advantages

- Solar energy is renewable and extremely abundant.
- It use has minimal environmental impact, although it uses large land areas compared to other renewables.

Disadvantages

- The availability of solar energy is limited by seasons, climate, and latitude.
- Solar energy requires energy storage or the use of supplementary sources of energy when sunlight is not available.
- Making PV cells requires large amounts of energy and polluting chemicals.

Mini-Activity 3-4 **Testing PV Cells**

Your teacher will provide you with a PV cell, bright light source, voltmeter, tape measure, and protractor. Design a procedure to investigate how changing the distance and angle of a light source affects the output energy of a PV cell. What do your results suggest about the way PV cells should be arranged in relation to the Sun? How could a PV cell system be modified to take into account the Sun's changing position in the sky over the course of each day?

Wind Energy

If you have ever opened an umbrella on a windy day, you have experienced the powerful energy of wind. Globally, this **wind energy** is the fastest growing source of alternative energy in the world. While the largest producers of wind energy are China, the United States, and Germany, Canada has not missed out on this trend. In 2012, Ontario generated more electricity using wind energy than coal for the first time in history. The Canadian Wind Energy Association predicts that wind power could supply 20% of Canada's energy needs by 2025. A typical Canadian wind farm is shown in **Figure 3.16**.

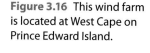

wind energy the energy of moving air

Figure 3.16 This wind farm is located at West Cape on Prince Edward Island.

Like many other energy resources, wind energy relies on turbines to transform kinetic energy into electrical energy. A typical wind turbine consists of a tower-mounted turbine and generator. The turbine has control features that can turn the rotating blades to compensate for different wind speeds. When wind speeds are very high, for example, the blades can be turned to prevent damage to the turbine. Wind turbines are most effective when they are placed in areas with significant wind speeds. The map in **Figure 3.17** highlights these areas in Canada.

Inquiry Lab 3C, Design an Alternative Energy Device, on page 85

Wind Speeds Across Canada

lowest ■ highest

Figure 3.17 This wind map shows the wind speed in different locations across Canada.

Advantages and Disadvantages of Wind Energy

Wind energy produces no greenhouse gases while operating and, in prime locations, has similar contruction costs per unit energy as fossil fuel plants. For these reasons, many people see wind power as the best alternative for reducing our reliance on fossil fuels.

However, wind energy is not without its problems and opponents. Winds suitable for generating energy blow only 15%–30% of the time, and no electricity is generated at other times. Therefore, wind installations need backup energy, which often comes from nonrenewable sources. Some people who live near wind farms find wind turbines unsightly and are concerned that the wind farms may lower property values. They also voice concern that the noise from wind turbines may cause negative long-term health effects resulting from stress and sleep deprivation. As well, wind turbines pose a risk to birds and bats.

Birds are killed when they collide directly with the turbines. Bats such as the one shown in **Figure 3.18** can also die this way. However, they also may be killed by the changes in air pressure created by the spinning blades. In these cases, the bats do not physically hit the blades. Rather, their internal organs are so damaged by the sudden change in pressure near the blades that they die. The data shown in **Table 3.2**, which were collected by researchers from the University of Calgary, support this idea as the cause of death for bats found near wind turbines.

Figure 3.18 Wind turbines can be harmful to bats such as this hoary bat (*L. cinereus*).

Inferring How could shutting down or slowing the turbines at certain times of day dramatically reduce bat losses?

Table 3.2 Injuries in Bats Killed at Wind Turbines in Alberta				
	Hoary Bat	**Silver-haired Bat**	**Other Bat Species**	**Total**
No external injuries	38%	55%	75%	46%
Internal organ damage and hemorrhage	90%	96%	100%	92%

Advantages
- Wind is a renewable energy resource.
- Its environmental impact is relatively low.
- Wind farms can be located in areas where they are less likely to disturb people, such as offshore. The land they occupy can also be used for other purposes, such as farming.

Disadvantages
- Wind is intermittent and its speed is variable.
- Wind turbines can kill birds and bats.
- Some people oppose wind farms due to concerns about appearance, reduced property values, and noise and other health concerns.

Mini-Activity 3-5 **Plan a Wind Farm**

Use the wind map of Canada in **Figure 3.17** and your knowledge of wind energy to write a proposal for a wind farm. In your proposal, explain the following:
- where you would build your wind farm and why

- how you would address unpredictable wind strength and timing
- how you would deal with any citizen or environmental concerns

Hydro Energy

Hydro energy uses the gravitational energy of falling water. When water falls over a height of land, it is accelerated by gravity, which produces energy. Dams are built to control this process, and special water turbines installed in the dam are attached to generators that produce electricity. In some cases, the energy of moving water can be transformed into electricity using a turbine and generator inserted into the water flow, but this is relatively rare compared to the use of dams. As a result of Canada's many large river systems, hydro energy supplies about 63% of the country's electricity demands.

hydro energy the energy of running or falling water

Hydroelectric Dams

A hydroelectric dam like the one in **Figure 3.19** is one of the largest engineered structures in the world. Hydroelectric dams can provide a reliable source of water for irrigation and other uses, such as recreation. They may also help prevent damaging spring or seasonal flooding.

Construction of dams can result in the formation of artificial lakes. These lakes may be many thousands of square kilometers in size. They can flood and destroy terrestrial habitat and create large aquatic habitats in its place. People living in the flooded area may be displaced from their homes. Organic material that decomposes as a result of this flooding also releases methane, a powerful greenhouse gas, and harmful mercury compounds. Additionally, hydroelectric dams can dramatically alter the natural flow of rivers. This can upset the river's ecological balance and destroy fish spawning habitat.

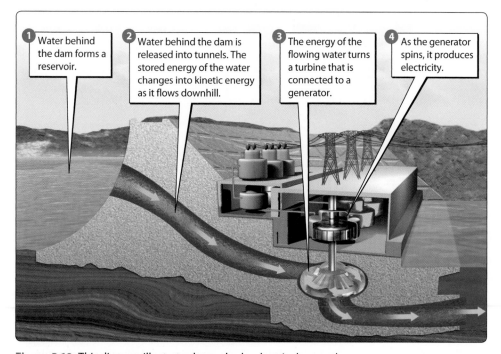

1. Water behind the dam forms a reservoir.

2. Water behind the dam is released into tunnels. The stored energy of the water changes into kinetic energy as it flows downhill.

3. The energy of the flowing water turns a turbine that is connected to a generator.

4. As the generator spins, it produces electricity.

Figure 3.19 This diagram illustrates how a hydroelectric dam works.

Pause and Reflect

4. List two advantages and two disadvantages of solar power.

5. Describe how hydroelectric power plants work.

6. Critical Thinking Use a point-form format to state at least two advantages and two disadvantages of hydro energy.

Other Renewable Energy Resources

Other renewable energy resources include geothermal energy, biomass energy, and tidal energy.

Geothermal Energy

One enormous renewable energy resource is literally buried beneath our feet—geothermal energy. **Geothermal energy** is thermal energy that is captured from Earth's interior. Much of this energy remains largely trapped by the insulating qualities of Earth's crust. However, where Earth's crust is thin and hot magma comes relatively close to the surface, high-temperature steam can be used to turn turbines to generate electricity, as shown in **Figure 3.20**. In other locations, surface water is injected deep underground, where it becomes superheated before returning to the surface. While the United States is the world's largest producer of geothermal electricity, western Canada also has the potential to develop significant geothermal energy. Canada is developing its first geothermal generating facility near Meager Mountain in British Columbia.

Geothermal power has many advantages. It is reliable, unlimited, and environmentally friendly. Geothermal energy tends to be economical in ideal settings such as this. However, these settings, where Earth's crust is thin, are few. While geothermal energy releases few pollutants, some underground water and steam contain harmful gases such as hydrogen sulfide and radioactive radon.

Figure 3.20 Geothermal power plants use thermal energy from Earth's interior to produce electricity.

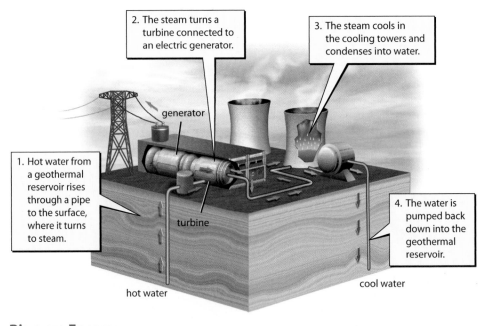

2. The steam turns a turbine connected to an electric generator.

3. The steam cools in the cooling towers and condenses into water.

generator

1. Hot water from a geothermal reservoir rises through a pipe to the surface, where it turns to steam.

4. The water is pumped back down into the geothermal reservoir.

turbine

hot water

cool water

Biomass Energy

Biomass energy refers to chemical energy in non–fossil fuel organic materials. These include wood and vegetation, plant oils, and organic wastes from municipalities, landfills, industry, forestry, and agriculture. These materials are burned, transforming chemical energy into thermal energy that is used directly to provide heat. The thermal energy can also be transformed into electrical energy.

Biomass is the major source of heating and cooking fuel in many developing countries, meeting up to 90% of energy needs in poorer countries. Other forms of biomass energy are also growing in popularity. Biogas is produced by fermentation of animal wastes. It provides an economical and renewable fuel for many farms. Biofuels such as ethanol also fuel vehicles to various extents.

Net carbon dioxide release from burning biomass is low. However, this renewable energy resource comes with other environmental problems. Using agricultural land for biofuel production is problematic. As less land is available to grow food, the cost of food may rise. This is already a problem in poorer nations where higher prices are more likely to result in malnutrition. Additionally, some nations cut down forests to grow biofuel crops. In tropical regions where rainforests remove massive amounts of carbon dioxide from the atmosphere, this contributes to climate change. Where forests are being cut down for wood fuel faster than they can regenerate, ecosystems are harmed and biomass production becomes self-limiting.

Tidal and Wave Energy

Ocean tides and waves contain large amounts of energy that can be harnessed to generate electricity. The west coasts of Canada, Scotland, South Africa, Australia, and the United States are considered among the best places to make use of wave energy, and research to develop suitable technologies is ongoing.

Tidal energy can be harnessed by turbines that convert the kinetic energy of ocean tides to electrical energy. For a tidal power station to be cost-effective, high and low tides must vary by at least 5 m. This occurs in about 40 places in the world, but only two power stations have been built to date. One of these is in the Bay of Fundy.

Tidal energy is renewable and reliable. It also generates little pollution. However, in addition to being limited to a few locations, ecosystem disturbance can be an issue. For example, the tidal generating facility across the mouth of the Annapolis River has increased riverbank erosion. Tidal turbines that cause less environmental damage are currently being tested. **Figure 3.21** shows one of these designs.

tidal energy energy of the regular movement of incoming and outgoing ocean tides

Figure 3.21 Tidal turbines operate on the same general principles as wind turbines. They must withstand powerful currents and continual exposure to corrosive salt water. This prototype turbine from Nova Scotia Power is also designed to minimize ecosystem disturbance.

Mini-Activity 3-6 — Assessing Canada's Energy Resources

In this activity, you will use your textbook, **Figure 3.10**, **Figure 3.17**, and online resources to create a large map of Canada that shows the locations of nonrenewable energy resources. Also include the best areas to use renewable energy resources across Canada. Your finished product should show a complete portrait of Canada's energy resources. Be sure to create a key to go with your map.

Summary

- Solar energy is produced and given off by the Sun. There are three main methods used to capture solar energy: passive solar collection, active solar collection, and photovoltaic cells.

- Wind energy produces very few greenhouse gases and, in prime locations, is very inexpensive compared to most other energy sources. However, some citizens think that wind turbines are unsightly and lower property values. As well, wind turbines pose a risk to birds and bats.

- Hydro energy refers to the energy of running or falling water. A hydroelectric dam is one of the largest engineered structures in the world.

- Geothermal energy is thermal energy that is captured from Earth's interior. Where Earth's crust is thin and hot magma comes relatively close to the surface, high-temperature steam can be used to turn turbines to generate electricity.

- Biomass energy refers to chemical energy in non–fossil fuel organic materials. These include wood and vegetation, plant oils, and organic wastes from municipalities, landfills, industry, forestry, and agriculture.

- Tidal energy is renewable, abundant, and reliable. It also generates little pollution. However, in addition to being limited to a few locations around the world, ecosystem disturbance can be an issue.

Review Questions

1. In 2012, for the first time, Ontario generated more electricity with wind power than with coal power. What are the major advantages of wind over coal? **K/U**

2. What are some of the advantages of installing wind farms offshore instead of on land? **T/I** **A**

3. As shown below, a number of Ontario schools have installed solar panels on their roofs. **T/I** **A**

   a) Would you be in favour of such a project at your school?

   b) What might be the educational benefits of having a school generate renewable energy?

4. Canadians with expertise in oil drilling and mining technologies are in demand around the world to help develop geothermal projects. Industry leaders have suggested that the Canadian government is not doing enough to promote geothermal energy here at home. Do you agree? Do you think geothermal energy should play a larger role in meeting Canada's future energy needs? Explain your reasoning. **A** **C**

5. Some solar collectors are mounted on towers that enable them to track the Sun. What is the advantage of such designs? **T/I**

6. When new reservoirs are formed, unavoidable flooding of large quantities of plants and other organic materials results in decomposition and the release of mercury compounds. This can lead to mercury contamination of food chains and, in particular, of valuable fish stocks. Over a period of 10 to 30 years, decomposition rates slow and mercury concentrations in the environment decline. **T/I** **A**

   a) How should we value the construction of a reservoir versus the damage caused by mercury pollution?

   b) If you depended on fish in the region for food and had to stop eating them, should you be compensated for your loss?

7. What source or sources of energy make Earth's core so hot? **K/U**

8. Trees and grasses, when grown sustainably for use as biomass fuels, have a very small carbon footprint. Do you think devoting large areas of land to growing these fuels is a good idea? What factors would you take into consideration for a particular area of land? **A**

9. Explain why a sustainable modern society could not exist without the use of renewable energy sources. **A** **C**

10. What could be done to help shift energy-generation practices toward a greater reliance on renewable energy sources? Think about both individual and societal actions. **A**

Skill Check

Initiating and Planning

Performing and Recording

Analyzing and Interpreting

Communicating

Materials

- access to online or library resources

Extraction of Nonrenewable Energy Resources

As you have read in this chapter, the extraction of fossil fuels and uranium has environmental, social, and economic consequences. In this Lab you will research more about one of these extraction processes and perform a risk-benefit analysis for the process.

Pre-Lab Questions

1. What are some of the ways in which fossil fuels and uranium are extracted?

2. Why is it important to be aware of the advantages and disadvantages of different extraction processes?

3. How will you perform a risk-benefit analysis for the process you chose to research?

Question

What are the advantages and disadvantages associated with certain extraction processes?

Procedure

1. Choose an extraction process from the list below.
   - strip mining for coal
   - mountain top removal
   - open-pit mining for coal
   - room-and-pillar mining for coal
   - offshore drilling for oil
   - hydraulic fracturing
   - underground or open-pit mining for uranium
   - milling to extract uranium from mined ore

2. Use the following questions as a guide as you research the process you chose.
   - How does the process occur?
   - What are the environmental effects of the process?
   - What social consequences does the process have, including those on human health?
   - What short-term and long-term economic impacts does the process have?
   - Are there advantages to the process in any of the following areas: environmental, social, economic?
   - Are there alternatives to the process that are better? If so, how are they better?

Analyze and Interpret

1. Summarize the process you chose to research. Include a diagram, sketch, animation, or video if appropriate.

2. Perform a risk-benefit analysis on the process.

Conclude and Communicate

3. Do you think this process should still be used? Why or why not?

Materials

- access to online and library resources

Considering a Controversy: The Alberta Oil Sands

Oil sands are made up of sand and shale particles that are coated in bitumen, a thick mixture of hydrocarbons. Canada has one of the world's largest and most accessible oil sands resources. Deposits in northern Alberta are estimated to be about 1.7 trillion barrels of oil, only about 10% of which is currently economically recoverable. In this Lab you will research more about the controversies surrounding the Alberta oil sands and perform a risk-benefit analysis.

Pre-Lab Questions

1. What are oil sands?
2. What are the different ways of extracting oil from oil sands?
3. How will you evaluate your sources for bias?

Question

What are the advantages and disadvantages associated with extracting oil from the Alberta oil sands?

Procedure

1. Use online and other available resources to research the Alberta oil sands. Use the questions below to guide your research. Remember to evaluate your sources for bias.
   - What are the different ways of extracting and processing oil recovered from the oil sands?
   - What are the environmental and health impacts of extracting oil from the oil sands?
   - What are the economic advantages of extracting oil from the oil sands? Are there any economic disadvantages?
   - What are the social and/or political controversies surrounding the oil sands?
2. Make a risk-benefit table to organize your ideas about this issue.

Analyze and Interpret

1. Summarize what is occurring at the Alberta oil sands.
2. Perform a risk-benefit analysis on the process.

Conclude and Communicate

3. Do you think oil should be extracted from the Alberta oil sands? Why or why not? Communicate your opinion using a medium of your choice. Ideas include writing an editorial for a newspaper or blog, making a video, or creating a web page.

Skill Check

Initiating and Planning

Performing and Recording

Analyzing and Interpreting

Communicating

Safety Precautions

- Use appropriate protective equipment such as aprons, goggles, and gloves as well as taking any other safety precautions that are stated in associated Material Safety Data Sheets (MSDS).
- Wash your hands with warm water and soap when you have completed the Lab.

Suggested Materials

- access to online or library resources
- black paint
- pop cans
- glass or Plexiglas®
- aquarium tubing
- aluminum foil
- foam insulation
- PV cells (from solar garden lights)
- turbine/fan blade
- small electric generator
- rechargeable batteries

Design an Alternative Energy Device

In this Lab your team will design, build, and test a device that uses a renewable energy source. A device that is able to use a renewable energy source must be able to perform a number of key tasks. The device must be able to absorb or "capture" the energy in its original form. If this energy is not already in the desired form, the device must be able to transform or convert the energy from one form into another. Lastly, the device must be able to use the energy to perform a task.

Pre-Lab Questions

1. How will the device capture sunlight (or wind) energy?

2. How will you test the device?

3. What safety issues must you consider when designing, building, and testing your device?

Question

How can you design and build a device that uses renewable energy?

Procedure

1. With your team members, brainstorm possible renewable energy devices that you may wish to build. Consider the following list of devices and think of other familiar devices that use renewable energy. Conduct online research to look for more ideas. Remember: your device must be able to perform a simple task using either sunlight or wind energy.

Possible devices:
- solar thermal collector (for heating air or water)
- solar (PV) battery charger
- solar-powered model car
- solar oven
- vertical-axis wind turbine
- wind-powered water pump
- wind-powered lamp

2. Once your team has decided on a device, make a labelled drawing of your design and a list of all the materials you will need to build it.

3. Submit your design drawing and materials list to your teacher for approval or suggestions.

4. Once you have a final approved design, build your device.

5. Test your completed device to determine how successfully it performs.

Analyze and Interpret

1. How was your device able to absorb or capture renewable energy?

2. Explain the energy transformations that took place with your device.

3. What was the final form of energy produced by the device? Describe how this form of energy was used to perform the desired task.

Conclude and Communicate

4. Was your device able to perform its task successfully? Why or why not?

5. What changes would you make if you were to build a "new and improved" model?

Chapter 3 SUMMARY

Section 3.1 Understanding Energy Resources

Different forms of nonrenewable and renewable energy resources are available.

Key Terms
energy resource
nonrenewable energy resource
renewable energy resource

Key Concepts
- An energy resource provides energy to bring about movement or change.
- Energy resources are classified based on their ability to be renewed. The two categories are nonrenewable and renewable.

- Nonrenewable energy resources are those that require thousands or millions of years to reform, if at all. Fossils fuels—coal, oil, and natural gas—are one type of nonrenewable energy resource. The other type of nonrenewable energy resource is uranium, which is used to generate electricity in nuclear power plants.
- Renewable energy resources are produced on a continual basis. Renewable energy resources include solar, wind, hydro, tidal, geothermal, and biomass.
- The use of energy resources by humans and human societies has changed in significant ways over time. Many energy-related discoveries and innovations have benefited society. All have taken their toll on human health and the environment.

Section 3.2 Nonrenewable Energy Resources

Nonrenewable energy resources include coal, natural gas, oil, and nuclear.

Key Terms
methane hydrates   radioactive material
hydraulic fracturing tailings
fission

Key Concepts
- Fossil fuels, including coal, oil, and natural gas, have properties that make them a convenient energy choice.
- Coal is abundant and relatively inexpensive to mine, however, coal mining often causes water pollution and land degradation. Regionally, burning coal is also a major

source of air pollutants. On a global scale, burning coal is the world's largest source of greenhouse gases.
- Extracting and processing oil from oil sands degrades large areas of land. When burned, oil produces a variety of pollutants. These contribute to the greenhouse effect and the formation of smog.
- The production of nuclear energy remains socially and politically controversial due to its potential effects on the environment. These effects occur during the entire life cycle of nuclear fuel, from mining and processing uranium to reactor operations to waste disposal.

Section 3.3 Renewable Energy Resources

Renewable energy resources include solar, wind, hydro, and geothermal.

Key Terms
solar energy   geothermal energy
wind energy   biomass energy
hydro energy   tidal energy

Key Concepts
- There are three main methods used to capture solar energy: passive solar collection, active solar collection, and photovoltaic cells.
- Wind energy produces very few greenhouse gases and, in prime locations, is very inexpensive compared to most other energy sources. However, some citizens also think that wind turbines are unsightly and lower property values.

- Hydro energy refers to the energy of running or falling water. A hydroelectric dam is one of the largest engineered structures in the world.
- Geothermal energy is thermal energy that is captured from Earth's interior. Where Earth's crust is thin and hot magma comes relatively close to the surface, high-temperature steam can be used to turn turbines to generate electricity.
- Biomass energy refers to chemical energy in non–fossil fuel organic materials. These include wood and vegetation, plant oils, and organic wastes from municipalities, landfills, industry, forestry, and agriculture.
- Tidal energy is renewable, abundant, and reliable. It also generates little pollution. However, in addition to being limited to a few locations around the world, ecosystem disturbance can be an issue.

Chapter 3   REVIEW

Knowledge and Understanding

Choose the letter of the best answer below.

1. Fossil fuels currently supply about what percentage of human energy demands?
   a) 20% d) 80%
   b) 40% e) 90%
   c) 60%

2. Which is considered a nonrenewable energy source?
   a) biomass
   b) geothermal
   c) wind
   d) natural gas
   e) hydro

3. Which environmental problem is not closely associated with the burning of fossil fuels?
   a) acid precipitation
   b) climate change
   c) the hole in the ozone layer
   d) water and air pollution
   e) increased acidity of the ocean

4. Which of the following statements is *false*?
   a) A nuclear reactor releases no greenhouse gases in operation.
   b) The mining of uranium produces relatively little waste.
   c) Some nuclear reactors have experienced "meltdowns" of their reactor cores.
   d) Nuclear reactors are costly to build and maintain.
   e) Disposal of radioactive nuclear waste is an unresolved problem.

5. Which would not be associated with a passive solar heating system?
   a) a large water tank or concrete interior walls
   b) pumps that circulate water through rooftop collectors
   c) large south-facing windows
   d) very well insulated walls
   e) design features to keep the home cool in summer

6. Photovoltaic cells are used to convert solar energy directly into
   a) thermal energy   d) electrical energy
   b) kinetic energy   e) chemical energy
   c) light energy

7. Which is a disadvantage of using wind energy to produce electricity?
   a) Wind is intermittent.
   b) Wind energy is inexpensive compared to other sources of energy.
   c) Wind is a renewable energy source.
   d) The land wind farms occupy can also be used for other purposes, such as farming.
   e) Wind farms can be located in areas where they are less likely to disturb people, such as offshore.

8. The construction of a dam can cause the flooding of very large land areas and create a large reservoir. The decomposition of organic material that has been flooded can result in the release of
   a) methane and ammonia
   b) methane and mercury
   c) carbon dioxide and mercury
   d) methane and lead
   e) lead and mercury

Answer the questions below.

9. In which nonrenewable resources is Canada particularly rich?

10. What are methane hydrates and where are they located?

11. What are two advantages and two disadvantages of using natural gas as a source of energy?

12. Compare and contrast passive solar collection and active solar collection.

13. What is the world's main energy source used to produce electricity? Why did Ontario decide to phase out the use of this source of energy?

14. List some of the hazards associated with
    a) coal mining
    b) uranium mining
    c) oil drilling

15. Explain how oil forms.

16. What was the first major renewable energy source used for long-distance transportation?

17. What physical property of flowing water allows it to carry far more energy than an equal volume of wind?

18. What is meant by the term nuclear *fission*?

19. Photovoltaic solar power plants are becoming increasingly common. How is this increase related to
    a) the cost of PV cells?
    b) concerns about the environment?

Chapter 3   REVIEW

20. Wind farms are also becoming a major energy source in Canada and around the world.
   a) What are the advantages of wind power?
   b) What are the disadvantages of wind power?

21. What are the advantages and disadvantages of tidal energy?

22. What is biomass energy?

Thinking and Investigation

23. Scientists only consider biomass a renewable resource when certain conditions are met. Explain what conditions are necessary for biomass to be renewable and sustainable on a long-term basis.

24. In many ways, geothermal energy is the cleanest, most environmentally friendly source of energy available. It is also extremely abundant. Why is it not more widely used?

25. Explain how changes in energy use over the past several hundred years have allowed humans to produce more and more food. Which change do you think was the most significant? Explain your choice.

26. Suppose the price of gasoline in your province were suddenly tripled.
   a) In terms of energy sources, what changes might occur?
   b) How might such an increase influence people's driving habits or choice in the purchase of a new vehicle?
   c) Do you think there would be any benefits to the environment? Explain your reasoning.

27. In 2012 a company installed 18 vertical wind turbines (shown below) on a large building in Oklahoma City. These turbines are very quiet, do not harm birds or bats, and are able to generate electricity that is used directly in the building.
   a) Would you favour the installation of similar wind turbines on office buildings throughout your province?
   b) What might be some of their advantages and disadvantages compared with more traditional wind farms?

28. Consider the advantages and disadvantages of biomass and solar energy. Describe a situation in which biomass energy would be a better energy choice for a community than solar energy. Describe a different situation in which solar energy would be the better choice.

29. Ontario's forests and farmland are capable of producing a very large quantity of biomass energy. What are the pros and cons of using a natural forest as a source of biomass compared to using farmland?

30. Over the next 20 years, Ontario's energy supply mix will change dramatically. The circle graphs below show one prediction of this transition.

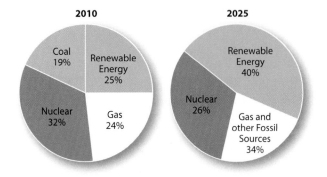

   a) Based on these values predict the overall impact on greenhouse gas emissions, air and water pollution, and energy costs.
   b) How might such a change influence job opportunities in the province?

Communication

31. You use energy from different sources many times each day of your life. When you use energy, you have an impact on the environment and you provide a benefit to yourself. In one or two paragraphs, describe the energy uses that you think are essential to your health and well-being and describe some the energy uses that could be avoided or reduced.

32. Supporters of nuclear power note that major accidents at nuclear power plants are extremely rare. Opponents of nuclear power note that even without major accidents, nuclear power is associated with environmentally damaging uranium mining, and there is still no permanent safe disposal method for high-level radioactive waste.
   a) Conduct online research of these two viewpoints and find at least two websites for each side of the issue.
   b) Describe the key points made by each side.
   c) Which do you think makes a stronger case?
   d) Do you consider nuclear power safe?

33. Examine the graph below.

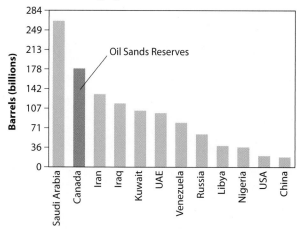

a) As of 2007, how many barrels of oil did Canada have in oil sands reserves?

b) Such large reserves often create tensions between people who wish to exploit the oil sands for economic advantage and people who believe the oil sands should not be used due to negative environmental effects, including climate change. Choose one side and argue whether oil from the oil sands should or should not be extracted.

34. The Three Gorges Dam on the Yangtze River in China is the largest hydroelectric dam in the world, with a generating capacity of 22 500 MW. The construction of the dam created a reservoir with a surface area of over 1000 square kilometers and required the relocation of 1.3 million people. The electricity produced by the dam is expected to pay for the entire cost of the project within 10 years, yet the project remains highly controversial. Conduct online research and summarize the major risks and benefits of this massive energy project.

35. Human society has changed in the past and continues to change today. Many of the most dramatic changes in human history have been associated with changes in how humans obtain and use energy sources. In three or four paragraphs, outline what you consider to be the key changes in energy use in the past and how these energy changes influenced human society.

36. **Did You Know?** Reread the quote by Thomas Edison on page 75. He made these comments in the 1930s. If Edison were alive today, what do you think he might say about society's use of energy resources? Explain your reasoning.

Application

37. In order to pump crude oil to the surface from a well, it is often necessary to also extract natural gas. If there is no gas pipeline at the well site the gas is often burned or "flared" off, as shown below. This process produces carbon dioxide and wastes a valuable energy resource.

a) Do you think it is acceptable to flare gas? Why or why not?

b) What suggestions could you make to reduce or eliminate this practice?

38. Some people do not like the appearance of wind turbines. Of course, many people also do not like the appearance of large industrial facilities, shopping malls, highways, or other infrastructure. To what extent should the appearance of energy installations such as wind turbines be a factor in their use? Conduct a simple survey among your friends and family members. How do they or would they feel about the appearance of these different structures if they were within view of their homes? Based on your own thoughts and your survey results, do you think that the appearance of wind turbines should influence whether or not they are approved for use in a particular location? Explain your reasoning.

> ## Pause and Reflect
>
> How could you incorporate what you have learned in this chapter into your daily actions or choices?

This photo shows bicycles that can be rented in Ottawa, Canada's capital city. By renting these bikes to get around rather than using a car, people use less energy and reduce their impact on the environment.

What programs are available in your area to help people use energy more sustainably?

Making Energy Use Choices

Opting to travel by bicycle rather than car is a choice that reduces individual energy consumption. Each day, people make a variety of choices that influence the amount of energy they consume. These choices often fall into the following general categories: food consumption, transportation, home and buildings, consumer goods, and entertainment and recreation.

1. In a small group, consider each of the above categories. For each category, brainstorm
   a) choices that consume relatively large amounts of energy
   b) choices that consume relatively small amounts of energy

   For example, water skiing and snowmobiling consume larger amounts of energy, while dancing and cross-country skiing consume very little.

2. Brainstorm and list other advantages of your low-energy choices. For instance, dancing helps you keep in shape and meet new people. It is relatively inexpensive and creates little if any pollution.

In this chapter, you will

- identify the causes and effects of global climate change
- explain the link between climate change and energy use
- explain the need for a sustainable energy system
- describe different tools and technologies that facilitate sustainable use of energy

Rethinking Global Energy Use

Global Climate Change

Chapter 3 discussed the advantages and disadvantages of different types of nonrenewable and renewable energy resources. Many of the disadvantages indicate that society's use of energy is creating harmful environmental conditions. These conditions affect the health of the planet and all living things that call it home, including humans. One of the most damaging and far-reaching of these conditions is global climate change. **Global climate change** refers to a long-term change in Earth's climate. The effects of this change are already being felt by millions of people. These effects are a major driver in a change toward more sustainable use of energy resources around the world.

global climate change a long-term change in Earth's climate

The Effects of Climate Change

Earth's surface temperature has increased by between 0.56°C and 0.92°C in the past 100 years. This small change may not seem like a big deal. However, it is important to keep in mind that Earth's climate is a system. This means that a change in one aspect of climate, such as temperature, can result in changes to other aspects, such as precipitation levels, wind patterns, and storm severity. Climate change is expected to affect conditions on Earth in many ways. Some of these changes are already evident, as you will read below. The flowchart in **Figure 4.1** explores some relationships between warming global temperatures and their effects around the world.

Figure 4.1 Warmer global temperatures are linked to changes in Earth's ecosystems.

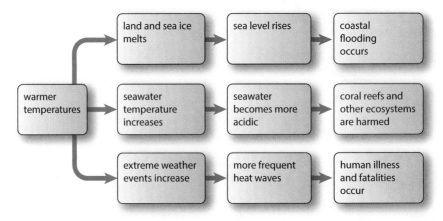

Melting Sea Ice—Warmer temperatures are causing sea ice to melt in the Arctic and Antarctic. Between 2005 and 2007, sea-ice levels in the Arctic Ocean decreased by 20%. This is the largest change in sea-ice levels recorded since scientists began taking these measurements in 1978. Many polar organisms that rely on sea ice for survival, such as polar bears, are negatively affected by this change. In the Arctic, decreasing sea-ice levels also affect Aboriginal people who hunt marine organisms as a food source.

Melting Land Ice—Warmer temperatures are melting glaciers around the world. **Figure 4.2**, on the next page, shows changes in the Bear Glacier in Alaska between 2002 and 2007. Glacial melting changes the volume and flow of rivers. These changes affect local flooding, sea level, and water available for human use, such as irrigation.

As temperatures continue to warm, permafrost (permanently frozen ground) in the higher latitudes is also at risk of melting. Permafrost covers one quarter of the northern hemisphere to a depth of 700 m. When it melts, it is expected to release large amounts of methane gas that will accelerate climate change.

Figure 4.2 Bear Glacier can be viewed from northern British Columbia. Photo (A) of the glacier was taken in 2002. Photo (B) shows the glacier in 2007. Between 2002 and 2007, the glacier had retreated by several kilometres and decreased significantly in height.

Rising Sea Level—Sea level is rising at a rate of about 2 mm to 3 mm per year. Already, two uninhabited islands in the Pacific Ocean have slipped beneath the waves. In other island nations, such as Samoa, residents are moving to higher ground as shorelines retreat inland. Salt water is also intruding into underground supplies of drinking water.

In addition to coastal flooding, sea-level rise destroys wetlands, mangroves, and salt marshes. These are important habitats for aquatic organisms and birds. Melting land ice is linked to this rise in sea level, and its effects will increase as the planet warms. For instance, the Antarctic Ice Sheet contains 90% of Earth's ice. Scientists predict that complete melting of the ice sheet will raise sea level by 50 m. As water warms, it also increases in volume, contributing to further sea-level rise.

Changing Ocean Chemistry—Sea surface temperature has increased over the last 100 years and is continuing to rise. As oceans warm, seawater absorbs more carbon dioxide from the air. When this gas dissolves in ocean water, the water becomes more acidic. This has a harmful effect on ocean ecosystems, as explained below.

Changing Ecosystems—Changes in temperature, ocean chemistry, and sea level are altering global ecosystems. So too are increased winds, flooding, and erosion brought about by more severe storms. When possible, organisms are relocating their habitats and moving out of these danger zones. Even plant growth patterns have shifted north over the last 30 years due to warmer temperatures. When a move is not possible, species may not be able to survive, which threatens biodiversity. The slow-growing coral reefs are one example of a temperature-sensitive organism that is unable to move. Corals, as well as shelled organisms, are also threatened by increased ocean acidity. This acidity dissolves calcium in the organisms' skeletons and shells.

Threatening Human Health—Extreme weather is predicted to increase with global temperature. Some people, including senior citizens and those with heart problems, are especially vulnerable to health problems and fatalities associated with more frequent heat waves. Water stress is another concern. Changes in rainfall can cause drought in some regions. This can lead to shortages of drinking and irrigation water. Dehydration, malnutrition, and even famine may result. Rising temperatures can also broaden the distribution of diseases like malaria, which are carried by organisms that thrive in warm regions. Certain air pollutants and air-borne pollen increase as temperatures warm.

Using Technology to Determine the Effects of Climate Change

Technology such as satellites and computers is helpful in determining the effects of climate change. For example, satellite data were used to create the image in **Figure 4.3**. The data have helped scientists understand how plant growth has been changing in response to warmer temperatures. Technology also helps scientists predict how climate change will affect Earth in the near future. Scientists use computers to model these changes. However, due to the complex interrelationships that make up Earth's climate system, these effects are not always easy to predict.

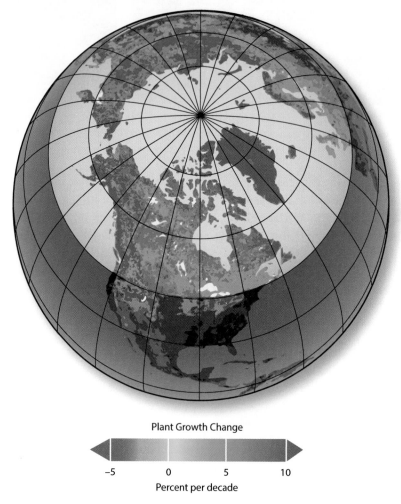

Plant Growth Change

−5   0   5   10

Percent per decade

Figure 4.3 This image was compiled by NASA from data recorded by satellites over 30 years. Due to changes in plant growth patterns, vegetation in the higher latitudes is now lusher and similar to that previously found 4 to 6 degrees of latitude farther south.

Mini-Activity 4-1 **Observing the Effects of Climate Change**

1. Use **Figure 4.3** to determine changes in plant growth per decade in the Canadian regions given below.
   a) Newfoundland and Labrador
   b) the Prairie Provinces
   c) southern Ontario
   d) northern British Columbia

2. Suggest one reason why plant growth change might be so different between Newfoundland and Labrador and the Prairie Provinces.

3. Temperatures are increasing more rapidly in the polar regions than in other regions on Earth. Describe any evidence you can find in **Figure 4.3** to support this fact.

Climate Change and Energy Use

Why does global climate change occur? The first step in answering this question is to understand the physical process that helps moderate Earth's temperature so that it can support life. This process is referred to as the **greenhouse effect**. Similarly, the atmospheric gases involved are called **greenhouse gases**.

greenhouse effect a process that absorbs outgoing infrared radiation in Earth's atmosphere

greenhouse gases gases that absorb infrared radiation in Earth's atmosphere

Understanding the Greenhouse Effect

In the greenhouse effect, greenhouse gases allow ultraviolet and visible light to pass through Earth's atmosphere, while absorbing infrared (heat) radiation. After solar radiation passes through the atmosphere, most of it is absorbed by land and water on Earth's surface. These surfaces re-radiate the energy they absorbed as infrared radiation, some of which remains trapped in the atmosphere by greenhouse gases. As a result, temperatures on Earth are much warmer than they would be if all the radiation escaped back into space. **Figure 4.4** shows how the greenhouse effect and greenhouse gases increase the temperature of Earth's surface and lower atmosphere.

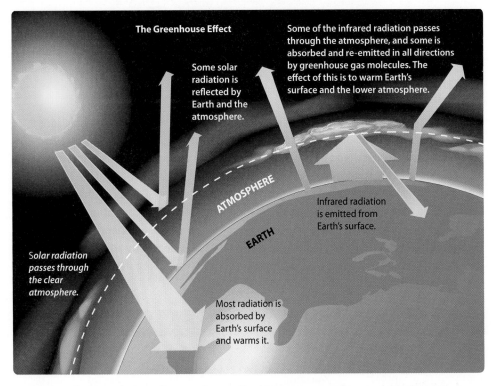

The Greenhouse Effect

Some solar radiation is reflected by Earth and the atmosphere.

Some of the infrared radiation passes through the atmosphere, and some is absorbed and re-emitted in all directions by greenhouse gas molecules. The effect of this is to warm Earth's surface and the lower atmosphere.

ATMOSPHERE

EARTH

Infrared radiation is emitted from Earth's surface.

Solar radiation passes through the clear atmosphere.

Most radiation is absorbed by Earth's surface and warms it.

Figure 4.4 The greenhouse effect moderates Earth's temperature. Average global temperature would be a chilly −18°C if greenhouse gases were not naturally found in the atmosphere.

Pause and Reflect

1. Describe two effects of climate change.
2. Explain the role that greenhouse gases play in the greenhouse effect.
3. **Critical Thinking** When considering the effects of global climate change, why is it important to keep in mind that Earth's climate is a system that interacts with other Earth systems?

Natural and Human-enhanced Climate Change

Climate change can be caused by either natural factors or human activity. Earth has experienced many periods of climate change due to natural processes such as natural variations in greenhouse gases, changes in ocean and and atmospheric circulation, and changes in Earth's orbit. For example, during the Cretaceous period, around 90 million years ago, high levels of volcanic activity released large amounts of carbon dioxide into the atmosphere, quadrupling the concentration of carbon dioxide in the atmosphere compared to today's values. Carbon dioxide is a significant greenhouse gas. As a result, global temperatures rose well above average. The graph in **Figure 4.5** shows that scientists estimate that the temperature of water at the bottom of the ocean during that time was about 19°C. Today water near the bottom of the ocean is around 3°C.

Figure 4.5 High levels of volcanic activity released greenhouse gases that warmed global temperature significantly during the late Cretaceous period. As a result, scientists estimate that the temperature of water at the bottom of the ocean was near 19°C.

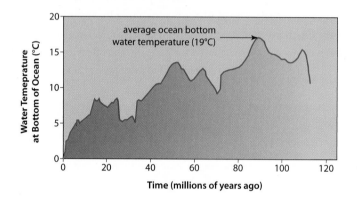

Unlike events in the past, current climate change is the result of human actions. There is a solid relationship between the recent increase in greenhouse gases (at least 70% in the last 40 years) and rapidly increasing global temperatures. This increase in greenhouse gases is mainly due to human activities that burn fossil fuels. These activities release carbon dioxide and other greenhouse gases, such as methane, chlorofluorocarbons, and nitrous oxide. **Table 4.1** shows the concentrations and sources of these gases. Other human activities related to energy use also enhance the natural greenhouse effect. These include the generation of hydroelectricity and deforestation for biomass. However, they are of minor importance when compared to warming due to the use of fossil fuels.

Table 4.1 Principal Greenhouse Gases				
Greenhouse Gases	**Pre-1750 Concentration (ppm)**	**2010 Concentration (ppm)**	**Contribution to Global Warming (%)**	**Principal Sources**
Carbon dioxide	280	388.5	64	• Burning fossil fuels • Deforestation
Methane	0.70	1.7–1.9	18	• Bacteria in wetlands, rice fields, and guts of livestock • Release of fossil fuels
Chlorofluorocarbons	0	0.000 85	12	• Release from foams, aerosols, refrigerants, and solvents
Nitrous oxide	0.270	0.322	6	• Fertilizers • Deforestation • Burning fossil fuels

Moving Toward a Sustainable Energy System

Society's growing awareness of the state of the environment is driving change on a global level. A shift is occurring in how energy is perceived, produced, and used. Many developed nations, including Canada, are rethinking their use of inexpensive, abundant, and easily accessible nonrenewable energy resources. In most cases, these energy resources have been fossil fuels.

Use and production of these resources have enabled developed nations to maintain high levels of productivity. Automated manufacturing, industry, and agriculture have fueled economic growth and helped developing nations attain a high standard of living. While the advantages of this approach have been many, the disadvantages are numerous as well. For instance, little, if any, consideration has been given to the environmental effects of development, such as global climate change. Ecosystem and human health have been an afterthought, and the availability of energy resources and sustainability for future generations has largely been ignored.

As a result of this greater awareness, individuals and governments have begun to see the need for a sustainable energy system. A **sustainable energy system** is a way of perceiving, producing, and using energy. The system has the following characteristics:

- the extraction, production, and use of energy have limited impact on environmental and human health

- there is less reliance on a decreasing supply of nonrenewable resources

- it ensures the availability of renewable and reliable energy resources for current and future generations

- it provides access to affordable energy for the entire global population

> **sustainable energy system** a system in which the perception, production, and use of energy ensure that energy is sustainable

Public demonstrations like the one shown in **Figure 4.6** are becoming common sights. Protestors hope that their actions will help facilitate a change to a sustainable energy system. However, as you will read in the following case study, such change is not always easy to bring about.

Figure 4.6 People around the world are beginning to realize the environmental consequences of energy production and use. This Inuit protest against climate change took place on Earth Day in 2005.

Mini-Activity 4-2   Voicing Your Concerns

Create a plan to raise awareness about the need for a sustainable energy system. For example, this could involve setting up an awareness group in your school, creating a poster campaign, or raising awareness through social media. At a minimum, your plan should

- summarize how you will organize your campaign

- describe who your campaign will reach and how it will reach them

- discuss how your campaign will raise awareness about this issue

- explain how you will judge the success of your campaign once it is over

If there is time, present your plan using a medium of your choice.

Case Study The Challenges of Changing to a Sustainable Energy System

As much as we understand the need for change, the transition to a sustainable energy system is not expected to be an easy one. Part of the challenge is that individuals and society do not just have to change how they act. They also have to change how they think. Changing opinions and habits can be difficult, especially if they are convenient. Additionally, energy systems are complex. How will we know that we are making the right choices? The questions below highlight some of the challenges that come with the transition to a sustainable energy system.

- Individuals, society, and government may not agree on the advantages and disadvantages of different renewable energy resources. Can we get all stakeholders to agree to a sustainable energy system? If so, what steps should we put in place to bring about this agreement?

- How can we make sure that information provided to stakeholders about energy resources is obtained and presented in an unbiased way?

- Canada is a large and diverse country. Should a sustainable energy system be the same for all of Canada, or should it differ for different regions?

- What types of energy resources should be included in a sustainable energy system? For instance, is there room for using fossil fuel or uranium-fueled nuclear energy in a sustainable energy system?

- Canada is a major energy exporter. How would a sustainable energy plan take into account our own energy needs as well as the energy resources we export? Should one take priority over the other?

- Energy can be expensive. How can we ensure that lower-income Canadians can afford access to reliable energy under a sustainable energy system?

- Most energy resources, including renewable resources, affect the environment in some way. How can we evaluate these effects in an unbiased way and prioritize them?

- We need energy now, but so will future generations. How should we balance the needs of today with those of tomorrow?

- It is estimated that 1.5 billion people on our planet have no access to electrical grids. Most of these people live in developing nations. What, if any, are our energy obligations to these nations?

- Developing renewable energy resources costs money. How much should we invest in these technologies? Who should pay for this research—the energy industry, the taxpayers, or both?

Research and Analyze

1. Why is making the transition to a sustainable energy system challenging?

2. Your teacher will assign your group one set of questions presented in this Case Study. Research more information about how to answer it. Gather and evaluate sources that represent the many sides of the issue.

Communicate

3. Use the information you and your group gathered to prepare for and carry out a debate, as instructed by your teacher.

Summary

- Global climate change refers to a long-term change in Earth's climate.

- The effects of global climate change include land ice and sea ice melting, seawater temperature increasing, and the possibility of extreme weather events becoming more frequent.

- The greenhouse effect is a natural process that traps outgoing infrared radiation in Earth's atmosphere. Greenhouse gases, such as carbon dioxide and methane, are gases that trap infrared radiation in Earth's atmosphere.

- The greenhouse effect moderates Earth's temperature. Average global temperature would be about –18°C if greenhouse gases were not naturally found in the atmosphere.

- Climate change can be caused by either natural factors or human activity. Unlike events in the past, current climate change is the result of human actions. There is a solid relationship between the recent increase in greenhouse gases and rapidly increasing global temperatures.

- This increase in greenhouse gases is mainly due to human activities that burn fossil fuels. These activities release carbon dioxide and other greenhouse gases, such as methane and nitrous oxide.

- Society's growing awareness of the state of the environment is driving change on a global level. A shift is occurring in how energy is perceived, produced, and used. As a result of this greater awareness, individuals and governments have begun to see the need for a sustainable energy system.

Review Questions

1. Describe the factors related to climate change that are causing sea level to rise. **K/U**

2. How is the melting of permafrost related to the release of greenhouse gases? **K/U**

3. Describe the ways in which climate change will affect human health. **K/U**

4. Explain how climate change may lead to food shortages. **T/I**

5. In what ways do you think technology has influenced our ability to model and predict the effects of climate change? **T/I**

6. Examine the graph below. Provide reasons to account for the dramatic jump in concentrations of each of these gases. **T/I** **C**

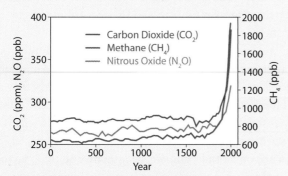

Source: United States Environmental Protection Agency

7. What forms of radiation are able to pass through the atmosphere without being absorbed? **K/U**

8. Explain why naturally occurring greenhouse gases are necessary for life to exist on Earth. **K/U**

9. Some people argue that climate change is not a concern because it is natural and that Earth's climate has changed in the past. What aspect of this argument is true? Explain your reasoning. What aspect is false? Explain your reasoning. **A** **C**

10. List three ways that society has benefited from the use of abundant but nonrenewable energy supplies. **K/U**

11. Examine the graph below, which shows trends in global energy supply. **A** **C**

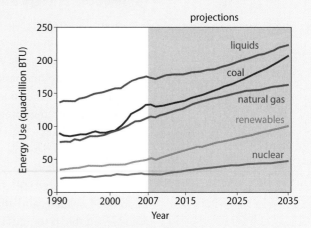

   a) Why do you think the change is so slow?

   b) Do you think the switch to renewables is happening fast enough to avoid serious climate change?

12. List the key characteristics of a sustainable energy system. **K/U**

Achieving a Sustainable Energy Future

Tools for a Sustainable Energy Future

Renewable energy resources, energy efficiency and technology, and energy conservation each play an important role in achieving a sustainable energy future.

Figure 4.7 shows these components as part of a sustainable energy pyramid. The top step consists of renewable energy resources. These resources tend to be expensive to produce on a large scale. However, a sustainable energy system is not possible without them. The middle step addresses energy efficiency. More efficient technology is affordable to many people in developed nations. However, it is more expensive and less accessible than energy conservation, which makes up the base of the pyramid. This is the most cost-effective step and the one that the most people can participate in.

Figure 4.7 This pyramid uses a hierarchy of steps to illustrate the sustainable use of energy.

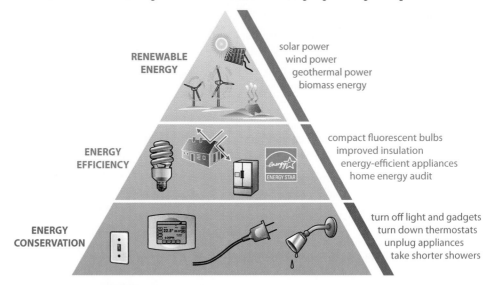

RENEWABLE ENERGY

solar power
wind power
geothermal power
biomass energy

ENERGY EFFICIENCY

compact fluorescent bulbs
improved insulation
energy-efficient appliances
home energy audit

ENERGY CONSERVATION

turn off light and gadgets
turn down thermostats
unplug appliances
take shorter showers

Renewable Energy Resources

As you read in section 4.1, renewable energy resources are important parts of a transition to a sustainable energy system. Arguably, they are the most important parts, since nonrenewable energy resources will one day run out. Renewable energy resources, on the other hand, can be renewed within 100 years and are not at risk of being used up over the course of human society. They are less harmful to ecosystem and human health than nonrenewable energy resources, and they contribute less to global climate change than fossil fuels.

Introducing renewable energy resources to achieve a sustainable energy future may be challenging at first. Today, nonrenewable resources such as fossil fuels and uranium-based nuclear energy still produce abundant amounts of energy. On top of this, they are established, reliable, and affordable energy resources. Although individuals can access some renewable resources at low cost, building new facilities that supply electricity to large numbers of consumers is expensive. Renewable energy can also be unreliable. Access and production often depends on latitude, climate, geography, and other factors. Further, some of the technology that transforms renewable energy resources into a usable form is still in its early stages of development. As a result of these limitations, some scientists believe it will be necessary to increase the use of renewable resources slowly while maintaining some use of nonrenewable resources. Such mixed use of energy resources could smooth the transition to a sustainable energy system.

Energy Efficiency

Energy efficiency is the next step on the sustainable energy pyramid. To fuel economic growth and maintain a high standard of living, we need to use energy. By using this energy more efficiently, we can continue to run machinery and manufacture materials and products at the same rate as before while using less energy. We can power our homes, schools, and businesses as always, but we cause less harm to the environment and emit fewer greenhouse gases.

How can energy efficiency actually be improved? The answer lies in the relationship between energy efficiency and technology. Recall that the first steam engine was patented in 1698. However, it was 100 years later that James Watt supplied the technology that reduced the energy wasted by this engine. This increased energy efficiency improved the cost-effectiveness of the steam engine so it could be used on an industrial scale. Today, we use devices that have a variety of energy efficiencies, as shown in Table 4.2. Some, like the incandescent light bulb shown in Figure 4.8A, are very inefficient. Others, such as the electric heater shown in Figure 4.8B, waste little if any energy. A transition to sustainable use of energy will focus on more energy-efficient devices and work to improve the efficiencies of machines that fall short of this mark.

Table 4.2 Efficiencies of Common Devices	
Device	**Energy Efficiency**
Incandescent bulb	5% (electrical to visible light energy)
Internal combustion engine	20% (chemical to mechanical energy)
Microwave oven	65% (electrical to thermal energy)
Electric motor	90% (electrical to mechanical energy)
Electric heater	100% (electrical to thermal energy)

 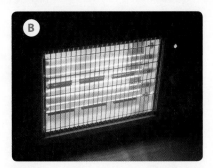

Figure 4.8 Incandescent light bulbs (A) are very inefficient. They convert only 5% of electrical energy they use into visible light energy. Electric heaters (B) are very efficient; they convert 100% of electical energy they use into thermal energy.

Pause and Reflect

4. Describe the three steps in the sustainable energy pyramid.

5. Explain why it might be challenging to increase our use of renewable energy resources.

6. Critical Thinking Many provinces offer subsidies that help homeowners pay for solar technology they install in their homes. Others, like Ontario, buy back excess electricity produced by the consumer. Do you think these programs will increase the number of people using this technology? Explain.

Energy Conservation

energy conservation
choices and changes in
behaviour that enable
people to use less energy
without sacrificing the
services energy provides

The base of the sustainable energy pyramid consists of energy conservation. **Energy conservation** refers to choices and changes in behaviour that enable people to use less energy. Although less energy is used, energy conservation does not mean going without the services energy provides. For example, a person can often conserve energy by changing how a particular task is performed. Choosing to ride a bicycle or walk rather than drive a car to a certain destination means that less energy is used to produce the same end result. Another example is hanging washed clothes on a line rather than drying them in a dryer.

Another way to conserve energy is to avoid wasting electricity. The simple act of turning off lights and other appliances when they are not being used conserves energy. Some devices, such as the computers, televisions, and others shown in **Figure 4.9**, also consume energy when they are turned off and in standby mode. Consumption of this *phantom power* can be reduced by turning these devices off using a power bar or by unplugging them. The above strategies can have a large impact on the amount of energy used on a daily basis. They are also great examples of how small changes by individuals can make an enormous difference. Energy conservation on a societal level often takes a long time to come about. But making small changes on an individual level and educating others about the need for these changes is the first step in this direction. Imagine how much energy can be saved if even one tenth of Canadians make one change, such as turning the thermostat down when they go out in winter.

Figure 4.9 Devices that draw phantom power include chargers, televisions, computers, microwave ovens, washing machines, and air conditioners.

Inferring Why do you think these devices continue to draw power even when they are technically turned off?

Reducing, re-using, and recycling, examples of which are shown in **Figure 4.10**, also play important roles in energy conservation. It takes a lot of energy to manufacture new products. Reducing purchases and re-using what you already have conserves energy. Recycling wastes also reduces the amount of energy used for production. It takes much less energy to manufacture a computer from recycled components and materials than to make one from completely new materials. But note that the processes involved in recycling do use energy. This means that reducing consumption and re-using products are still better options in terms of energy conservation.

reduce, re-use, recycle—**see section 7.1 on page 220**

Figure 4.10 Reducing, re-using, and recycling are often overlooked forms of energy conservation

Describing *How is each person in this visual helping to conserve energy?*

Re-use

Reduce

Recycle

> **Did You Know?**
> "Energy conservation is the foundation of energy independence."
> *– Thomas Allen, American politician*

Pause and Reflect

7. Define *energy conservation* in your own words using a specific example.

8. What is phantom power and how does it waste energy?

9. **Critical Thinking** Explain why choosing to drink from a water fountain rather than purchasing a bottle of water helps conserve energy.

Mini-Activity 4-3   Complete a Phantom Power Inventory

Devices that draw phantom power are found in most homes, schools, and businesses. Complete a phantom power inventory in your home or school. Create a table or other graphic organizer to record your inventory. Your inventory should include the following information for each device:
- type of device
- location of device
- number of devices in home or school

When you have completed your inventory, compare your results with those of your classmates. Do you think phantom power is a significant energy waster in homes and schools? Explain your reasoning.

Sustainable Energy Use for Homes and Buildings

The tools you just read about can facilitate sustainable use of energy in many different situations. Most people spend a lot of time indoors in homes, schools, and other buildings, so this is a good place to start.

Conserving Energy in Buildings

Below are a few simple actions people can carry out to conserve energy in buildings.

- turn off lights when not in use
- use natural lighting whenever possible
- wash a full load of clothes using cold water
- run the dishwasher only when full
- take showers instead of baths, and take shorter ones
- close curtains in summer during the day to reduce air-conditioner use
- unplug appliances that use phantom power
- adjust the thermostat when a building is unoccupied

InquiryLab 4A, Household Energy Consumption, on page 114

Improving Energy Efficiency Indoors

Current technological advances provide many easy, low-cost ways to improve energy efficiency in buildings. For instance, many light bulbs currently on the market are much more energy efficient than incandescent (traditional tungsten) bulbs. With an incandescent light bulb, only 5% of the electrical energy consumed by the bulb is converted into light. The remaining 95% is released as heat. As shown in **Figure 4.11**, compact fluorescent lights (CFLs) and light emitting diodes (LEDs) are more efficient. In an effort to reduce energy use and greenhouse gas emissions, the Canadian government has been phasing out the use of incandescent light bulbs over several years.

Figure 4.11 Halogen, CFL, and LED bulbs are more efficient than incandescent bulbs. However, light bulbs are not very energy efficient in general. Even LED bulbs, the most efficient technology, only convert 25% of the energy they use into light.

Incandescent: 5% efficiency   Halogen: 10% efficiency   CFL: 20% efficiency   LED: 25% efficiency

---

Mini-Activity 4-4   **Comparing Lighting Options**

Design an experimental procedure that allows you to compare the energy consumption, quality of light, and time to reach maximum brightness of several types of light bulbs. The following tips will help you get started:

- The intensity of light is measured in units called lumens. Make sure that each light bulb you use emits the same number of lumens.

- Wattage is the rate at which the bulb uses electrical energy. Multiplying the wattage by time gives the amount of energy used.

- Light quality can refer to the colour or warmth (yellow tone) of light.

Have your teacher review your design. Carry out your experiment if directed to do so by your teacher.

Energy Star and EnerGuide Labelling

Many governments and other organizations have created rating and identification systems to help consumers purchase products that use energy more sustainably. Two such systems are EnerGuide and ENERGY STAR®.

EnerGuide—Energuide is a labelling system created by the federal government. It provides comparative information about the energy consumption for major appliances and heating and cooling equipment. EnerGuide labels, such as the one in **Figure 4.12A**, display a number of key pieces of information. In terms of energy use, this label states the estimated annual energy consumption of the appliance. It then compares this rate with the annual consumption of other appliances of the same type. For room air conditioning units, the EER or energy efficiency rating is also given. The higher the EER value is, the greater the cooling efficiency.

ENERGY STAR®—ENERGY STAR is an international labelling system symbol of premium energy efficiency. It identifies products that meet or exceed strict energy efficiency criteria. Any product with an ENERGY STAR label symbol like the one in **Figure 4.12B** is highly efficient. The ENERGY STAR Most Efficient designation identifies the most energy efficient of all ENERGY STAR rated qualified products. For example, ENERGY STAR qualified natural gas furnaces labelled this way in Canada are more than 97% efficient.

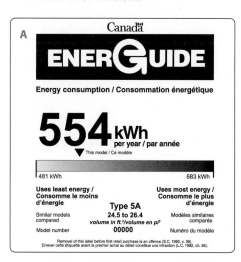

Figure 4.12 An EnerGuide label (**A**) gives consumers specific information about an appliance and how much energy it consumes. Appliances with the ENERGY STAR designation (**B**) are extremely energy efficient.

Mini-Activity 4-5   Analyzing an EnerGuide Label

Your teacher will provide you with an EnerGuide label for a certain appliance. In your notebook, create a table for the appliance to record the following:

- type of appliance
- purpose of appliance
- an explanation of the appliance's EnerGuide label

When you have completed your table, answer the questions below.

1. What factors other than energy efficiency do people consider when buying an appliance? How important are each of these to you in comparison to the EnerGuide rating when purchasing this type of appliance?

2. Based on your analysis, would you buy this appliance? Why or why not?

3. How much do you think EnerGuide ratings influence purchasing decisions made by the general public? Explain your reasoning.

Using Renewable Energy Resources

Some buildings in Canada use electricity that is produced by renewable energy resources such as solar, hydro, or wind energy. However, even where this is not the case, it may be possible to install small generators that use these energy resources. For example, Abbotsford Middle School in British Columbia, shown in **Figure 4.13**, uses renewable energy to run the computer lab. As a result, running the lab produces no greenhouse gases. **Figure 4.14** shows more ways that technology can promote the sustainable use of energy in buildings.

Figure 4.13 The computer lab in Abbotsford Middle School is powered by solar panels, a wind turbine, and a people-powered electric generator that is hooked up to a bicycle.

Applying *Suggest one way that renewable energy could be used in your school.*

Figure 4.14 There are many ways to promote sustainable energy use indoors.

A very effective way to reduce energy is to take full advantage of natural lighting. Buildings can be designed with well-placed windows and skylights to help illuminate rooms and interior spaces. Solar energy that passes through windows can also help heat the building. However, windows are also a major source of heat loss. To reduce this problem, some energy-efficient windows have thin films on their interior that reflect infrared radiation back into the room. The windows also consist of several panes of glass. Spaces between the windowpanes may be filled with an inert gas, usually krypton or argon. These gases have excellent insulation properties. Window frames are also constructed to reduce conduction and leaking of warm air.

Energy-efficient appliances can help reduce energy use in homes. Energy-efficient washing machines use about 20% less energy and 35% less water. A larger tub capacity also means that more clothes can be washed per load. These appliances tend to have an Energy Star label.

The average Swedish home uses one third of the energy of an average Canadian home. Swedes, like Canadians, experience long cold winters. However, their homes are designed to conserve energy better. For instance, effective insulation can reduce the loss of thermal energy when temperatures are colder. It also helps keep buildings cooler during the warm summer months. Common types include fibreglass batts, polystyrene foam boards, and loose cellulose-based materials. More innovative materials such as straw bales are also found in some buildings. The effectiveness of insulation is given by its **R-value**. The higher the R-value, the better the insulation prevents thermal energy from entering or leaving a building.

R-value a measure of the effectiveness of insulation

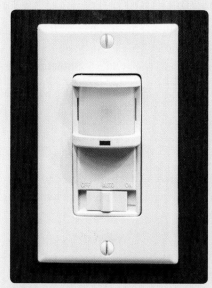

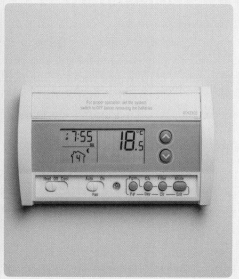

Some newer buildings have motion-sensing light switches that control lighting. Lights come on when someone enters a room and turn off when the sensor detects that the room is empty.

A programmable thermostat allows a person to precisely control the temperature inside a building by setting different temperatures during different portions of the day. For example, a home thermostat can be set to a lower temperature during the night, when people are sleeping.

Inquiry Lab 4B, Which Insulation Is Most Effective?, on page 115

Using solar collectors to heat water is the most effective way to reduce energy consumption for this purpose. The next best choice is to use an on-demand hot-water heater. Although most modern hot-water heaters are extremely well insulated, they are designed to keep a large tank of water hot consistently. However, the actual demand for hot water totals only minutes a day. On-demand hot-water heaters address this problem by heating water only when it is needed. These tankless heaters provide a continuous supply of hot water as long as there is demand and shut off as soon as demand stops. This can reduce energy use to heat water by 30%.

Mini-Activity 4-6 Researching R-values

Use online and other available resources to answer the following questions about R-values and insulation.

1. What is the difference between R-values and RSI values?

2. Create a table that compares the R-values or RSI values of at least three types of insulation. Which product is the best insulator? How do you know?

3. Canada's home building industry has established an R-2000 Standard for energy efficiency, air tightness, and other environmental criteria. What requirements must be met in order for a house to qualify as an R-2000 home? Do you think all new homes should be built to the R-2000 Standard? Explain your reasoning.

Sustainable Energy Use for Transportation

The transportation sector is Canada's largest contributor to climate change. It accounts for about 20% of all of Canada's greenhouse gas emissions. This sector includes all public and private passenger, freight, and off-road travel. The greatest opportunity for conservation is in passenger travel—both public and private—since it is responsible for over half of all transportation emissions. Key variables influencing the amount of energy an individual uses for transportation are discussed below.

Frequency and Distance Travelled

How often and far an individual travels is usually a personal decision. One can often avoid a trip, combine multiple errands, or shop closer to home. If possible, living near school or work can reduce the need to commute.

Vehicle Efficiency

Vehicles with conventional internal combustion engines are among the least energy-efficient forms of transportation. Engines in motor vehicles are only 18% to 20% efficient at converting chemical energy into mechanical energy. Jet engines in aircraft are more efficient, but these gains are offset by energy losses resulting from high-speed travel.

Cost of Fuel

The price of fuel also has a significant effect on energy consumption. As fuel prices rise, people are more likely to purchase more fuel-efficient vehicles and to drive less.

Mode of Transportation

Choosing to walk, bicycle, car pool, or use public transit saves energy and has less environmental impact than driving alone. Many cities encourage such choices with bicycle, car pool, and bus lanes and improvements to public transportation. Many cities also promote bicycle and car sharing programs, as shown in **Figure 4.15**. Although shared cars may be driven by one person, drivers are less likely to use a car to run short errands, and therefore they emit fewer greenhouse gases than those with their own vehicles.

Figure 4.15 Car share programs are found in most major Canadian cites.

Evaluating *Would you join a car share program? Why or why not?*

Number of Passengers

Perhaps the most influential factor of all is the number of passengers per vehicle. When multi-passenger vehicles like buses are at near-full capacity, they emit fewer greenhouse gases per person than other motorized vehicles. Private cars can also be relatively efficient when transporting three or more people. Private cars with single drivers are the greatest emitters per person.

Advances in Motor Vehicle Technology

Over the past few decades, motor vehicle technology has advanced greatly toward using energy sustainably. Consumers can choose from hybrid, pure electric, and hybrid electric options when purchasing a personal motor vehicle.

Hybrid Vehicles—These motor vehicles derive all of their energy from gasoline, but they are much more efficient than conventional vehicles. They have a regenerative braking system that captures energy that is otherwise lost during braking. The energy is stored as chemical energy in a large battery, as shown in **Figure 4.16**. The recovered energy powers an electric motor that, in combination with the gas engine, accelerates the vehicle back up to speed.

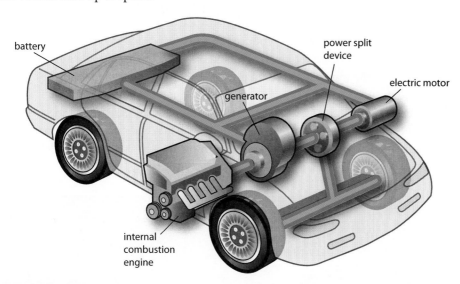

battery

power split device

generator

electric motor

internal combustion engine

Figure 4.16 A hybrid motor vehicle uses both gasoline and electrical energy.

Pure Electric Vehicles—These vehicles do not use fossil fuels. Instead, they are powered entirely by chemical energy stored in batteries. Because the batteries must be recharged from an external power supply, pure electric vehicles can only travel a limited distance between charges. On the positive side, these vehicles produce no pollutants when they run. Thus, they have the potential to significantly reduce smog in large cities. However, it is important to realize that electric vehicles are only as environmentally friendly as the energy resource that is used to charge their batteries. When a renewable energy resource such as solar or wind energy is used, these vehicles have minimal greenhouse gas emissions. But if the electricity is generated by a coal-fired power plant, electric vehicles become linked to more greenhouse gas emissions and more impact on the environment.

Hybrid-Electric Vehicles—Hybrid-electric vehicles are able to run in pure electric mode, but they also have a gasoline engine. The latter can be used when travelling beyond the battery-powered range of the car.

Pause and Reflect

10. Identify three factors that influence the amount of energy an individual uses for transportation.

11. In general, how does the number of passengers influence the amount of greenhouse gases emitted per person by motorized vehicles?

12. Critical Thinking Hybrid, pure electric, and hybrid-electric vehicles all promote sustainable use of energy to some extent. Which would you be most likely to purchase? Why?

Hydrogen Fuel Cell Technology

One challenge electrically powered vehicles face is the limited storage capacity of their batteries. Fuels like gasoline and propane store more energy per mass than the best batteries currently available. However, they are nonrenewable and produce greenhouse gases and other pollutants. Some vehicles, such as buses, are currently adopting a technology that deals successfully with this problem—hydrogen fuel cells. A *hydrogen fuel cell* is shown in **Figure 4.17**. Hydrogen fuel cells convert chemical energy that is in pressurized hydrogen directly into electrical energy. During this process, the hydrogen combines with oxygen molecules in the air to produce water as an end product.

A key advantage of a fuel cell is that unlike a battery that must be recharged, it can operate continuously as long as hydrogen and oxygen (from the air) are available. Hydrogen gas is also a renewable fuel if the energy used to generate it comes from a renewable resource. (Hydrogen is produced in a process called electrolysis, where electricity breaks water down into hydrogen and oxygen.) Nevertheless, a number of factors stand in the way of widespread use of fuel cells. Hydrogen gas is highly explosive and must be stored safely. As well, the extremely small hydrogen gas molecules move through many solid materials. This makes the gas difficult to store. Finally, fuel cells are expensive to manufacture. In Canada, hydrogen fuel cells are mainly used in public vehicles. For instance, many buses used by the British Columbia transit system (shown in **Figure 4.18**) are powered by fuel cells.

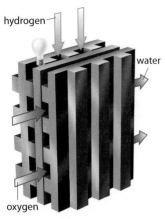

Figure 4.17 Hydrogen fuel cells create an electric current that can run a motor in a vehicle. Hydrogen gas and oxygen enter the fuel cell. Water leaves the fuel cell as an end product.

Figure 4.18 Canada is a recognized leader in fuel cell technology and the use of hydrogen powered vehicles. This bus in British Columbia is powered by hydrogen fuel cells.

Mini-Activity 4-7   The Pros and Cons of Fuel Cells

As you have read, using hydrogen fuel cells has advantages and disadvantages. Use online and other resources to find out more about these pros and cons, as well as how advances in technology are helping overcome certain disadvantages. Then, use your research findings to create an advertisement promoting hydrogen fuel cells for use in buses in your town or region.

Your Energy Future

Canada has one of the highest rates of per person energy consumption in the world. It also ranks near the bottom of developed countries when it comes to efficient energy use. While these statistics are nothing to applaud, they do leave a lot of room for improvement. In Chapters 2 and 3, you learned about Earth's energy resources, their advantages and disadvantages, and various approaches to enhancing energy efficiency and conservation. The final piece in the sustainable energy puzzle is you. What role will you play in a sustainable energy future? How can your choices and actions help reduce your own energy consumption and encourage others to do the same? Like the road sign in **Figure 4.19**, the questions below can help guide your actions and place you on the path to a sustainable future.

Figure 4.19 A sustainable future involves making wise personal choices.

Food—energy used to grow, process, deliver, and prepare food

- What kind of food should I consume? It generally requires more energy to produce meat products than plant-based foods.

- How can I avoid over-packaged and processed foods and beverages? Food packaging and processing consume energy.

- How can I avoid wasting food? When food is wasted, the energy used to produce and transport this food is wasted as well.

Transportation—energy used for personal travel

- How often do I have to travel? Often, unnecessary travel can be avoided.

- What mode of transportation should I choose? Choosing to walk, bicycle, or use public transit reduces energy consumption. Car sharing may be an alternative.

Buildings—energy used to power, heat, and cool homes and other buildings

- How can I conserve energy at home? Adding insulation and using more energy-efficient appliances can reduce home energy use.

- Can I encourage the use of renewable energy resources in my home, school, or community? It may be possible to purchase electricity generated by renewable energy resources or to install small solar or wind power devices.

Consumer Goods—energy used to manufacture and use consumer goods

- Do I really need to purchase a particular item? Can a product be re-used or recycled? Reducing, re-using, and recycling save energy.

- How can I buy more energy-efficient products such as appliances? EnerGuide or ENERGY STAR labels can help establish product energy efficiency.

- How can I encourage using energy sustainably in my community? There are many ways to get involved, from supporting recycling programs to encouraging people to donate used clothing.

Entertainment and Recreation—energy used for fun and exercise

- Can I choose entertainment that is close to home? Travelling long distances to a rock concert consumes more energy than going to see a local band.

- How can I use less energy for recreation? Activities that do not consume fuel, like canoeing, are just as fun as those that consume fuel, such as motor boating.

Mini-Activity 4-8 **Personal Sustainable Energy Use Plan**

Use the questions listed above, as well as your own ideas, to create a personal plan for using energy sustainably.

If there is time, share your ideas using a medium of your choice.

Case Study   Smart Meters and Smart Grids

Electricity requirements change based on the time of day, season, weather conditions, and economic activity. On a hot summer day many homes, malls, and office buildings switch on the air conditioning. This dramatically increases the amount of electricity required and results in *peak demand*. In contrast, on a cool summer night when most people are sleeping the demand for electricity is very low. On weekends and holidays demand is lower because many businesses are closed.

The supply of energy does not always match the demand. Nuclear power stations always operate at the same capacity, unless they are shut down for maintenance. Hydroelectric and natural gas electricity production can be "turned up and down" as needed by changing the volume of water driving the turbines or the amount of gas being burned. But renewable energy sources depend on weather conditions. Wind turbines and solar panel systems only produce energy when there is wind and sun to power them.

Ontario uses nuclear power to meet most of the "base load" demand for electricity. When demands are high the province turns to hydroelectric and gas-generated power. This means that peak demand coincides with the highest costs and the greatest production of greenhouse gases due to the burning of natural gas.

Smart Meters

To help reduce peak demand and greenhouse gas emissions while still meeting our electricity needs, *smart meters* have been installed in Ontario homes and businesses. These smart meters monitor electricity consumption by time of use. Utility companies offer lower prices for electricity used during "off-peak" and "mid-peak" hours. Consumers can choose to shift some of their energy use to non-peak times to save money and help the environment. For example, high energy use appliances such as dishwashers

Smart Meter

or dryers can be run in the evening or overnight instead of during the day.

Smart Grids

Once energy is generated it must be delivered over great distances. This is accomplished by a network of transmission lines and equipment called the *power grid*. The power grid is similar to a network of highways and smaller roads. It must be built to handle present-day capacity but also plan for future expansion. Today, planners are designing *smart grids* that use sophisticated sensors, digital communication, and automated systems to manage electricity supply and demand. A smart grid can respond to changing energy needs, can integrate localized energy supplies (like wind and solar power), and is less vulnerable to major power failures.

Analyze and Conclude

1. Predict when demand for electricity in a 24-hour period is highest during the summer. Explain your reasoning. Make the same prediction for winter and explain your reasoning. (Hint: Think about daily activities, seasonal temperatures, etc.)

2. The circle graph shows summer electricity rates in 2012 in Ontario. Think about the times most energy is used in your household. For example, when are major appliances run? Suggest a new schedule for some of these activities to take advantage of lower energy rates in off-peak times.

Summer Electricity Rate per Kilowatt Hour

MIDNIGHT

P.M.   A.M.

NOON

6.3¢	off-peak
9.9¢	mid-peak
11.8¢	on-peak

Communicate

3. Will smart meters motivate people to use less energy or will they just change when they use energy? Write a short (250 word) essay expressing your opinion.

Summary

- Renewable energy resources, energy efficiency and technology, and energy conservation each play an important role in achieving a sustainable energy future.

- Introducing renewable energy resources to achieve a sustainable energy future may be challenging at first. Today, nonrenewable resources such as fossil fuels and uranium-based nuclear energy still produce abundant amounts of energy.

- Energy conservation refers to people using less energy by changing their choices and behaviour.

- Sustainable use of energy in homes and buildings includes conserving energy and increasing energy efficiency, such as by using more efficient light bulbs and appliances.

- Sustainable use of energy for transportation includes increasing the energy efficiency of vehicles and aircraft and using public transit, walking, or bicycling to reach a destination.

- Your choices and actions will play a role in a sustainable energy future. Your choices and actions can help reduce your own energy consumption and encourage others to do the same.

Review Questions

1. Compare and contrast the concepts of *energy conservation* and *energy efficiency*. **K/U**

2. What is *phantom power* and how can it be avoided? **K/U**

3. Compact fluorescent and LED light bulbs waste over half the energy they consume. Why, then, are they considered energy efficient choices? **T/I**

4. ENERGY STAR and EnerGuide labels are used to provide important information about energy efficiency. **K/U**

   a) Why do we use two systems instead of just one?
   b) How do they differ?

5. List four actions that reduce energy consumption in a home. **K/U** **A**

   a) Rank them in order from most beneficial to least beneficial.
   b) Which of the actions do you think would be easiest to perform?
   c) Which would be most difficult?

6. The World Wildlife Fund and other environmental organizations promote an annual event called *Earth Hour*. During this event, people around the world are encouraged to turn off their non-essential lights for one hour to raise awareness about the need to take action on climate change. Each year millions of people around the world participate in the event. **A** **C**

   a) Do you think this is a valuable event? Why or why not?
   b) Did you or would you participate in this event? Why or why not?
   c) How might it be enhanced?

7. The use of pure electric vehicles has many potential advantages over gasoline and diesel powered vehicles. **K/U** **T/I**

   a) Outline these advantages.
   b) What factors are currently preventing more widespread use of electric vehicles?
   c) Why is the source of the electricity used to recharge these vehicles of concern to environmentalists and climate scientists?

8. A key problem with a switch to hydrogen fuel cells is the lack of a distribution network. Businesses will not create a network of fueling stations across the province and country unless there is a market for the hydrogen. Similarly, people cannot be expected to purchase hydrogen fuel cell cars unless they know they will be able to obtain fuel for them when they travel. Brainstorm ways to overcome this problem. What roles could governments and/or the hydrogen fuel industry play in addressing this problem? **T/I** **A**

9. For each of the following, describe one action you could take to reduce energy consumption when: **K/U**

   a) consuming food
   b) using daily transportation
   c) designing a home
   d) choosing to engage in a recreational activity

10. Some people think drying clothes on a line is an eyesore that should not be allowed in residential areas. Others think it should be encouraged as an important way to conserve energy. With whom do you agree, and why? **C**

Skill Check

Initiating and Planning

Performing and Recording

Analyzing and Interpreting

Communicating

Materials

- record of household electricity use over 12 months
- record of household natural gas use over 12 months (if applicable)
- record of any other source of fuel for your household over 12 months (if applicable)

Household Energy Consumption

In this Lab, you will investigate the energy consumed in your home during the last year and the costs of using that energy. Based on your findings, you will suggest ways that your household can conserve energy. Note: If you do not have access to this information for your home, your teacher will provide you with data to use.

Pre-Lab Questions

1. What are the different types of fuel that are used as a source of energy in your home?
2. What type of energy is used to heat your home during winter? What type of energy does your hot-water tank use?
3. What appliance in your home uses the most electrical power? (Energy consumption by large appliances is available in the user's guide that came with the appliance or online.)

Question

What are the yearly energy consumption and associated costs at home?

Procedure

1. Ask a parent or guardian to provide you with a record of the following information (typically shown on the bills sent by the companies that provide the electricity or fuel).
   - the amount of electricity used and the cost of that electricity for each month of the last 12 months
   - the amount of fuels such as natural gas, propane, heating oil, and wood used in each month over the last 12 months, and the total cost of each fuel
2. Draw a bar graph that shows the consumption of electricity in each month over the last 12 months. Draw similar bar graphs for the other fuels used in your home for each month over the last 12 months.
3. Draw a pie graph that represents total energy costs for your household. Each sector of the pie graph should represent the percentage of total cost in a 12-month period for consuming a particular type of energy.

Analyze and Interpret

1. Summarize the trends you saw in energy consumption and costs.
2. How do energy consumption and costs in winter compare with those in summer? If your home uses different sources of energy, which energy source consumption shows the greatest difference between summer and winter months? Explain why.

Conclude and Communicate

3. Most electric companies charge different rates at different times of the day. On-peak hours are more expensive than off-peak hours. When are off-peak and on-peak times of the day? Why do companies have the different rates?
4. Suggest three ways that your family can conserve energy and reduce the energy bills.

Skill Check

Initiating and Planning

Performing and Recording

Analyzing and Interpreting

Communicating

Safety Precautions

- Use appropriate protective equipment such as aprons, goggles, and gloves as well as taking any other safety precautions that are stated in associated Material Safety Data Sheets (MSDS).
- Some forms of insulation can cause irritation. Avoid contact with skin and wear eye protection. Avoid inhaling insulation fibres.
- Wash your hands with warm water and soap after handling insulation.

Suggested Materials

- different types of insulation such as straw, foam, fibreglass, or blown cellulose
- reference material and/or computer with Internet access
- ice
- water
- kettle
- thermometers
- suitable containers
- timer
- metric ruler
- graph paper or graphing software

Which Insulation Is Most Effective?

Many different types of insulation are available today. Each type has its own advantages and disadvantages. In this Lab, you will plan and conduct an investigation to evaluate the effectiveness of two or more different insulation materials.

Pre-Lab Questions

1. What variables should you consider as you plan your procedure?
2. What will you use as a control in your procedure?
3. What safety precautions should you take as you carry out your investigation?

Question

How do different materials compare in their ability to insulate?

Procedure

1. Choose two or more types of insulation to investigate from the selection provided by your teacher.
2. Research each type of insulation to find out its R-value.
3. Prepare a procedure to determine the effectiveness of different insulation materials. Consider how you will
   - test the effectiveness of the insulation
   - set up a control
   - measure your variables
   - record and assess your results
   - include safety precautions, such as how you will safely handle the materials, in your procedure
4. Ask your teacher to review your procedure when you have completed it.
5. Make changes to your procedure as suggested by your teacher.
6. Once your teacher has approved your final procedure, carry out your investigation with appropriate supervision as determined by your teacher.

Analyze and Interpret

1. Briefly summarize your observations.
2. Even though you were comparing different types of insulation, why was it important to set up a control for this investigation?
3. Were your results what you expected based on the R-value of each type of insulation? Explain.

Conclude and Communicate

4. Describe at least two changes you could make to improve your procedure if you completed your investigation again. Explain your reasoning.

Inquire Further

5. Heating replaces thermal energy that is lost from a home. Some super-insulated homes have such a slow rate of energy loss that they require little ongoing heating. Conduct online research to learn how super-insulated homes are designed and how they reduce energy use.

Chapter 4   SUMMARY

Global climate change is linked to several effects on ecosystems around the world.

Key Terms
global climate change
greenhouse effect
greenhouse gases
sustainable energy system

Key Concepts
- Global climate change refers to a long-term change in Earth's climate.
- The effects of global climate change include land ice and sea ice melting, seawater temperature increasing, and the possibility of extreme weather events becoming more frequent.
- The greenhouse effect is a natural process that traps outgoing infrared radiation in Earth's atmosphere. Greenhouse gases, such as carbon dioxide and methane, are gases that trap infrared radiation in Earth's atmosphere.

- The greenhouse effect moderates Earth's temperature. Average global temperature would be about −18°C if greenhouse gases were not naturally found in the atmosphere.
- Climate change can be caused by either natural factors or human activity. Unlike events in the past, current climate change is the result of human actions. There is a solid relationship between the recent increase in greenhouse gases and rapidly increasing global temperatures.
- This increase in greenhouse gases is mainly due to human activities that burn fossil fuels. These activities release carbon dioxide and other greenhouse gases, such as methane and nitrous oxide.
- Society's growing awareness of the state of the environment is driving change on a global level. A shift is occurring in how energy is perceived, produced, and used. As a result of this greater awareness, individuals and governments have begun to see the need for a sustainable energy system.

Sustainable energy use includes conservation, increased efficiency, and use of renewable resources.

Key Terms
energy conservation
R-value

Key Concepts
- Renewable energy resources, energy efficiency and technology, and energy conservation each play an important role in achieving a sustainable energy future.
- Introducing renewable energy resources to achieve a sustainable energy future may be challenging at first. Today, nonrenewable resources such as fossil fuels and uranium-based nuclear energy still produce abundant amounts of energy.

- Energy conservation refers to people using less energy by changing their choices and behaviour.
- Sustainable use of energy in homes and buildings includes conserving energy and increasing energy efficiency, such as by using more efficient light bulbs and appliances.
- Sustainable use of energy for transportation includes increasing the energy efficiency of vehicles and aircraft and using public transit, walking, or bicycling to reach a destination.
- Your choices and actions will play a role in a sustainable energy future. Your choices and actions can help reduce your own energy consumption and encourage others to do the same.

Chapter 4 REVIEW

Knowledge and Understanding

Choose the letter of the best answer below.

1. Which are predicted consequences of climate change?
   a) more frequent storms
   b) more severe droughts
   c) rising sea level
   d) melting permafrost
   e) all of the above

2. Which form of radiation is readily absorbed by the atmosphere?
   a) radio waves
   b) visible light
   c) ultraviolet radiation
   d) infrared radiation
   e) none of the above

3. Which is causing the oceans to become more acidic?
   a) a decrease in water temperature
   b) increased absorption of carbon dioxide
   c) the release of methane into the atmosphere
   d) the release of nitrous oxides into the atmosphere
   e) acidic water entering the ocean from melting glaciers

4. Which are the principal greenhouse gases produced by human activity in order from most significant to least significant?
   a) carbon dioxide, methane, nitrous oxide, chlorofluorocarbons
   b) carbon dioxide, methane, chlorofluorocarbons, nitrous oxide
   c) carbon dioxide, nitrous oxide, chlorofluorocarbons, methane
   d) carbon dioxide, nitrous oxide, chlorofluorocarbons, ozone
   e) carbon dioxide, methane, ozone, nitrous oxide

5. A sustainable energy system is characterized by
   a) dependence on renewable energy sources
   b) ensuring long term availability of energy supplies
   c) ensuring energy is affordable for all
   d) b and c only
   e) all of the above

6. *Phantom power* refers to
   a) thermal energy released during combustion.
   b) invisible infrared radiation.
   c) energy wasted by devices in standby mode.
   d) waste energy given off when an inefficient device such as light bulb is being used.
   e) power lost by electrical transmission lines.

7. Which is NOT a method of conserving energy in the home?
   a) using a programmable thermostat
   b) closing curtains during the day in summer
   c) using natural light whenever possible
   d) unplugging appliances that use phantom power
   e) washing a number of small loads of laundry rather than a large single load

8. Which statement best describes a hydrogen fuel cell?
   a) It burns hydrogen gas to produce energy.
   b) It converts chemical energy to electrical energy.
   c) It uses electrical energy to produce hydrogen gas.
   d) It is used to produce hydrogen gas from water.
   e) It is similar to a battery and must be "recharged."

Answer the questions below.

9. Why is ocean acidification considered a very serious environmental problem?

10. How can rising sea levels negatively affect drinking water supplies?

11. How are land-based ecosystems being affected by climate change?

12. Climate change is resulting is a relatively modest overall increase in average global temperatures. How can such a change have serious consequences for human health?

13. What is the key difference between a hybrid vehicle and a pure electric vehicle?

14. How can eating locally grown food contribute to a sustainable energy system?

15. Describe the types of information provided on an EnerGuide label.

16. List the four main types of light bulbs and compare their electrical efficiencies.

17. Describe the energy-conserving features of high-efficiency windows.

18. Explain how improving the insulation of a home can help reduce energy costs during the summer.

19. What is an "on-demand" hot water heater? Why is it more energy efficient than a typical hot water tank?

20. What is regenerative braking and how does it contribute to energy efficiency?

21. List two recreational activities that use large amounts of energy and two that use very limited amounts of energy. How do these activities compare in terms of "enjoyment" value?

Thinking and Investigation

22. Do you think climate change has the potential to affect your own health and/or the health of your friends and family members? Why or why not?

23. Why is the natural greenhouse effect considered beneficial while the release of greenhouse gases by humans is considered harmful?

24. If Earth's climate has changed many times in the past, why are scientists concerned about the changing climate now? Explain your reasoning.

25. Taking public transit, walking, or using a bicycle are examples of ways to reduce the amount of carbon dioxide entering the atmosphere. Brainstorm some ways that these greener modes of transportation could be encouraged in your own community. What do you think are the main barriers to their use?

26. One of the most effective ways to reduce the consumption of a commodity is to raise its price. Do you think raising the price of gasoline and other fossil fuels would be a good way of reducing their consumption and promoting the use of renewable energy alternatives? Explain your reasoning.

27. How does the use of passive solar energy have the potential to lower energy costs for lighting and heating of homes?

28. Explain some of the advantages and disadvantages of using hydrogen fuel cells to power vehicles compared to using pure electric vehicles.

29. Many large urban centres now have dedicated high-occupancy vehicle (HOV) lanes, as shown in the photo below. These lanes may be restricted to vehicles carrying at least two or three passengers. What are the advantages of these lanes for drivers? How do these lanes benefit the environment?

30. How can the use of car-share programs reduce environmental impacts if the people who use them are still driving cars?

31. Two people driving identical pure electric vehicles for the same distance each year can have very different environmental impacts depending on the source of energy they use to charge their car batteries. Explain why this is the case.

32. Canada has among the highest rates of energy consumption per person in the world. Why do you think this is the case?

33. Many scientists and environmentalists are opposed to the consumption of water from plastic bottles even though most bottled water is sold in bottles made out of recyclable plastic, like the ones shown below. How does consuming bottled water waste energy and affect the environment? What are the alternatives to bottled water?

Communication

34. Conduct online research to investigate the effects of climate change that are already influencing people living in the Canadian Arctic. Write a one-page summary to report your findings.

35. People living in different countries have very different rates of per capita energy consumption. In general, developed countries use the most energy while developing countries use the least. However, the rate of energy use is increasing in developing countries. How do you think this information should influence how much responsibility different countries have in moving to a sustainable energy system?

36. Suppose you have been asked to give a talk to a group of Grade 6 students on the topic of "Making Wise Energy Choices."

   a) Describe how you could use the energy pyramid diagram in **Figure 4.7** to explain how energy conservation, energy efficiency, and renewable energy choices can benefit the environment and people's daily lives.

   b) What examples could you use for students in Grade 6?

Application

37. Climate change will affect people and the environment in many different ways. Which impacts do you think will be most serious for each of the following?

   a) a farmer growing crops

   b) people living in a large city on the coast

   c) people working in the forestry industry

   d) you

38. Examine the map below, which shows the potential impact of climate change on droughts by the end of this century.

   a) Which parts of the world are at greatest risk?

   b) How might increased droughts influence food availability and prices?

   c) How might increased droughts affect human health?

Potential for Drought by the End of This Century

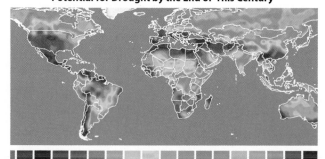

| High risk | Moderate risk | Low risk |
| (drier than current conditions) | (similar to current conditions) | (wetter than current conditions) |

39. An overwhelming number of the world's scientists and all of the world's major scientific organizations believe that climate change is extremely serious and is caused by human activities. Yet, many people do not agree with them. What biases do you think people and organizations might have that cause them to disagree with or not accept the findings of scientific experts?

40. A transition to a sustainable energy system that relies on renewable energy sources will be expensive. However, the negative effects of climate change will also be expensive and in many cases irreversible. Compare and contrast these two sets of "costs." How will these costs affect individuals, communities, and entire countries?

41. The cost of hybrid cars can be prohibitive for many people. Identify two actions that could be taken to help increase the availability and use of hybrid cars by all people of all economic backgrounds.

42. Compact fluorescent light bulbs (CFLs) contain a small amount of toxic mercury that can escape into the environment if the bulbs are not properly recycled. In early 2013, a large Canadian home supply retail chain that sells these bulbs discontinued their CFL recycling program.

   a) Do you think this action is acceptable? Why or why not?

   b) Should stores that sell CFLs be required to recycle them?

   c) Would you choose to purchase CFLs at a store that recycles them over a store that does not? Why?

43. Describe a particular choice you could make for each of the following categories that would benefit the environment *and* directly benefit your health or some other aspect of your life:

   a) a food or diet choice

   b) a recreational activity

   c) where you choose to live in relation to where you go to work or school

44. Did You Know? Reread the quote from Thomas Allen, on page 103. Do you agree that energy conservation is the foundation of energy independence? Why or why not? If energy conservation is not the foundation of energy independence, what is?

> ## Pause and Reflect
>
> How could you incorporate what you have learned in this chapter into your daily actions or choices?

Canadians in Environmental Science

Vibha Singh: Spreading the Word On Renewable Energy

The campaign to educate the architects, engineers, construction specialists, and business leaders of tomorrow in the possibilities of sustainable energy is what York University's Vibha Singh is all about. At the university's Faculty of Environmental Studies, she acts as team leader developing and maintaining the International Renewable Energy Learning Portal, an online resource supported by the International Renewable Energy Agency. The portal gives students access to libraries, webinars, e-learning lectures, and other materials to help deepen their understanding of renewable energy. Vibha also co-ordinates workshops on the use of RETScreen, a free software tool by Natural Resources Canada that helps decision makers assess the possibilities of renewable energy projects.

She began her career as an architect, completing her Bachelor of Architecture degree at Aligarh Muslim University in India. "Both my parents are teachers," she says. "My father teaches science and my mother teaches art. I guess I have inherited some of their qualities and architecture allows me to make best use of them." Vibha completed her degree with a design for an Arts and Crafts Village in New Delhi, and her interest in building led to completion of Construction Management and Technology courses at North Lake College, Dallas. There, she was inspired by her professor's enthusiasm for sustainable architecture that is "in harmony with nature ... that involves maximum use of natural light, minimal use of energy for heating/cooling, minimal use of water, and so on to create a balance between the built and the natural environment." This inspiration led her to enroll in graduate school and receive her master's degree in Environmental Studies (Planning) from York University.

While the commitment to renewable energy has been more advanced in places like Europe, or India, where some cities have made solar panels and water heaters mandatory, Vibha feels it is the way of the future here in Canada. "This is a long journey," she says, "but we're on a good path, and I feel that it's going to become popular, with many projects adopting renewable energy, because in the long run it saves money."

Environmental Science at Work

Focus on Energy Use and Conservation

Oil and Natural Gas Engineer

Energy Auditor

Powerline Technician

Clean Energy Researcher

Energy Use and Conservation

Nuclear Mining Engineer

Emerging Energy Researcher

Emerging energy researchers are committed to developing environmentally sustainable sources of alternative energy that are technically and financially viable. They may work on wind energy, solar energy, or geothermal energy. They may work for universities, government departments, or in industrial research.

Renewable Energy Technician

Renewable energy technicians are specialists in alternative energy systems such as solar panels, geothermal heat pump systems, or wind turbines. They may assist with the installation, monitoring, and maintenance of equipment or perform energy audits on businesses or homes in order to suggest ways to make energy use more efficient.

Energy Management and Sustainable Building Technologist

Energy management and sustainable building technologists apply the principles of energy efficiency and conservation in commercial and institutional buildings. These technologists are focused on best practices for managing, optimizing, and distributing energy in large buildings and building complexes.

For Your Consideration

1. What other jobs and careers do you know or can you think of that involve the fields of energy resources and sustainable energy systems that you explored in this unit?

2. Research a job or career that interests you. What essential knowledge, skills, and aptitudes are needed? What are the working conditions like? What attracts you to this job or career?

Compare Energy Sources

Deciding on the best source of energy in a particular setting depends on many variables. Costs, environmental impacts, availability, reliability, and convenience are just a few of the many factors. One major distinction is whether an energy source is renewable or nonrenewable.

Question

Is a renewable or nonrenewable energy source the best choice in a given situation?

Initiate and Plan

1. As a group, decide on a scenario in which you will compare two sources of energy. Some scenarios you may choose from are below. You may also conduct research to create your own scenario.
   - Scenario 1: A small family wants to generate electricity for their cottage in the Muskoka region of Ontario. They have very little wind so they want to choose between using solar energy or a gasoline-powered electric generator.
   - Scenario 2: A 20-home housing development is being constructed on the southern coast of Nova Scotia. It is an area with fairly steady winds, so the builder is choosing between wind power with land-based turbines or buying electricity generated by a coal-fired power plant.
   - Scenario 3: A mining operation in a remote northern Ontario location needs a source of electricity to run their processing facilities. They could purchase electricity generated by nuclear power or build a small hydroelectric dam on a local river.
2. Decide on factors you will consider and how you will rank and judge their importance. In a group, brainstorm answers to the following questions.
   - What key factors should be considered when choosing an energy source for a particular situation?
   - How can one factor be weighed against another?
3. Prepare a plan for comparing the two sources of energy in your group's scenario based on steps 1 and 2. Have your teacher approve your plan before proceeding.

Perform and Record

4. Make a list of the major advantages and disadvantages of each energy source.
5. Identify factors that might be particularly important in the scenario you chose.
6. Decide on which factors you think should be given the greatest weight, and rank them from most important to least important. Explain your reasoning.

Analyze and Interpret

1. Compare each energy source based on the factors you chose. Assign each energy source a score for each factor. For example, an energy source might receive a +2 for environmental impacts but a "–1" for cost.
2. Many people think the financial costs of an energy source should always be a top priority. Do you agree? Why or why not?
3. Some advantages and disadvantages are immediate (such as the initial cost of construction) while others are realized over a much longer period of time (such as gradual pollution of an ecosystem). Summarize how you weighed short-term versus long-term factors.

Communicate Your Findings

4. Tally your results and reach a conclusion.
5. Prepare a recommendation that could be used to inform people involved in the situation you chose. Use a format of your choice, keeping your audience in mind.

Assessment Checklist

Review your project when you complete it. Did you ...

- ☑ **K/U** choose a scenario to assess the best energy source for that situation?
- ☑ **T/I** decide on factors you will use and how the importance of each is ranked?
- ☑ **A** develop a method of assigning a score for each energy source?
- ☑ **T/I** reach a conclusion about the best energy source to use for the chosen scenario?
- ☑ **A** make a recommendation based on several factors and the given situation?
- ☑ **C** choose an appropriate presentation format, keeping your audience in mind?

An Issue to Analyze

A Power Plant for Kossmannopolis

The map below shows the city of Kossmannopolis (population: 65 000) and surrounding area. The map shows the layout of the city, the shape of the valley, and two of the closest suburbs: McLarenvale and Martinapt. The Pritch River winds through the valley and supplies water to the area. The city has a system of floodgates to cope with the spring floods that often affect the valley. The valley floor is very fertile and grows some of the province's best watermelon. The slopes of the valley are used mostly for marmot ranching, though there are a few ranches that use the slopes to raise alpacas. A ski hill has opened on the side of the valley by McLarenvale. Since opening it has seen moderate activity during the winter. Other tourism and recreational opportunities are found in and around most of the communities.

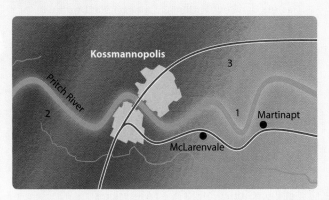

Issue

The energy needs of Kossmannopolis are growing, and the people have decided to build a local power plant to promote future growth. The city requires a power source that will supply 3 kW to every person. Anything above this can be sold to neighbouring communities as surplus. The Sigmunmax Power Company is contracted to build a facility for the city. There are many different power plants that will work for Kossmannopolis. The infrastructure is in place and the resources are available. All that is required now is for a plan to be accepted by the town council.

Initiate and Plan

1. Your teacher will provide details to your group about which proposal, stance, or role your group will take regarding the plan for a new power plant.

Perform and Record

2. Use the information on this page, including the map, as well as Internet and library resources to conduct research as needed to learn more about the proposal, stance, or role your group is taking.

3. Organize the results of your research to prepare your argument for a mock town council meeting. If your group is serving as the town council, prepare for the meeting as directed by your teacher.

4. Your teacher will briefly explain each group's proposal, stance, or role. Prepare questions to ask each group after their presentation at the meeting.

5. Make your argument before the town council, keeping your audience in mind.

Analyze and Interpret

6. What was the final decision of the town council? What was their reasoning?

7. Do you agree with the decision? Why or why not?

Communicate Your Findings

8. Summarize the argument you made to the town council in the form of a report. Consider including statistics, illustrations, or diagrams as needed.

Assessment Checklist

Review your project when you complete it. Did you ...

- ☑ **T/I** conduct research to learn more about the proposal, stance, or role your group took?

- ☑ **K/U** organize the results of your research in order to prepare your argument?

- ☑ **A** prepare questions to ask each group after their presentation?

- ☑ **C** make your argument before the town council, keeping your audience in mind?

- ☑ **C** explain why you agreed or disagreed with the town council's decision?

- ☑ **C** summarize the argument you made before the town council in the form of a report?

Unit 2 REVIEW

Knowledge and Understanding

Choose the letter of the best answer below.

1. Which has seen the greatest increase over the past 100 years?
   a) energy use for transportation
   b) energy use for machinery (including agricultural and industrial)
   c) energy use for food production
   d) energy use for heating
   e) energy use for lighting

2. Which best describes petroleum or crude oil?
   a) a fossil fuel that forms from deep sea sediments
   b) a fossil fuel formed from ancient plant debris
   c) a fossil fuel released by "fracking"
   d) the fossil fuel that releases the most carbon dioxide when burned
   e) the fossil fuel extracted from Earth by strip mining

3. Which statement regarding fracking is *false*?
   a) It involves the high-pressure injection of water into shale formations.
   b) It is a method used to extract natural gas from beneath Earth's surface.
   c) It may be responsible for some earthquakes.
   d) It may be responsible for contamination of drinking water supplies.
   e) It is now widely used and no longer controversial.

4. Which statement regarding nuclear power is *false*?
   a) The release of energy occurs when atoms undergo fission.
   b) Many people remain concerned about the potential of a serious nuclear accident.
   c) Nuclear power is a nonrenewable energy source.
   d) High levels of radioactive waste are buried deep below Earth's surface.
   e) Nuclear power produces less greenhouse gases than electricity generated by fossil fuels.

5. Which statement is *true*?
   a) Most biomass energy is used in wealthy countries with large forestry industries.
   b) Geothermal energy is considered environmentally friendly but nonrenewable.
   c) Many of the best locations for wind turbines are over or near large bodies of water.
   d) An advantage of hydroelectric power is that it creates no water pollution.
   e) Tidal energy is environmentally friendly but it is unreliable and very limited in supply.

Answer the questions below.

6. Provide definitions and use examples to clearly distinguish between each of the following sets of terms.
   a) *renewable resources* and *nonrenewable resources*
   b) *passive solar* and *active solar*
   c) *energy efficiency* and *energy conservation*

7. List the key characteristics of a sustainable energy system.

8. Describe two environmental impacts of each of the following energy sources.
   a) fossil fuels
   b) nuclear power
   c) hydroelectric dams
   d) biomass

9. The choice for location of large-scale electric generating facilities is influenced by a number of factors. For each of the following, list one or more factors that could influence where a plant would be located.
   a) coal-powered generating station
   b) a wind farm
   c) a geothermal generating station
   d) a solar farm

10. Briefly explain how each of the following can be used to reduce energy consumption.
    a) programmable thermostat
    b) Smart Meter
    c) EnerGuide/ENERGY STAR systems
    d) home energy audit

Thinking and Investigation

11. Consider the advantages and disadvantages of different solar energy systems.
    a) Why are passive systems often used for heating a home, but active systems are used for hot water heating?
    b) Why must a home or community that uses solar power have a way of storing energy or an alternative energy source?
    c) Suggest at least one advantage and one disadvantage of generating electricity using photovoltaic cells.

12. People were aware of coal and petroleum many centuries before these fossil fuels were used as major sources of energy by industrialized societies. Use this idea to explain the role that technology plays in determining the use of an energy source.

13. The production of crops, such as corn, requires high energy inputs in the form of fertilizers and the fuels used for planting, harvesting, and processes. As a result the net energy gains can be low. Examine the table below.

Biomass Fuel Efficiency

Fuel	Inputs (GJ/ha)	Outputs (GJ/ha)	Net Energy Ratio
Corn ethanol	75.0	93.8	1.2
Soy ethanol	15.0	28.9	1.9

a) Which crop is the most efficient source of biofuel? Explain your reasoning.

b) Which crop produces the most net energy per hectare (ha)? Explain your reasoning.

Communication

14. There are many reasons for choosing one source of energy over another. Examine the graph below, which shows past and projected future energy use. Draw a new graph with your own version of what you think *should* be the trend from 2007 to 2035. Write a supporting paragraph that explains your reasons for any changes.

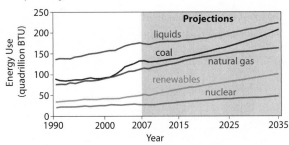

15. Suppose a friend tells you that she is interested in helping the environment but that she does not think there is anything she can do that would make a real difference. Create a one-page information pamphlet that shows the actions she can take and choices she can make in daily life to reduce energy consumption and contribute to a sustainable energy future.

16. History has shown that when the price of fossil fuels, such as gasoline, increases, consumers respond by using less fuel and purchasing more efficient vehicles. One way to increase fuel costs is to limit their supply. Canada has the world's second-largest supply of petroleum in the form of oil sands. Do you think Canada should limit the development of these resources in order to reduce fossil fuel consumption? Present your answer in the form of a short letter to the editor of a local newspaper.

Application

17. Wind turbines can pose a threat to birds. It is also true that millions of birds are killed each year by other human-related factors, as shown in the table below.

Causes of Human Related Bird Fatalities

Cause of Human Related Bird Fatality	Number per 10 000 Fatalities
Buildings/windows	5820
High-tension lines	1370
Cats	1060
Vehicles	850
Pesticides	710
Communication towers	50
Wind turbines	< 1

a) What is the leading cause of human-related bird fatalities?

b) How do deaths from wind turbines compare to the answer from part a)?

c) By placing the wind turbines away from major bird "flyways," most deaths can be avoided. Do you think the number of bird deaths should be considered acceptable given the overall environmental benefits of wind power? Explain your reasoning.

18. Suppose an oil company discovers a deposit containing 25 billion barrels of crude oil, which is enough oil to supply the entire world for one month. Assume that this oil is discovered in an extremely fragile environment and drilling and extraction would permanently damage a very large ecosystem. Also assume that at current prices this oil would be worth about 2 trillion dollars.

a) Do you think extracting one month's supply of oil would justify the damage to the ecosystem? Why or why not?

b) Who should decide whether or not this oil should be extracted? Explain your reasoning.

19. Ontario's Feed-in Tariff program pays producers of renewable energy a higher price for their energy than producers of nonrenewable energy. This is being done as an incentive to encourage the development of more environmentally sustainable energy resources.

a) Why might some people be critical of this program?

b) Why might others consider it a very positive step?

UNIT 3 Sustainable Agriculture and Forestry

Which ecosystems, both natural and human-made, are represented by the green areas in the photo?

Why is it important to manage agricultural and forest ecosystems sustainably?

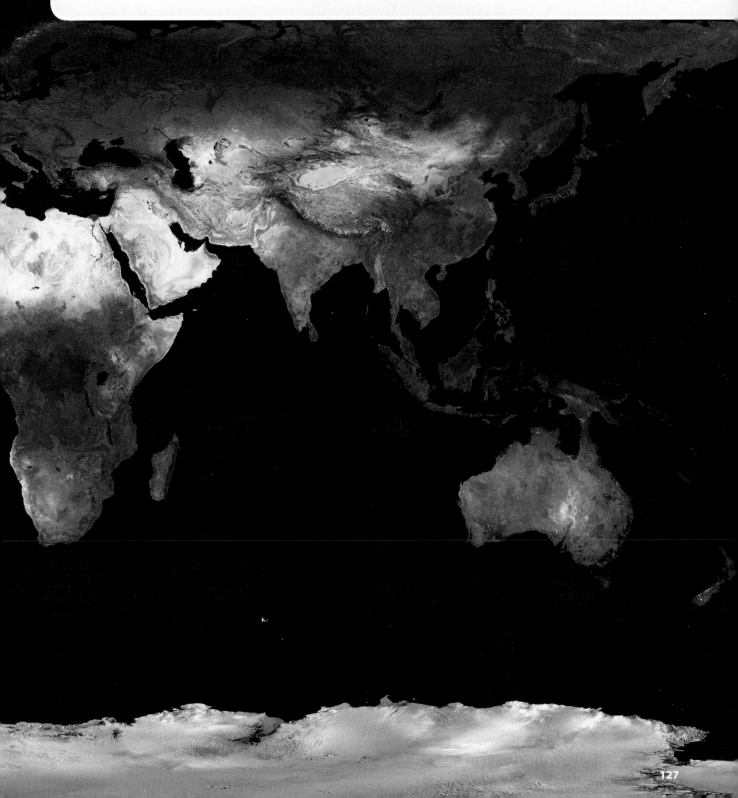

BIG IDEAS

- Modern agricultural and forestry practices can have positive and negative consequences for the economy, human health, and the sustainability of ecosystems, both local and global.

Overall Expectations

- Evaluate the impact of agricultural and forestry practices on human health, the economy, and the environment.

- Investigate conditions for plant growth and environmentally sustainable methods to promote growth.

- Demonstrate an understanding of conditions required for plant growth and environmentally sustainable practices used to promote growth.

Unit Contents

Chapter 5

Sustainable Food Production

Chapter 6

Sustainable Forestry

Topic 1: Sources of Food

Most of our food comes from two main systems: agriculture, which includes crop production and livestock production, and fisheries and aquaculture. Agricultural crops include grains, such as wheat, rye, rice, teff, and quinoa, as well as fruits, vegetables, nuts, herbs, and spices. Livestock, including cattle, hogs, goats, sheep, and chickens, are raised on rangelands, pastures, and feedlots. Protein, in the form of meat, eggs, and dairy, comes from livestock. Fisheries and aquaculture provide protein from fish and seafood that is either wild-caught from marine and freshwater ecosystems or raised in pens.

Topic 2: Food Production

Our early ancestors obtained food from nature by hunting and gathering. The development of agriculture involved manipulating the natural environment to produce the kinds of foods humans want, and it allowed for an increase in human population. Evidence suggests that agriculture began around 9500 B.C.E., with the cultivation of wheat. Between 9500 B.C.E. and 0 C.E., agriculture continued to develop with the cultivation of barley, rice, cotton, sorghum, and maize and the domestication of cattle, chickens, and horses. During this time period, there is also evidence that people used irrigation to water crops and ploughs to work soil. Between 0 C.E. and 1900 C.E., more crops were cultivated, including coffee, and steel ploughs, milking machines, and gasoline-powered tractors were introduced.

In many developing countries today, agriculture is still very labour intensive. Through manual labour or the use of animals, soil is *tilled*, seeds are planted, and crops are harvested , as shown in **Figure 1**. In other areas of the world, including North America, Australia, and much of South America and Europe, food is presently produced through mechanized agriculture. In mechanized agriculture, machines and fossil-fuel energy replace the energy previously supplied by human and/or animal muscles. Mechanized agriculture has substituted the energy stored in petroleum products for the labour of humans and/or animals. As shown in **Figure 2**, mechanized agriculture also involves the use of fertilizers and pesticides that are produced from petroleum products.

Producing food to feed a world population of 7 billion can have major effects on the environment if not done sustainably. For example, large areas of forests and prairies have been cleared worldwide so that land can be used for agriculture. Cows, from which we get beef and milk, are one of the biggest producers of methane gas, a greenhouse gas that contributes to global warming. Nutrients from fertilizer and animal wastes affect aquatic ecosystems when they reach lakes and oceans through run-off. The overharvesting of certain fish and seafood species has affected food chains in aquatic ecosystems and led to a loss of *biodiversity*. Knowing where your food is from and how it is produced is an important part of being an informed citizen and consumer.

Figure 1 Some aspects of agriculture are still as labour intensive today as they have been in the past.

Figure 2 The majority of food today is produced through mechanized agriculture and industrial fishing. The system is based on the use of fossil fuels and other petroleum-based products to grow, harvest, and process crops and raise and process livestock. Food may be transported thousands of kilometres from where it is originally produced to reach grocery store shelves or restaurants.

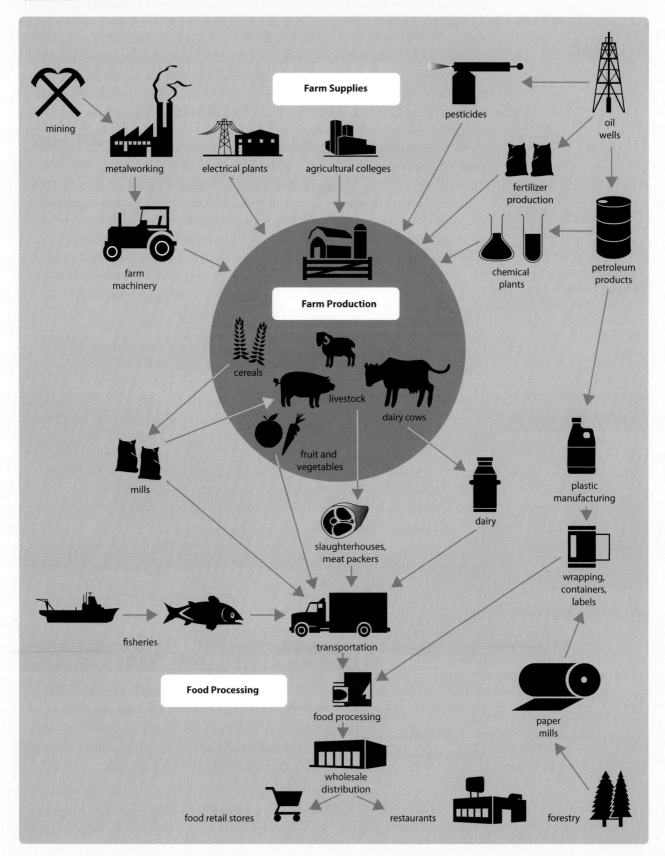

Topic 3: The Green Revolution

Throughout the 1950s, 1960s, and 1970s, the introduction of new plant varieties and farming methods resulted in increased agricultural production worldwide. This has been called the *Green Revolution*. Both the developed world, which uses highly mechanized farming methods, and the developing world, where labour-intensive farming is typical, have benefited from these advances, and food production has increased significantly. Research continues to provide improved varieties of crops, more efficient irrigation, better farming methods, more effective use of agricultural chemicals, and more efficient machines. As shown in **Figure 3**, yields per hectare of land being farmed have increased over much of the world, particularly in the developed world, which includes Canada.

These advances have also had some negative effects. For example, many modern varieties of plants require fertilizer and pesticides that the traditional varieties they replaced did not need. In addition, many of the crops require higher amounts of water, thus increasing the demand for irrigation. The increased profits from increased yields have also attracted large corporations to the business of farming. Many smaller family farmers have sold their land to big-business interests. As a result, much of the agriculture that feeds most of Earth's population today is practised by a fairly small number of multinational companies all over the world. These companies grow much of the fruits, vegetables, and grains that are used to make the millions of food products you see at the supermarket.

Figure 3 The increased crop yields in many parts of the world are the result of a combination of factors, including the development of high-yielding varieties, changed agricultural methods, the application of fertilizers and pesticides, and more efficient machinery.

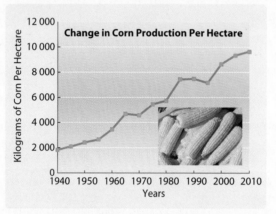

Another result of the Green Revolution is that of the thousands of edible plants and animals in the world, only about 12 types of seeds and grains, 3 root crops, 20 fruits and vegetables, 6 mammals, 2 domesticated birds, and a few fish and other marine life make up almost all of the food humans eat. **Table 1** shows the worldwide yield of some of these food sources.

Table 1 Important Food Sources and Yields			
Food Source	**2011 Yield (millions of tonnes)**	**Food Source**	**2011 Yield (millions of tonnes)**
Wheat	704	Sugar cane	1794
Rice (paddy)	722	Cow's milk	606
Maize (corn)	883	Eggs	65
Potatoes	374	Beef (cattle)	62
Soybeans	260	Pork (hogs)	108
Cassava	252	Poultry (chicken)	89
Bananas	106	Fish and seafood	90*
Tomatoes	159	Aquaculture	55*

Source: Food and Agriculture Organization 2011
* These data are from 2010.

Topic 4: Soil Formation

A combination of physical, chemical, and biological events acting over time is responsible for the formation of soil. Soil building begins with the fragmentation of the *parent material*, which consists of ancient layers of rock or more recent geologic deposits from lava flows or glacial activity. As shown in Table 2, the kind and amount of soil that develops depends on five main factors: the kind of parent material present, climate, the slope of the land, the plants and animals present, and the time involved. Soils and their layers differ from one another, depending on how and when they are formed.

Table 2 Soil Forming Factors	
Factor	**Description**
Parent material	The material in which soils form is called parent material. Some soils come directly from the breakdown of underlying rocks. These soils have the same general chemistry as the original rocks. Most soils form from materials that have been moved in from other places by wind, water, or glaciers.
Climate	Soils vary depending on the climate. Seasonal and daily changes in temperature affect moisture, biological activity, rates of chemical reactions, and kinds of vegetation. Wind redistributes sand and other particles, especially in dry regions. The amount, intensity, timing, and kind of precipitation influence soil formation.
Topography	Slope can affect the moisture and temperature of soil. Steep slopes facing the Sun are warmer. Soils on steep slopes may be eroded and lose their topsoil, making the soils thinner than the more nearly level soils. Deeper, darker-coloured soils are often found on land at the bottom of a slope.
Living organisms	Plants, animals, micro-organisms, and humans affect soil formation. Animals and micro-organisms mix soils and form burrows and pores. Plant roots open channels in the soils. Micro-organisms affect chemical exchanges between roots and soil. Humans can mix the soil so extensively that the soil material is again considered parent material.
Time	Time for all of the above factors to interact with the soil is also a factor. Soil formation processes are continuous. For example, recently deposited material, such as the deposition from a flood, shows no features from previous soil formation. The previous soil surface and underlying layers become buried.

Topic 5: Secondary Succession

Secondary succession is the recolonization of an area after an ecological disturbance in which soil has remained intact. For example, soil, which contains nutrients and organic matter, usually survives disturbances such as forest fires, floods, and agricultural activity.

Often, the seeds and roots of plants remain buried in the soil, as do the spores of ferns and mosses. As shown in **Figure 4**, secondary succession includes changes in the composition and number of species over time. The stages of succession may occur over weeks in an area recovering from a flood. In other areas, such as a new forest, succession may continue for 150 years.

Figure 4 Secondary succession is a series of changes that leads to a mature community.

| Annual plants | Shrubs | Grasses/ herbs | Pines | Young oak/ hickory | Pines die, oak/hickory mature | Mature oak/hickory forest |

| 0 | 1–2 years | 3–4 years | 4–15 years | 5–15 years | 10–30 years | 50–75 years |

Sustainable Food Production

These photographs show different food products that are grown or harvested across the country. Shown here are cattle raised in Saskatchewan, wheat grown in Ontario, crabs fished in British Columbia, and salmon raised in pens in New Brunswick.

What food products are grown, raised, or harvested in your area?

Where Does Food Come From?

If your answer to the question above is "the store," you are not alone. Many people who live in developed countries such as Canada spend most of their lives in cities. As a result, they lack an appreciation for the sources of the food they depend on, who grows or raises it, and how it gets to them.

1. Make a list of the different foods you ate over the past two to three days.

2. If you ate a food that was processed, such as pasta, bread, crackers, or potato chips, think about what grain or vegetable was used to make that food.

3. For each food item you listed, identify how it was produced. Is it an agricultural crop that is grown, such as wheat, rye, or a fruit or vegetable? Does it come from an animal source, such as a cow, pig, chicken, or goat? Did it come from an aquatic ecosystem, such as a lake or the ocean?

4. What do you think it takes to grow or raise the plants and animals that supply you with the nutrients and energy that keep you alive? Who do you think grows or raises them?

5. How could knowing where food really comes from be valuable to society—especially people who live in big cities?

In this chapter, you will

- identify the importance of soil and describe its properties
- describe methods used in sustainable agriculture
- explain how livestock can be raised sustainably
- explain the impact of fishing and aquaculture industries on the environment

Plants and Soil

Basic Requirements of Plants

In sections 5.2 and 5.3 of this chapter, you will learn about the methods used to grow the plants we depend on for food. These include grains to make cereal, bread, and pasta; fruits and vegetables; and seeds and nuts. Before this discussion takes place, however, it is necessary to understand more about what plants need to survive, as well as the nature and properties of the life-supporting material that make it possible for plants to grow: soil.

Like all living things, plants need certain factors and conditions in order to survive in the environment where they grow and live. **Figure 5.1** shows the main factors and conditions that plants need to grow, reproduce, and carry out their other life functions.

Figure 5.1 Plants need sunlight, gases from the atmosphere, water, nutrients and minerals, and space to thrive in their environment.

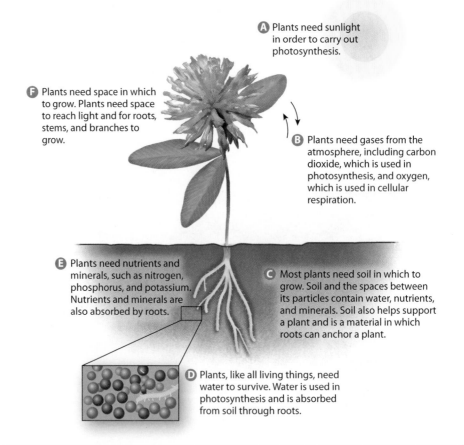

A Plants need sunlight in order to carry out photosynthesis.

F Plants need space in which to grow. Plants need space to reach light and for roots, stems, and branches to grow.

B Plants need gases from the atmosphere, including carbon dioxide, which is used in photosynthesis, and oxygen, which is used in cellular respiration.

E Plants need nutrients and minerals, such as nitrogen, phosphorus, and potassium. Nutrients and minerals are also absorbed by roots.

C Most plants need soil in which to grow. Soil and the spaces between its particles contain water, nutrients, and minerals. Soil also helps support a plant and is a material in which roots can anchor a plant.

D Plants, like all living things, need water to survive. Water is used in photosynthesis and is absorbed from soil through roots.

Pause and Reflect

1. List the factors plants need to survive.

2. Choose one of the factors you listed in question 1 and explain why it is needed for plant survival.

3. Critical Thinking Choose one environmental issue such as air pollution, water pollution, or global climate change and explain how it may affect a plant's ability to survive in its environment.

Components of Soil

It is common to refer casually to soil as "dirt," but this description is far from the truth. **Soil** is a loose covering of weathered (broken-down) rock particles enriched with decaying organic matter. Soil overlies the bedrock of Earth's surface. Plants anchor their roots in soil, and obtain water and nutrients from it. Soil is the product of thousands of years of activity that includes the combined action of wind, rain, the natural movements of Earth's surface, the life processes of billions of organisms that live on and in it, and the chemical compounds that result from the death and decay of those organisms. All the life-supporting regions on Earth—the forests, grasslands, and even certain parts of deserts—cannot and would not exist without soil.

Soil is a mixture of four main components: mineral grains from weathered rock, air, water, and organic material. The organic material in soil includes living organisms such as worms, beetles, and bacteria, as well as the decaying remains of these and other organisms. Different soils have different proportions of the four soil components. **Figure 5.2** shows the proportions of a soil that is considered to be well suited for growing agricultural crops.

soil a mixture of mineral grains, air, water, and organic material that support plant life

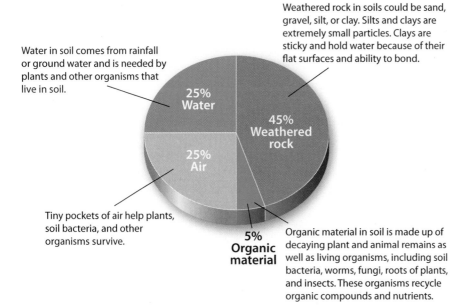

Water in soil comes from rainfall or ground water and is needed by plants and other organisms that live in soil.

Weathered rock in soils could be sand, gravel, silt, or clay. Silts and clays are extremely small particles. Clays are sticky and hold water because of their flat surfaces and ability to bond.

25% Water

45% Weathered rock

25% Air

5% Organic material

Tiny pockets of air help plants, soil bacteria, and other organisms survive.

Organic material in soil is made up of decaying plant and animal remains as well as living organisms, including soil bacteria, worms, fungi, roots of plants, and insects. These organisms recycle organic compounds and nutrients.

Figure 5.2 The four main components of soil are weathered rock, water, air, and organic material.

Inferring Where do you think the air spaces in soil come from?

Mini-Activity 5-1 **Modelling the Plant-supporting Soil Layer**

Carry out the following thought experiment. Your teacher might also demonstrate it for you.

- Imagine Earth as an apple cut into four equal wedges. One of these wedges represents all the land on Earth. What do the other three wedges represent?
- If you cut the land wedge in half, one of the pieces represents the part of Earth where everyone lives. What could the other piece represent?
- If you cut the land wedge into four smaller segments, one of these represents all the land on Earth that can be used to grow crops. What could the others represent?

- Suppose you peeled off the thin layer of skin on this last, tiny piece. This thin, fragile layer of skin represents the part of the soil in which all plants grow and which supplies all our food.

Why would a farmer want to understand the composition and properties of soil? Why might someone who lives in an urban (city) environment benefit from understanding soil?

Layers of Soil

soil profile the series of horizontal layers in soil

topsoil an upper layer of soil that contains nutrients and organic material

A **soil profile** is a series of horizontal layers in soil that differ in chemical make-up, physical properties, particle size, and amount of organic matter. **Figure 5.3** shows a typical soil profile. Notice that just below the layer of surface litter is a layer of topsoil. **Topsoil** is an upper layer of soil that contains nutrients and organic material; this is the layer of soil in which plants have most of their roots.

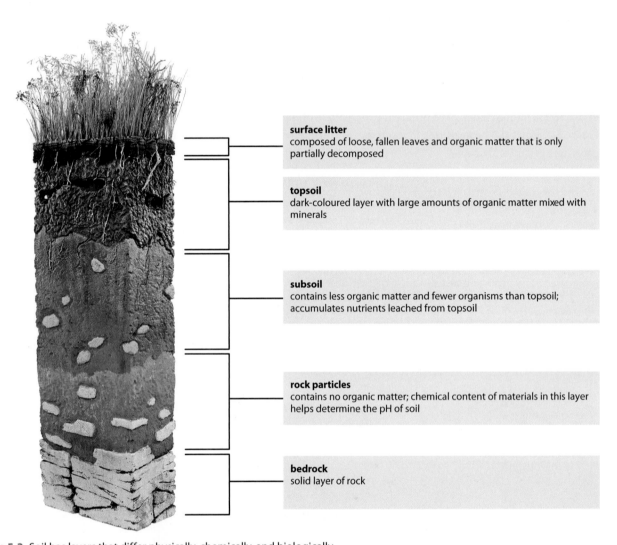

surface litter
composed of loose, fallen leaves and organic matter that is only partially decomposed

topsoil
dark-coloured layer with large amounts of organic matter mixed with minerals

subsoil
contains less organic matter and fewer organisms than topsoil; accumulates nutrients leached from topsoil

rock particles
contains no organic matter; chemical content of materials in this layer helps determine the pH of soil

bedrock
solid layer of rock

Figure 5.3 Soil has layers that differ physically, chemically, and biologically.

Inferring *Why do you think topsoil is considered an important natural resource?*

Humus: The Organic Component of Soil

humus organic material resulting from the breakdown of plant and animal remains

The organic material in soil that results from the decay of plant and animal remains is **humus**. Humus is an important part of soil that builds up on the surface and eventually becomes mixed with the top layers of mineral particles. Humus contains nutrients that are taken up by plants from soil. Humus increases the water-holding ability and the acidity of soil, which makes it easier for plants to absorb the nutrients. Humus also tends to make other soil particles stick together and helps create a loose, crumbly soil that allows water to soak in and air to be incorporated. A good soil for agricultural use will crumble and has spaces for air and water.

Living Organisms in Soil

As shown in **Figure 5.4**, inhabitants of soil include burrowing animals such as worms and insects, soil bacteria, fungi, and the roots of plants. One of the most important burrowing animals is the earthworm. As earthworms move through soil, they mix organic and inorganic material in soil. This mixing increases the amount of nutrients available to plants. Soil drainage is also improved by the burrowing of earthworms and other soil animals, such as mites and pill bugs.

Bacteria and fungi are important in the decay and recycling of materials. Their chemical activities change complex organic materials into simpler forms that plants can use as nutrients. For example, some of these micro-organisms can change the nitrogen contained in organic matter into nitrogen compounds that can be used by plants. When the roots of plants die and break down, they release organic matter and nutrients into the soil and provide channels for water and air.

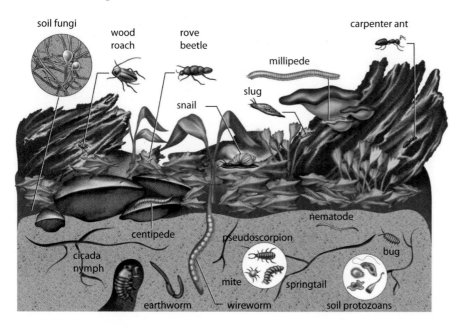

Figure 5.4 Beneath the surface of soil, countless organisms live in a sunless world. They form a network of food chains within soil and open up spaces for air and water to move through soil.

Soil Properties: Texture

Soil texture is determined by the size of the mineral particles in the soil. Particles between 0.05 and 2.0 mm in diameter are classified as sand. Silt particles range from 0.002 to 0.05 mm in diameter. The smallest particles are clay, which are less than 0.002 mm in diameter. Rarely does soil consist of a single-size particle. As shown in **Figure 5.5**, various particles are mixed in many different combinations, resulting in many different soil classifications. The ideal soil for agriculture is *loam*, which combines large spaces for air and water drainage with the ability of clay particles to hold nutrients and water.

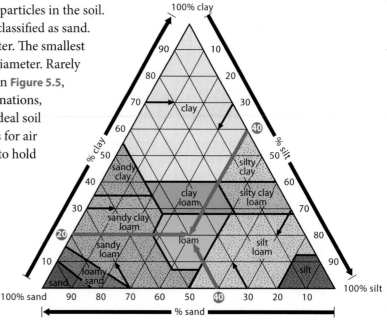

Figure 5.5 Soil texture depends on the percentage of clay, silt, and sand particles in the soil. A loam soil has the best texture for most crops.

Describing *What is the percentage of sand, silt, and clay in loam?*

Soil Properties: Porosity

The porosity of soil refers to the size and number of spaces among particles of soil. The importance of these spaces is that they allow water and air to move through soil. Different types of soil have different degrees of porosity.

As shown in **Figure 5.6A**, soil that is made up of particles of different sizes has spaces for both water and air. The soil in **Figure 5.6B** is made up of small particles that are all about the same size. This soil has less space for air. Since plant roots need both air and water, the soil in **Figure 5.6A** would be better able to support agricultural crops than the soil in **Figure 5.6B**. The amount of water and air in soil is also important for determining the numbers and kinds of organisms that live in the soil.

Inquiry Lab 5A, Characteristics of Various Soil Samples, on page 164

Figure 5.6 (A) The different sizes of soil particles allow air and water to move through soil. **(B)** The soil particles in this example are more uniform in size, leaving less space for air among the particles.

Explaining Why is it important that soil contain air?

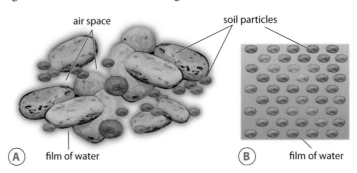

air space   soil particles

(A)   film of water   (B)   film of water

Soil Properties: pH

pH measures how acidic or basic a substance is. A substance that has a pH of less than 7 is acidic, one that has a pH of 7 is neutral, and one that has a pH higher than 7 is basic. Most plants grow well in soil with a pH between 6 and 7, although some plants, such as blueberries and potatoes, grow well in more acidic soil.

The pH of soil depends on the amount of rainfall and the amount of organic material the soil contains. High annual rainfall makes soil more acidic. The rainwater dissolves and removes minerals such as calcium, magnesium, and potassium from soil, leaving more acidic material behind. In addition, the breakdown of organic matter tends to increase the acidity of soil.

Soil pH is important because it affects how easily nutrients can be removed from soil, and this affects the kinds of plants that will grow in the soil. In turn, the kinds of plants that grow in soil affect the amount of organic matter that farmers have to add to soil. Also, since plants need the nutrients calcium, magnesium, and potassium, their loss from soil reduces the fertility of soil. Aluminum dissolves more easily in soil that is highly acidic, and soil that is high in aluminum is toxic to many plants.

Mini-Activity 5-2 **Filtering Water through Soil**

- Put a piece of filter paper in a funnel, and place the funnel in a beaker or flask.

Your teacher will provide each group with a different soil sample.

- Obtain a soil sample from your teacher. Place the soil in the funnel.

- Pour 100 mL of water over the soil. Time how long it takes for the water to drain through the soil. Then measure how much water drained through.

- Describe the appearance of the water that drained through the soil. Compare your observations with those of other groups. How did the texture of the soil affect the variables you measured or described? How do you think what you observed is related to the issue of nutrients, pesticides, or other chemicals in surface water run-off from farms?

Reviewing Section 5.1

Summary

- Plants need certain factors and conditions, including light, water, and soil, in order to survive in their environment.
- Soil is a thin covering over the land made up of a mixture of minerals, organic material, living organisms, air, and water that together support the growth of plant life.
- A soil profile is a series of horizontal layers in soil that differ in chemical make-up, physical properties, particle size, and the amount of organic matter they contain.
- Humus is the organic material in soil that results from the decay of plant and animal remains.
- Inhabitants of soil include burrowing animals, such as worms and insects, soil bacteria, fungi, and the roots of plants. Bacteria and fungi are important in the decay and recycling of materials.
- Soil texture is determined by the size of the mineral particles in the soil.
- The porosity of soil refers to the size and number of spaces among particles of soil. Different types of soil have different degrees of porosity.
- Soil pH affects how easily nutrients can be removed from soil, which affects the kinds of plants that will grow.

Review Questions

1. Explain why plants need space to survive. K/U
2. From which source do plants get nutrients and minerals? K/U
3. List the four components of soil. Identify the origin of each component and describe its function. K/U
4. Identify and describe the layers in a soil profile. K/U
5. Create a flowchart or concept map to show the relationships among (a) the role of humus in soil, (b) the role of soil in plant growth, and (c) the role of plants in producing humus. C
6. Identify three organisms that live in soil and what each organism does to improve the quality of soil. K/U
7. Why is loam considered a good soil texture for growing agricultural crops? K/U
8. How is the size of the particles in soil related to its porosity? K/U
9. Soils that contain a high percentage of sand drain water very quickly. Explain why sandy soil may not be a good soil in which to grow agricultural crops. T/I A
10. The quality of soil depends on many environmental conditions, including climate, topography, and the types of particles it contains. Predict how each event or condition would affect the quality of soil in which an agricultural crop is growing. T/I A

    a) too much rain
    b) extreme cold
    c) a steep slope
11. Soil has been described as "the ultimate resource." Develop an argument to support this statement. Develop a counter-argument in favour of some other resource that could reasonably be regarded as the "ultimate" one. C T/I

12. The table below contains information about the function of certain nutrients in plants and what happens if there is not enough of the nutrient. A T/I

Nutrient	Function	Not Enough
Nitrogen	Gives plants their dark green colour	Plants have yellowing leaves and stunted growth
Phosphorus	Helps plants develop roots, buds, and seeds	Plants grow slowly; young leaves may be greyish, older leaves may be reddish
Potassium	Builds strength and disease resistance, and improves quality of plant seeds	Plants are stunted and the edges of older leaves turn brown and die
Magnesium	Needed for photosynthesis	Leaves are yellow with green veins
Calcium	Helps develop healthy cell walls	Plants may be unable to grow new leaves, stems, or roots

a) Which nutrient is the plant in the photo lacking? Explain your reasoning.

b) You are starting a garden. You notice that some plants are not able to grow new leaves, some have stunted growth, and the older leaves are turning brown and dying. Do you live in an area with high rainfall? Explain your reasoning.

Methods of Mechanized Agriculture

Mechanized Agriculture and Crop Production

agriculture the practice of raising plants and livestock for food or other human needs

Agriculture is the practice of growing and raising plants or livestock for food and other human needs. Mechanized agriculture refers to the use of machines that use fossil fuels to supply the energy they need to work. Mechanized agriculture requires large areas of flat land for machines to work efficiently. **Figure 5.7** shows an example of the type of farming that is mostly associated with mechanized agriculture: monocultures.

Monocultures: Advantages and Disadvantages

monoculture the growth of a single crop, usually on a large area of land

A **monoculture** is the growth of a single crop on a large area of land. Corn, cotton, wheat, rice, and soybeans are examples of monoculture crops. Since the seeds used to grow each crop are all the same species, a monoculture has low genetic and biological diversity. Monocultures have several advantages that make them attractive to farmers. The process of ploughing, irrigating, fertilizing, applying pesticides, and sowing seeds is simpler and less expensive than it would be for growing multiple crop species together at the same time. This makes caring for and harvesting the crop simpler and less expensive as well.

Figure 5.7 Soybeans, like the ones shown here, are often grown as monoculture crops.

Inferring Why is mechanized agriculture well-suited to the development of monocultures?

The increased crop yields made possible by monocultures carry with them several disadvantages as well. For example, growing the same crop in the same soil year after year removes soil nutrients that then must be replaced. As a result, monocultures require the use of synthetic nutrient-supplying fertilizers. Because monocultures lack biodiversity, they are more vulnerable to pests and diseases that are specific to the crop that is growing. To prevent losses of their crops, and therefore income, farmers of monocultures must use pesticides to help reduce pests and the damage they can cause.

The Use and Effects of Synthetic Fertilizers

When a crop is harvested, the nutrients that have been absorbed by the mature plants are removed from the field with the crop. Since many of those nutrients come from the soil, and since plants cannot grow without them, they must be replaced if another crop is to be grown in the same field. The three soil nutrients that are most often depleted are nitrogen, phosphorus, and potassium. Other nutrients needed in smaller amounts include boron, manganese, and zinc. Farmers who use monocultures replace these nutrients with synthetic fertilizers, which are products of the petrochemical industry.

Scientists estimate that 25% of the world's crop yield can be directly attributed to the use of synthetic fertilizers. The use of synthetic fertilizers increased greatly in the last few decades of the 20th century. As the world population increases, scientists predict that fertilizer use will increase even more.

A chief concern about the widespread use of synthetic fertilizers is its impact on aquatic ecosystems. The action of rainwater can dissolve and carry fertilizers in rivers, lakes, and other bodies of water. When the concentrations of nutrients such as phosphorus and nitrogen become too high, algae reproduce in large numbers very quickly. This population explosion of algae is often called an *algal bloom*, shown in **Figure 5.8A**. Eventually the algae die, and bacteria in the water begin to decompose them. During this process, the bacteria use large amounts of oxygen, thus reducing the amount of oxygen in the water. Low oxygen levels cause fish and other aquatic organisms to die quickly and in large numbers. This is often referred to as a *fish kill*, shown in **Figure 5.8B**.

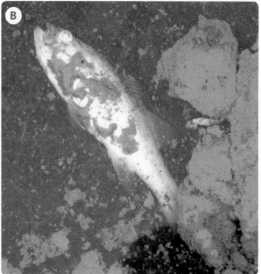

Figure 5.8 (A) An explosion in the population of algae can result when nutrient levels in aquatic ecosystems become too high. **(B)** Bacteria that decompose algae deplete oxygen levels in the water. Fish and many other aquatic organisms cannot survive without oxygen.

Pause and Reflect

4. What is mechanized agriculture?

5. Why do monocultures depend on the use of synthetic fertilizers?

6. Critical Thinking How do you think the use of synthetic fertilizers could affect a person who catches and sells fish for a living?

Use of Synthetic Pesticides

Many types of organisms can affect the health and growth of agricultural crops. A **pesticide** is any chemical used to kill or control populations of unwanted organisms. In agriculture these unwanted organisms, generally referred to as "pests," are most often plants (weeds), insects, or fungi. Synthetic pesticides are those that are made by humans, as opposed to naturally occurring pesticides that are produced by plants to defend themselves against bacteria, fungi, and animal pests.

The use of synthetic pesticides prevents the loss of crops and food supplies to pests or infection. The cost of applying synthetic pesticides is balanced by the savings that result from the reduced need to tend the fields and the increased yields, which result in greater profits for farmers.

Effects of Synthetic Pesticides on Ecosystems

Although the use of pesticides by farmers prevents loss of crops and food supplies to pests or infection, pesticide use can have negative effects on the environment.

- Pesticides are designed for one or several specifically targeted pest species. Since pesticides are poisons, however, other species also may be affected unintentionally. These organisms are often called non-target organisms. Many pesticides are sprayed broadly and destroy populations of beneficial insects as well as pests. Non-target organisms affected by pesticides commonly include insects, fish, birds, plants, and soil organisms.

- Over time, some pest populations become resistant to pesticides, as shown in **Figure 5.9**. In any given population of pests, there are some individuals who are naturally resistant to a pesticide due to their genetic make-up. These individuals survive when the pesticide is applied. When they reproduce, they pass along the genes that allow them to be resistant to the pesticide to their offspring. Some species of insect pests can produce a new generation each month. With each new generation, more and more members of the population have the genes for resistance. In these cases, within five years, 99% of individuals in the population are able to survive exposure to the pesticide.

Figure 5.9 Many pests have developed resistance to pesticides. Because insecticides were the first pesticides to be used widely, insects show resistance earlier than other pest populations. More recently, plant pathogens and weeds are also becoming resistant to pesticides.

bioaccumulation—**see section 10.1 on page 313**

- Many pesticides break down into substances that are less harmful within days or weeks of applying them. However, some pesticides do not break down easily and remain highly toxic. Because they stay in the environment for years, even decades, and move freely through air, water, and soil, they often show up far from the point of original application. Some of these compounds have been discovered far from any possible source and long after they were most likely used. Polar bears, for example, have been shown to have concentrations of certain pesticide compounds 3 billion times greater than the seawater around them.

Case Study   Effects of Pesticides on Human Health

The effects of pesticides on human health can be divided into two categories: acute effects and chronic effects. Acute effects include poisoning and illnesses caused by high doses and accidental exposures. Chronic effects result from long-term exposure to low levels of pesticides in the air, water, or food. Chronic effects can include cancers, nervous system disorders, and decreased fertility in both men and women.

The World Health Organization (WHO) estimates that 25 million people are poisoned by pesticides each year, and 20 000 die as a result. Most illness and death comes from occupational exposures in developing countries, where people use pesticides without wearing protective clothing and often without a proper understanding of the health risks.

When properly applied, most pesticides pose little danger to the person using them. In developed countries such as Canada, the Pest Management Regulatory Agency (PMRA) is responsible for pesticide regulation. Under the Pest Control Products Act, pesticides are strictly regulated to make sure they pose minimal risk to human health and the environment. Through this Act, Health Canada registers pesticides after a thorough science-based evaluation that ensures any risks are acceptable. As well, pesticides are re-evaluated every 15 years to ensure they meet scientific standards and support sustainable pest management. In many parts of the developing world, on the other hand, regulations for the use of pesticides may be poorly enforced.

For most people, the most critical health problem is unintentional exposure to very small quantities of pesticides. Studies of farmers who are exposed to pesticides over many years show that they have higher levels of certain kinds of cancers compared to the general public. There are also questions about the effects of chronic, low-level exposures to pesticide residues in food or through contaminants in the environment.

Working with plants and pesticides in a greenhouse, this scientist is wearing proper protective gear, including protective clothing, gloves, and a respirator.

In U.S. studies of a wide range of foods between 1994 and 2000, 73% of conventionally grown food had residue from at least one pesticide and were six times as likely as organic foods to contain multiple pesticide residues.

Research and Analyze

1. Do research on pesticide drift. What is it? How does it affect human health? Choose one actual incident and explain what happened and how people were affected. What steps can be taken to reduce or eliminate pesticide drift?

2. Do research about bioaccumulation. What evidence is there of bioaccumulation in the tissues of fish, birds, and mammals? What evidence is there of bioaccumulation of pesticides in the tissues of humans? What are some possible health effects?

Communicate

3. Explain why is it important that agencies such as PMRA regulate pesticides.

Irrigation

irrigation adding water to an agricultural field to allow certain crops to grow where the lack of water would normally prevent their cultivation

Crop **irrigation** accounts for 70% of water consumption around the world. Agricultural water use varies from one place to another due to variations in rainfall, soil type, and farming methods. In Canada, about 8% of water is used for crops. In India, over 90% of water is used for agricultural purposes. However, in Kuwait, where water is especially scarce, only 4% is used for crops. Just as the amount of water used to irrigate crops varies from one area of the world to another, the method of irrigation also varies. **Figure 5.10** shows several methods of irrigation used by farmers.

Figure 5.10 Irrigation Methods

Surface Irrigation

Surface irrigation (also called flood irrigation) supplies water to crops by having the water flow over the field in canals or ditches. The land has a small downhill slope so that the water flows from the source into the fields. This type of irrigation can be inefficient. As much as half of the water can be lost through evaporation. Much of the rest runs off before plants can absorb it.

Spray irrigation uses a sprinkler system to spray water into the air above the plants. Sprinkler systems can also be inefficient. As much as 35% of the water can evaporate in the air before it ever reaches the soil.

Spray Irrigation

Drip irrigation, also called trickle irrigation, uses a series of pipes with strategically placed openings to deliver water directly to the roots of plants. Although this method conserves water, it requires an extensive network of pipes.

Drip Irrigation

The amount of water used for irrigation and for livestock to drink continues to increase throughout the world. Future agricultural demand for water will depend on factors such as the following:

- the cost of water for irrigation
- the demand for agricultural products
- government policies
- the development of new technology
- competition for water from a growing human population

Pause and Reflect

7. What is a pesticide?

8. What are two effects that synthetic pesticides can have on ecosystems?

9. Critical Thinking If you were a farmer, what factors would you consider when deciding how to irrigate your crops? Explain your reasoning.

Genetically Modified Crops

Genetic engineering involves taking a small section of DNA called a gene from one organism and inserting it into another organism. The DNA can be taken from any source—even an entirely different species. An organism with a genetic make-up (DNA) that has been altered by scientists in this way is a *genetically modified organism (GMO)*.

In agriculture, two kinds of genetically modified organisms have received attention. One involves inserting genes from a specific type of bacterium. The bacterial genes produce a material that destroys the gut of insects that eat it. Crops that have been genetically modified to contain these bacterial genes are called *Bt* crops. (*Bt* is short for the type of bacterium used, *Bacillus thuringiensis*.) It is likely that some of the food you eat and clothes you wear were grown from Bt crops, because these commonly include soybeans, potatoes, corn, and cotton.

A second kind of genetic engineering involves inserting a gene for herbicide resistance into the DNA of crop plants such as corn, soybeans, and canola. By planting a herbicide-resistant crop, a farmer can plant the crop with very little preparation of the field to rid it of weeds. When both the crop and the weeds begin to grow, the field is sprayed with a specific herbicide that will kill the weeds but not harm the crop because the crop plants contain genes that allow it to resist the herbicide's effects.

Table 5.1 lists several kinds of traits that have been modified by genetic engineering and the kinds of crops involved. Worldwide, 70% of soybeans, 49% of cotton, 26% of corn, and 21% of canola are genetically modified.

Table 5.1 Some Genetically Modified Crops and Their Altered Traits	
Modified Trait	**Crop**
Insect resistance	Tomato, potato, corn, cotton
Herbicide resistance	Corn, canola, cotton, flax, alfalfa, sugar beet, sugar cane, rice, radish
Insect and herbicide resistance	Corn, cotton
Virus resistance	Zucchini, papaya

Mini-Activity 5-3 Benefits and Risks of Genetically Modified Crops

Research more about the benefits and risks of genetically modified (GM) crops.

- How do GM crops benefit farmers?
- How have they benefited the world's food supply?
- What are some of the risks associated with GM crops?
- What happens if pests become resistant to GM crops?
- How can superweeds be created and how can they negatively affect ecosystems?

- What is known about the long-term effects on human health of consuming foods that contain GM ingredients?

Present the results of your research in a format approved by your teacher. Hold a class debate about the benefits and risks of GM crops. You may consider having classmates take on different roles during the debate, including farmer, consumer, manufacturer, and scientist.

Summary

- Agriculture is the practice of raising plants or livestock for food and other human needs.
- A monoculture is a single crop that is grown on a large area of land.
- Synthetic fertilizers replace soil nutrients removed by plants.
- Farmers use synthetic pesticides to prevent loss of crops and food supplies to pests or infection.

- Worldwide, agriculture accounts for the greatest use of water. Crop irrigation accounts for 70% of water consumption.
- Genetic engineering has allowed scientists to insert specific pieces of DNA into the genetic make-up of organisms. An organism with a genetic make-up (DNA) that has been altered is called a genetically modified organism.

Review Questions

1. Use a T-chart to organize the advantages and disadvantages of monocultures. K/U C

2. Why are synthetic fertilizers used? K/U

3. Make a flowchart to organize the events that occur if too many nutrients enter an aquatic ecosystem. C

4. Suppose you are an official for the Pest Management Regulatory Agency (PMRA) who is going to recommend whether an agricultural pesticide can remain on the market or should be banned. T/I A

   a) What are some facts you would need to make your recommendation?

   b) How would you go about finding the facts?

   c) Who would you want to talk to about the pesticide before making your recommendation? Why?

5. Identify the type of irrigation shown below. Choose one other type of irrigation and compare it to the type shown below. K/U

6. Do non-farmers have an interest in how water is used for irrigation? Why or why not? Under what conditions should the general public be involved in making these decisions with farmers who are directly involved? T/I A

7. Explain how crops can be genetically modified to be insect resistant. K/U

8. What are the advantages of herbicide-resistant crops? What problems could result from the repeated planting and growth of herbicide-resistant crops? K/U A

9. In one study, a population of houseflies regularly exposed to the pesticide Cyfluthrin was removed from the poultry CAFO where they lived and taken to a lab. There, the population was divided into three groups, each group was exposed to a specific dose of Cyfluthrin, and the rate of mortality was recorded for each dose. Also in the lab, a population of the same species of housefly that had never been exposed to Cyfluthrin was exposed to a standard dose of the pesticide. The mortality rate for this population was also recorded. The results of the experiment are shown in the data table below. T/I A

Population	Dosage Level of Cyfluthrin (ng/cm^2)	Mortality Rate (%)
Houseflies from poultry facility – Group 1	8.3 (standard dose)	0
Houseflies from poultry facility – Group 2	83	10
Houseflies from poultry facility – Group 3	830	62
Houseflies from regular population	8.3 (standard dose)	100

   a) Explain what happened to the three groups of houseflies from the poultry facility after they were exposed to Cyfluthrin in the lab. What is the significance of the dosage level the flies were exposed to?

   b) What happened to the houseflies from a regular population when they were exposed to the standard dose of Cyfluthrin?

   c) What do the results of this experiment show? Why are they important to both farmers and non-farmers?

Sustainable Agriculture

The shift to mechanized agriculture and monocultures has dramatically increased the amount of food that farmers are able to grow. Despite the gains that result from growing food on such large scales, there have been negative consequences as well. With respect to the environment, these include loss of topsoil, pollution of air and water, and an increased need for synthetic chemicals. In economic terms, these consequences include increasing costs of materials and energy. In social terms, these consequences include lowered standards for living and working conditions for people who work on the farm or in the production facilities.

Today, many people—in science, business, government, and communities large and small—believe that the current practices of mechanized agriculture are not sustainable in the long term. **Sustainable agriculture** is the practice of growing and producing food in ways that meet the needs of the present while also enhancing the health of farmland, human ecosystems, and natural ecosystems so that the needs of people years into the future can also be met. Sustainable agriculture addresses environmental, economic, and social needs throughout the process of food production.

sustainable agriculture producing food to meet the needs of the present without compromising the ability of future generations to meet their needs

Polyculture

One of the main differences between mechanized agriculture and sustainable agriculture is the use of **polycultures**. Unlike monocultures, polycultures contain a diversity of crops grown on the same plot of land, as shown in **Figure 5.11**. Polycultures mimic natural ecosystems, allowing for increased biodiversity not just of the crops planted, but also the species that visit the crops. Polycultures help conserve topsoil, while at the same time reduce the need for water, fertilizers, and pesticides.

polyculture an agricultural practice in which diverse species are raised in the same area

Figure 5.11 This polyculture includes day lilies, beans, corn, and raspberries.

In a two-year study done by scientists in China, farmers grew a mixture of rice varieties in the same fields. Some of the varieties were resistant to a fungus that spreads easily in rice plants. By mixing the disease-resistant rice with the types that are vulnerable to the fungus, farmers increased crop yield by 89% and reduced infection from the fungus by 94% compared to rice plants grown in a monoculture. By the end of the study, farmers could grow rice in polycultures without any fungicide.

Companion planting is planting two or more plant species close to each other so that some benefit, such as pest control, nutrient absorption, or higher yield, occurs. Companion planting is a type of polyculture. One example of companion planting used by many Aboriginal peoples is the "three sisters": corn, beans, and squash. The corn provides a good structure on which the beans can grow up; the beans return nitrogen to the soil, which benefits the squash; and the squash provides dense leaf coverage and ground cover, which helps keep weeds from growing.

companion planting planting two or more plant species close to each other so that some benefit occurs

Soil Conservation

Conserving soil and maintaining soil fertility in a sustainable manner are important to all farmers no matter which agricultural methods they practise. Soil conservation involves both reducing soil erosion and maintaining soil fertility. One of the best ways to conserve soil is to reduce the loss of topsoil due to erosion. Strip cropping, shown in **Figure 5.12**, involves planting alternating strips of one type of crop with another crop that totally covers the soil, called a *cover crop*. Examples of cover crops include clover, rye, and alfalfa. A cover crop helps reduce soil erosion and water run-off. Another way to reduce erosion is to leave the stalks, stems, and leaves of plants on the land after a crop is harvested. Some farmers plant cover crops right after harvesting to save topsoil.

Figure 5.12 In strip cropping, farmers plant one type of crop, such as corn, soybeans, cotton, or sugar beets, in alternating strips with a cover crop.

Soil Fertility

Retaining topsoil is the best way to maintain soil fertility. Restoring the nutrients that are lost as crops grow and are harvested is another important part of keeping soil fertile. Mechanized farming relies on synthetic fertilizers to restore lost nutrients. Now farmers are looking at more sustainable ways to maintain soil fertility.

One method involves crop rotation. **Crop rotation** is the practice of growing different crops at different times on the same land in order to preserve nutrients in the soil. A usual rotation sequence is for farmers to plant a crop that removes nutrients from soil, such as corn or cotton, one year. The next year they plant the same areas with legumes, such as beans or peas. Legumes add nitrogen back into the soil.

crop rotation the practice of growing different crops at different times on the same land

Organic Fertilizers

Farmers who are practising sustainability are less likely to use synthetic fertilizers when they need to replace lost soil nutrients. Instead they may choose from a variety of organic fertilizers. These include compost, mulch, and green manure. **Compost** is produced when micro-organisms in soil break down organic matter, such as leaves, food wastes, and crop residues, in the presence of oxygen. Aside from adding nutrients to soil, compost can also help reduce soil erosion and help control pests. **Mulch** is a general term for a protective ground cover than can include manure, wood chips, straw, seaweed, leaves, and other organic products. Like compost, mulch adds nutrients to soil. It also helps to reduce soil erosion and maintain soil temperature and moisture.

compost produced when micro-organisms in soil break down organic matter in the presence of oxygen

mulch general term for protective ground cover

Replacing synthetic fertilizers with green manure can help reduce the amount of nutrients in surface water run-off, while still nourishing soil. *Green manure* is produced by growing vegetation such as grasses or legumes on a field and then ploughing it into the topsoil at a later time. This increases the amount of nutrients and organic matter in the soil for the commercial crop that will be planted next.

Inquiry Lab 5B, Characteristics of Compost Samples, on page 166

Integrated Pest Management

Integrated pest management (IPM) is a system that uses biological organisms, chemical substances, and crop rotation to control pest populations. IPM is often used in mechanized agricultural systems, and it is considered an important aspect of sustainable agriculture as well.

The goal of IPM is to keep pest populations at a low enough level so that the farmer does not experience economically unacceptable losses. Chemical pesticides are used only when other options have failed to control pest populations. IPM uses biological methods, such as natural predators, parasites, or disease-causing organisms, to help keep pest populations in check. As shown in **Figure 5.13**, one type of parasitic wasp invades the eggs of the European corn borer, a damaging insect that feeds on sweet corn, peppers, snap beans, and apples. The wasp can be used as part of an IPM system to combat the European corn borer.

Rotating crops, changing planting times, and using resistant varieties of crops are also part of an IPM system. Steps in IPM include identifying pests, monitoring their population levels, and deciding when action must be taken to control the pest. Deciding which control method to use and evaluating its success are also part of IPM.

> **integrated pest management (IPM)** a system that uses biological organisms, chemical substances, and crop rotation to help keep pest populations under control

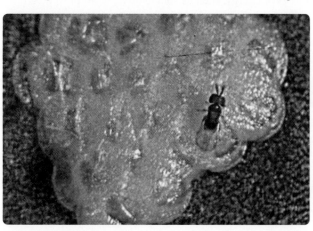

Figure 5.13 This parasitic wasp is invading the eggs of a European corn borer. The wasp is an example of biological pest control.

Pause and Reflect

10. What is sustainable agriculture?

11. How does strip cropping help maintain soil fertility?

12. Critical Thinking Why do you think many farmers, no matter which agricultural methods they practise, use integrated pest management?

Mini-Activity 5-4   Biological Pest Control

Choose one of the examples of biological pest control in the list on the right and research the ecological and economic advantages and disadvantages of its use. Write a short summary of when and how the species is used to fight pests. If you were a grower, would you choose to use the method you researched as part of your IPM? Why or why not?

- using nematodes to control crop damage from grubs
- using ladybird beetles (ladybugs) to control aphids
- spraying *Bacillus thuringiensis* (*Bt*) to control the tomato hornworm, corn rootworm, or cabbage loopers
- spraying neem (an extract from the *Azadirachta indica* tree) to control various crop pests and diseases

Irrigation Practices Used in Sustainable Agriculture

drip irrigation—**see** section 5.2 on page 144

Drip irrigation delivers water directly to the roots of plants. Since specific amounts of water are released just above plant roots, nearly all of the water is used by the plants. A 90% efficiency rate is possible with a well-designed and maintained drip irrigation system. The use of drip irrigation systems can also reduce fungal diseases, because the amount of water on plant leaves is less than the amount when other irrigation systems are used. One drawback of drip irrigation is that the initial cost is more than for other irrigation systems.

wastewater—**see** section 10.2 on page 318

In some parts of the world, recycled water is used to irrigate crops. Recycled water is water from wastewater treatment plants that has been treated so it is safe to use. The use of recycled water for irrigation benefits the environment by reducing the amount of water drawn from the ground, wetlands, and other ecosystems. In Canada, recycled water is used for irrigation on a small scale and is applied to non-food crops.

Renewable Energy Resources and Sustainable Agriculture

Another aim of sustainable agriculture is to reduce the use of fossil fuels and replace them with renewable sources of energy, such as solar power, wind power, or hydropower. The farm shown in **Figure 5.14** uses solar power to help reduce the use of fossil fuels. Some farms obtain all of their energy from renewable sources. Some farmers may also use the natural gas produced from farm waste as a renewable energy source.

Figure 5.14 This family farm has installed solar panels on the roofs of its outbuildings.

Mini-Activity 5-5   Buying Local

Farmers are not the only participants in sustainable agriculture—consumers are as well. One factor that has become an important part of sustainable agriculture is buying food that is grown in nearby locations. Many grocery stores now provide information about the source of foods, especially fruits and vegetables. Some studies show that buying foods from local sources reduces the amount of fossil fuels used to transport food.

- Research more information about how far food in a local grocery store has travelled to reach the shelf.

- What other factors should a consumer consider when determining whether buying locally grown food is more sustainable than buying food shipped a longer distance?

- How does buying local support sustainable agriculture? How does it benefit the environment? How does it benefit local culture or society?

- Identify any disadvantages or controversies associated with buying food grown locally.

Organic Farming

In many countries, including Canada, the United States, France, and Australia, food or other products that are labelled organic meet strict standards defined by the government. An **organic product** is one that has been produced without the use of synthetic fertilizers or pesticides, hormones, antibiotics, synthetic additives, genetically modified ingredients, or irradiation.

In Canada, about 2% of farms are certified organic farms. About 40% of those organic farms are in Saskatchewan, Manitoba, and Alberta. In 2011, the largest growth in organic farming occurred in Ontario and Québec. Aside from producing food or other products according to organic standards, organic farmers also use sustainable practices to increase biodiversity, maintain soil fertility, reduce soil erosion, conserve energy, and conserve water with innovative irrigation practices.

Typically, it takes three to five years for a farmer in Canada to make the change from conventional to organic farming profitable. Organic foods, which carry a label like the one shown in **Figure 5.15**, are becoming increasingly accepted as the benefits of sustainable agriculture become more widely understood. Besides farmers' markets and natural food stores, many large supermarkets now carry organic produce and other foods.

organic product a product that has been produced according to standards defined by government or independent agencies

Figure 5.15 Labels like the one shown here identify products such as fruits, vegetables, grains, dairy products, eggs, and meats as certified organic.

Pause and Reflect

13. What are the benefits of using recycled water to irrigate crops?

14. Why do farmers save seeds?

15. Critical Thinking Why do you think a consumer would be motivated to buy an organic product?

Mini-Activity 5-6   What's in a Label?

There are many different claims on food labels today, such as certified organic, non-GMO ingredients, all natural, free-range, cage-free, free-run, raised without the use of antibiotics, raised without the use of hormones, and grain-fed. These claims and others may appear on plant and animal products.

- Examine some of the claims that appear on food labels in your local grocery store or examples provided by your teacher.

- Research what these claims mean. Is there a defined standard for each of the claims? If not, what does the claim really mean?

- How would you know if a claim is misleading or not?

- Which agency or agencies oversee the regulation of claims on food labels in Canada?

Case Study   Native and Heritage Varieties

The fruits and vegetables you see in the grocery store are dominated by a small number of *hybrid* varieties. Hybrids, which are created by crossing two plant lines, are bred to enhance certain characteristics such as size, appearance, or longer shelf life. However, hybrids may be less flavourful and less nutritious. Hybrid crops often cannot produce fertile seeds. This means that farmers must purchase new seeds each year from the commercial seed market, much of which is controlled by large corporations.

Advantages of Native and Heritage Varieties

An alternative to growing hybrids is to plant native and heritage varieties of crop plants. Heritage (or heirloom) varieties are usually defined as being at least 50 years old, and are often linked to a particular region. Both native species and heritage varieties are non-patented, naturally pollinated plants that produce fertile seeds.

Native and local heritage varieties are adapted to a region's environmental conditions. These adaptations give them increased resistance to pests and disease, which in turn means they require less maintenance and irrigation. Native and heritage varieties that are well chosen for a site grow better with less investment of money and resources.

Genetic diversity refers to the variation naturally found among individuals. This diversity is important because when environmental conditions change, there is more chance that some individuals will survive. Maintaining more plant varieties in the gene pool through continuing to grow native and heritage crops gives us greater genetic diversity to draw on if changing conditions or drastic events ever threaten agricultural productivity.

Seed Libraries and Seed Banks

To help maintain genetic diversity, mechanisms for distributing and preserving seeds have been established. *Seed libraries* provide seeds and information about heritage plant varieties. Members grow these varieties, and then promote further seed sharing via seed swaps. *Seed banks*, which are not generally open to the public, preserve seeds in case some major disaster destroys other seed reserves. Some seeds can be kept for decades by drying and storing them in cold conditions. There are currently about 6 million seed samples stored in seed banks worldwide. However, this represents only a fraction of the world's biodiversity. A combination of seed banking, seed sharing, and the increased planting of native and heritage crops may be the best way to preserve this diversity for the future.

Inside the Svalbard Global Seed Vault in Norway, seeds are stored in boxes in freezing temperatures.

Analyze and Conclude

1. In a group, brainstorm how you could encourage people to plant and purchase heritage crops in your community. Give at least three detailed ideas.

2. With a partner, analyze the pros and cons of either hybrids or heritage crops. Record your points in a PMI chart.

Communicate

3. Suppose Canada was proposing to establish a seed bank. Write a rough draft of a letter you could send to your local member of Parliament to express your support or lack of support for this proposal.

Summary

- Sustainable agriculture is the practice of producing food to meet the needs of the present without compromising the ability of future generations to meet their needs.

- Practices used in sustainable agriculture include polyculture, planting cover crops, using organic fertilizers, using integrated pest management, using green manure, and reducing reliance on fossil fuels.

- Conserving soil and maintaining soil fertility in a sustainable manner are important to all farmers no matter which type of agricultural methods they practise.

- Soil conservation involves using a variety of ways to reduce soil erosion and restore soil fertility.

- Integrated pest management (IPM) is a system that uses biological organisms, chemical substances, and crop rotation to help keep pest populations under control.

- A specific type of farming that uses sustainable agriculture practices is organic farming. In many countries, including Canada, the United States, France, and Australia, food or other products that are labelled organic meet strict standards defined by the government.

Review Questions

1. Use a Venn diagram to compare and contrast sustainable agriculture practices and conventional mechanized agriculture practices. K/U C

2. Explain why companion planting is a type of polyculture. K/U

3. Study the data in the table below. T/I A

Soil Cover and Soil Erosion

Crop System	Average Annual Soil Loss (tons/hectare)	Rainfall Run-Off (%)
Bare soil (no crop)	41.0	30
Corn, planted continuously	19.7	29
Wheat, planted continuously	10.1	23
Rotation: corn, wheat, clover	2.7	14

a) Which crop system results in the least amount of soil loss annually?

b) Which crop system results in the least amount of rainfall run-off?

c) Why is it important to reduce rainfall run-off from farms?

4. Make a table to show the similarities and difference among compost, mulch, and green manure. K/U C

5. What is IPM and how is it used to control pests? K/U

6. Which of the three types of irrigation discussed in section 2 is best suited to the practices of sustainable agriculture? Explain your reasoning. K/U

7. As a consumer, would you consider buying food that is certified organic, even if it may be more expensive than the same non-organic product? Why or why not? A C

8. In Indonesia, after years of farmers applying pesticides to control brown planthoppers (an insect that destroys rice crops), the insects developed resistance to the pesticides. In 1986, the government banned the use of 56 of 57 pesticides, forcing farmers to allow natural predators to combat the pests and spraying only when absolutely necessary with chemicals specific to planthoppers. The table below compares data from when farmers used chemicals to manage the pests to when they switched to biological control. T/I A

Alternative Pest-control Strategies

	Pest Management Using Chemicals	Pest Management Using Biological Control
Number of times pesticide used in rice season	4.5 applications	0.5 applications
Cost to farmers per hectare	7.5 rupiah (local currency)	2.5 rupiah
Cost to government per hectare	27.5 rupiah	2.5 rupiah
Rice yield per hectare	6 tons	7.5 tons

a) How did the change in pest control affect costs to farmers and the government?

b) How did the change affect yield?

c) Would you agree with the Indonesian government, which declared the program a success? Explain.

9. What criteria should be used to determine whether farmers should use sustainable practices? How would your response differ if you were a farmer, a farmer's neighbour, someone downstream of a farm, or someone far from farming regions? A C

Livestock Production

Methods of Raising Livestock

Meat from cows, hogs, chickens, and other animals is an important source of protein, iron, fats, and other nutrients that give us the energy to lead productive lives. Dairy products are also a key protein source: globally, we consume more than twice as much dairy as meat. Meat and dairy consumption have quadrupled since the 1970s, with China representing about 40% of that increased demand.

About 50% of livestock are raised by grazing on rangelands and enclosed pastures, as shown in **Figure 5.16A**. Rangelands are ecosystems dominated by grasses, wildflowers, and shrubs on which livestock roam and feed. Enclosed pastures are fenced areas covered by grasses and legumes such as alfalfa and clover, on which livestock graze. The other 50% of livestock are raised in confined animal feeding operations (CAFOs) and feedlots. CAFOs, shown in **Figure 5.16B**, are energy-intensive industrialized systems where animals are housed and fed for rapid growth. Animals are confined to giant enclosures with up to 10 000 hogs or 1 million chickens in a barn complex, or 100 000 cattle in a feedlot.

Figure 5.16 (A) Rangelands can be found in several different ecosystems, including grasslands, woodlands, and deserts. **(B)** These chickens are housed and fed within this enclosed coop.

Advantages of Rangelands, Enclosed Pastures, and CAFOs

The most important advantage of raising livestock using these methods is that they increase meat and dairy production. Since the late 1960s, meat production has doubled globally. In Canada, both rangelands and pastures are an important part of the grazing industry, chiefly for beef cattle. Rangelands and seeded pastures provide food for livestock. They also provide ecosystem services including providing habitat for wildlife, reducing soil erosion, and regulating the flow and quality of water. Agriculture and Agri-Food Canada helps to maintain rangelands and pastures so that they remain in a healthy state.

Aside from increased production, CAFOs can reduce the risk of overgrazing on rangelands and yield higher profits. Due to the lower cost associated with efficient housing of animals, the cost of meat, milk, and eggs is lower. CAFOs also provide employment at the local level.

Disadvantages of Rangelands, Enclosed Pastures, and CAFOs

- Rangelands can be overgrazed. *Overgrazing* occurs when livestock are allowed to eat so much of the grasses and other plants on rangelands that the ecological health of the habitat is damaged. Overgrazing can lead to a decrease in important plants on which animals graze, an increase in weeds, and an increase in soil erosion. When rangelands are overgrazed in dry areas, a series of ecological and climate change events can lead to **desertification**, as shown in **Figure 5.17**.

- Allowing livestock to graze on rangelands also can affect biodiversity. To increase the productivity of rangelands, management techniques may specifically eliminate certain species of plants that are poisonous or not useful as food to grazing animals. In some cases, the populations of native animals are reduced if they are a threat to livestock through predation or by spreading disease. In addition, the selective eating habits of livestock tend to reduce certain species of native plants and encourage others.

- CAFOs require specially prepared mixtures of soy, corn, and animal protein that maximize animals' growth rate. These mixtures are not the normal food that animals would eat. These foods require large inputs of energy, mostly from fossil fuels, to produce. As a result of this and other factors, it takes about 16 times as much fossil fuel energy to produce a kilogram of beef from a CAFO than it takes to produce a kilogram of vegetables or rice.

- CAFOs fatten animals quickly and efficiently, but create enormous amounts of waste and expose livestock to unhealthy living conditions. The animal waste contributes nutrients to surface water run-off, which can pollute aquatic ecosystems. Bacteria in the animal waste can also pollute both surface water and ground water.

- CAFOs require the constant use of antibiotics, which are mixed in the animals' daily feed. The antibiotics are needed, because the chances of developing infections and diseases are so high due to having so many animals living in such close quarters together. In the United States, the amount of antibiotics added to animal feed each year is about eight times the amount of antibiotics used to treat human illnesses.

desertification the conversion of arid and semi-arid lands into deserts by inappropriate farming practices or overgrazing

Figure 5.17 Overgrazing contributed to the desertification and soil erosion shown here.

Pause and Reflect

16. What are the different ways in which livestock can be raised?

17. What is overgrazing?

18. Critical Thinking What types of problems do you think could result from using antibiotics in raising livestock?

Mini-Activity 5-7 — Livestock and Greenhouse Gases

The United Nations Food and Agriculture Organization estimates that livestock produce 20% of the world's greenhouse gases. This is more than is produced by transportation. Do research to find answers to the following questions.

- Which greenhouse gases do livestock release?

- Does the way in which the livestock are raised affect how much of a particular greenhouse gas they release?

- How much do these gases contribute to global climate change?

- What are some solutions to reduce the amount of greenhouse gases released by livestock?

Present the results of your research in a format approved by your teacher.

Case Study   What's in Livestock Feed?

In recent years, scientists, livestock producers, and consumers have become concerned about some of the substances that are added to livestock feed. These include antibiotics, growth hormones, and animal by-products. Each of these substances is associated with its own set of concerns.

Raising large numbers of animals in confined animal feeding operations (CAFOs) can allow disease to spread quickly among the livestock. To combat this, low levels of antibiotics are added to animal feed on a long-term basis. The widespread use of antibiotics in livestock operations has contributed to many pathogens becoming resistant to antibiotics. Studies conducted by Health Canada show that antibiotic-resistant strains of bacteria, including *Salmonella*, commonly occur in Canadian livestock. The constant use of antibiotics on livestock can also make pathogens more resistant to antibiotics that are used to treat human illnesses. This has made it difficult for some types of infections in humans to be treated.

Another substance that is added to the feed of livestock, particularly beef cattle, is growth hormones. Growth hormones help animals grow faster so that they can be sold sooner. Some types of growth hormones increase lean tissue growth and reduce fat. This results in a healthier product that is sold at a lower cost to the consumer. Concerns about the use of hormones in livestock are related to their impact on human health and whether they may cause cancer.

Between 1993 and 2010, almost 200 000 cases of bovine spongiform encephalopathy (BSE) were confirmed in the United Kingdom. BSE, also referred to as "mad cow disease," is caused by a mutated protein that damages the central nervous system of cows and eventually kills them. The likely source was cattle feed that contained animal by-products infected with BSE. There is strong evidence that consumption of meat containing the protein that causes BSE causes a similar neurological disease in humans, called Creutzfeldt-Jakob disease (CJD). As shown in the graph, by 2012 there were 23 confirmed cases of BSE in North America, 19 in Canada, and 4 in the United States. As a result of the BSE outbreak in the United Kingdom and the effect it had on human health, Canada updated laws concerning animal feed in 2007. The updated law prohibits most proteins, including those potentially infected with BSE, from all animal feeds, pet foods, and fertilizers.

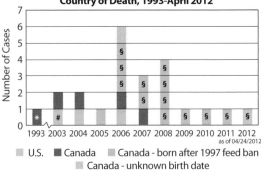

BSE Cases in North America, by Year and Country of Death, 1993-April 2012

Number of Cases

1993 2003 2004 2005 2006 2007 2008 2009 2010 2011 2012
as of 04/24/2012

■ U.S.   ■ Canada   ■ Canada - born after 1997 feed ban
■ Canada - unknown birth date

*Imported UK to Canada   #Imported Canada to US
§Born after March 1, 1999

Source: Centers for Disease Control

Research and Analyze

1. Research more about the long-term effects the use of antibiotics in animal feed has on human health. What is the significance of antibiotic-resistant strains of bacteria to human health? How important is this issue to public health? Is it possible to raise livestock without antibiotics and still make a profit? How?

2. Is there a correlation between the use of growth hormones in beef and dairy cattle and cancer in humans? Do research and summarize your results.

3. Research more about the use of animal by-products in livestock feed and the incidence of BSE and CJD. Why should people be concerned about this issue? Do you think the updated laws will protect cattle and humans from becoming infected with the proteins that cause BSE and CJD? Why or why not?

Communicate

4. Choose one of the issues you researched and design an information campaign about it. Include the facts about the issue, as well as your opinion about its importance and solutions to the problem.

Raising Livestock Sustainably

- Rotational grazing, shown in **Figure 5.18**, mimics the effects of wild herds. **Rotational grazing** involves confining animals to a small area of pasture for a short time, often only a day or two, before shifting them to a new location. Forcing livestock to eat everything equally, to trample the ground thoroughly, and to fertilize heavily with their manure before moving on helps keep weeds in check.

Figure 5.18 In rotational grazing, livestock graze in a small pasture area for a short time, and then are moved to a different area of pasture.

- Smart pasture operations (SPOs) offer an economically practical alternative to CAFOs. SPOs feed animals such as beef and dairy cattle a grass diet, which requires less maintenance and energy than feeding them corn and soybean. As well, eating grass eliminates health problems that result when the animals are fed an unnatural diet of grain. SPOs are smaller than CAFOs, which means that the waste produced by the animals can be managed more efficiently than in CAFOs. SPOs are also less crowded, which leads to less use of antibiotics.

- Organically raised livestock are another sustainable alternative to CAFOs. Meat, dairy products, or eggs that are certified organic means that the animals were raised without growth hormones or antibiotics. They were not fed anything that contained antibiotics, growth hormones, genetically modified ingredients, animal by-products, fertilizers, or pesticides.

- Another sustainable alternative for raising livestock involves a polyculture system. In Argentina some large farms allow cattle and chickens to graze on grassland for five years, and then the farmers switch to growing grain. Using this system, farmers can grow grain on the naturally fertilized soil for three seasons in a row without adding any fertilizer. Then, they switch back to grazing livestock.

Mini-Activity 5-8 — Cutting Back on the Consumption of Meat

Some people believe that reducing the amount of meat and dairy products in our diet is another sustainable solution. Globally, over one third of grains grown each year are used as livestock feed. We could feed about eight times as many people by eating those grains directly, rather than converting them to animal protein.

- Research more about how much consuming meat and dairy products contributes to a person's ecological footprint, and how making dietary changes can increase sustainability.

- Look up the differences among the terms *vegetarian*, *lacto-ovo vegetarian*, *semi-vegetarian*, *flexitarian*, and *vegan*.

- Find out about the sources of protein consumed by vegetarians and vegans.

Hold a debate about whether people in developed countries such as Canada should consider reducing or eliminating animal proteins from their diet.

Summary

- Meat from cows, hogs, chickens, and other animals is an important source of protein, iron, fats, and other nutrients that give us the energy to lead productive lives. Dairy products are also a key protein source.

- About 50% of livestock is raised by grazing on rangelands and enclosed pastures. The other 50% is raised in confined animal feeding operations and feedlots.

- Rangelands, enclosed pastures, and CAFOs have benefits and risks.

- *Overgrazing* occurs when livestock are allowed to eat so much of the grasses and other plants on rangelands that the ecological health of the habitat is damaged.

- Confined animal feeding operations require large inputs of energy and the constant use of antibiotics; they create enormous amounts of waste.

- Methods of raising livestock sustainably include rotational grazing, raising wild species, small pasture operations, raising livestock organically, and using polyculture systems.

Review Questions

1. Identify and describe the three main ways livestock are raised. **K/U**

2. Make a table to show the advantages and disadvantages of rangelands. **K/U** **C**

3. Study the photo below and explain why CAFOs add antibiotics to animal feed. **K/U**

4. Identify and describe two systems that are alternatives to CAFOs. **K/U**

5. Suppose you have just received 10 hectares of good pasture. This area is equal to about the size of 20 soccer fields. You think you would like to keep some cows on the land and sell their milk to make money. How many cows will give you the most profit? For example, if you keep 10 cows, each cow will have one hectare of pasture on which to graze. If you keep 50 cows, you will get five times more milk, but each cow will have only 0.2 hectare of pasture. You do not want to spend money on extra food. What questions must you ask before you can calculate the optimum number of cows that could graze on your land for an indefinite period? Assume that you want to manage the pasture and cows sustainably, with the least amount of inputs as possible. **T/I** **A**

6. What are the benefits and risks of confined animal feeding operations? **K/U**

7. Suppose you are a farmer who wants to start a CAFO. What conditions would make this a good strategy for you, and what factors would you consider in weighing its costs and benefits? What would you say to neighbours who wish to impose restrictions on how you run the operation? **A** **C**

8. What is rotational grazing? What are its benefits? **K/U**

9. The diagram below shows the number of kilograms of grain needed to produce one kilogram of bread or one kilogram of weight gain in an animal. **A** **C**

a) Which source of animal protein is the most efficient to produce?

b) Which source of animal protein is the least efficient to produce?

c) After considering the information in the diagram, would you make any changes to your dietary choices? Why or why not?

Fisheries and Aquaculture

Fish and Seafood Are Major Sources of Protein

We currently harvest about 95 million tonnes of wild fish and seafood every year. (Seafood includes lobsters, shrimp, crabs, clams, and scallops.) Of this amount, we eat about two thirds directly. The remaining third is used as feed in aquaculture operations. **Aquaculture** is commonly called "fish farming," and it is the breeding, raising, and harvesting of animals in specially designed aquatic environments.

aquaculture the breeding, raising, and harvesting of animals in specially designed aquatic environments

Fish and seafood are the main source of animal protein for about 1.5 billion people in developing countries. In developing countries, people eat mainly locally caught fish. In developed countries, industrial-scale fishing provides most fish and seafood. The development of freezer technology on ocean-going ships allowed annual catches to increase every year by about 4% between 1950 and 1988.

As shown in **Figure 5.19**, aquaculture is becoming increasingly important as a source of fish and seafood production. Salmon farming has been particularly successful. Norway, Chile, Scotland, and Canada are the leading countries in salmon production. New Brunswick produces about 40% of all Canadian salmon.

The production of salmon from aquaculture has increased rapidly. In 1988, less than 20% of the salmon sold were from aquaculture, compared to over 65% in 2004. During the same time period, the production of wild-caught salmon has been relatively constant.

World capture fisheries and aquaculture production

— Aquaculture production
— Capture production

Source: Food and Agriculture Organization of the United Nations

Figure 5.19 The number of fish captured rose steadily until about 1987. Since then, the amount has stayed almost constant. Aquaculture has continued to increase, so total production continues to increase.

Inferring Why do you think that the number of captured fish has remained about the same since 1987?

Overfishing of Marine Fisheries

The United Nations estimates that 70% of the world's marine (ocean) fisheries are being **overexploited** or are in danger of being overexploited. A 2009 study by Dr. Boris Worm of Dalhousie University and Dr. Ray Hilborn from the University of Washington found that 63% of the fish stocks they analyzed worldwide were declining in numbers. They concluded that exploitation would need to be reduced to avoid the collapse of vulnerable fish stocks. In marine ecosystems in which efforts were made to limit overfishing, they found improvement—a sign that the efforts were working.

overexploit to harvest so much of a resource that its existence is threatened

Aside from scientific studies, another sign that marine resources are being overexploited is the change in the kinds of fish being caught. The commercial fishing industry has been trying to market fish species that were previously viewed as unacceptable to the consumer. Examples of "newly discovered" fish in this category are monkfish and orange roughy.

Effects of Overexploitation on Sustainability

Thought Lab 5C, Overfishing and Moratorium of Cod Fishery, on page 167

Because wild fish in the open ocean are a shared resource belonging to no single country, the competition to catch them is high. Rising numbers of boats, along with more efficient technology, allow for the exploitation of a shrinking resource. Boats as big as ocean liners travel thousands of kilometres, using global satellite positioning equipment (GPS), sonar, spotter planes, and other technology to locate and catch large numbers of fish and seafood. This helps meet the growing demand for these sources of protein. **Figure 5.20** shows three main industrial fishing techniques and how they affect the sustainability of ocean ecosystems.

Figure 5.20 Different fishing methods can affect the sustainability of marine ecosystems in different ways.

Longline fishing boats set cables up to 130 km long with hooks every 2 m to catch fish such as tuna, halibut, and swordfish. In the process, longlines also catch unintended animals such as sea birds, turtles, sharks, and dolphins. These unintended captures are called *by-catch*. The loss of marine organisms as by-catch can have significant effects on local populations as well as endangered species. Researchers in Costa Rica believe that longline fishing led to a 60% reduction in the shark population in Costa Rican waters between 1999 and 2009. Another study found that about 250 000 loggerhead and 60 000 leatherback turtles, both endangered species, are accidentally caught in longlines every year.

Trawlers drag heavy nets across the ocean bottom to catch fish and seafood, including cod, flounder, shrimp, and scallops. As the heavy net is dragged across the ocean floor, it damages large areas of habitat. Organisms that live on the ocean floor, such as corals, sponges, and other fish, are killed. As well, clouds of sediment stirred up from the enormous net remain in the water long after the trawler has passed. Some scientists suggest that trawling is more damaging to marine ecosystems than any other type of industrial fishing.

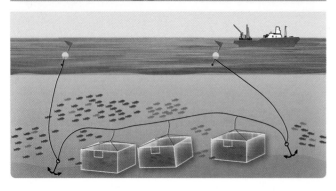

Pots and traps are wire or wooden cages used to catch seafood such as crab, shrimp, and lobster, as well as fish such as cod and Chilean sea bass. Pots and traps are set out along the ocean bottom—usually attached to a line with floating buoys at the surface. By-catch includes small-sized individuals of the target species. Habitat damage can occur when pots or traps are dragged along the bottom when harvested. Traps that are abandoned still trap fish and seafood and can lead to increased death rates in a given area. In the Gulf of Mexico, abandoned traps had enough of an impact on the blue crab fishery that efforts were made to remove the traps.

Pause and Reflect

19. What is aquaculture?

20. What are some of the disadvantages of longline fishing?

21. Critical Thinking How do you think overfishing affects the ecological, economic, and social sustainability of a local area?

Case Study   How Can Aquaculture Affect Natural Ecosystems?

In many aquaculture operations, fish and seafood are raised in pens that are close to shore in marine environments, like the ones shown here. Raising fish in concentrated settings can affect the natural ecosystems in which they are located. For example, increased nutrients are released into the surrounding water from uneaten food and the wastes produced by the fish. This can cause local algal blooms that affect the surrounding natural ecosystem. Shellfish, such as clams and oysters, are especially vulnerable when oxygen levels in the water drop. Net pens anchored near shore also allow the spread of diseases, antibiotics, and other pollutants into surrounding ecosystems.

Sea lice, parasites that infect both farmed and wild fish, have affected salmon raised in aquaculture pens around the world. Sea lice are also a threat to wild salmon where salmon fishing is an important part of the economy. Scientists estimate that in the early 2000s, sea lice associated with salmon farms near Vancouver Island accounted for 90% of the deaths of juvenile wild salmon after deaths from other known causes were counted. A new approach to treating farmed salmon for sea lice has reduced the impact of sea lice on wild populations in the area. There are still concerns, though, that the sea lice are becoming resistant to the parasite treatments.

Another way that near-shore aquaculture operations may affect natural ecosystems is through escaped fish. For example, many of the salmon species raised in farms are not native to the waters in which they are raised. Sooner or later, some of these fish escape. When Atlantic salmon from aquaculture pens escape into the Pacific Ocean, they compete for food with wild Pacific salmon. In other areas, escaped fish may compete with wild individuals for mates, disturb habitat, or become invasive.

Currently, about 60% of all aquaculture production takes place in freshwater ecosystems, and production is growing rapidly. Aquaculture of freshwater species typically involves the construction of ponds, which allows for the

Many aquaculture pens are located close to shore in marine environments. Fish or seafood raised in these pens are fed and managed like livestock.

close management of the fish, shrimp, or other species. The environmental impacts of freshwater aquaculture are similar to those of aquaculture in marine systems. An excess of nutrients from fish wastes can pollute local bodies of water, and the escape of non-native species may harm native species. In addition, freshwater aquaculture involves the conversion of land to a new use. Often the lands involved are mangrove swamps or other wetlands that many people believe should be protected. Mangrove swamps and other wetlands are important habitats for juvenile wild fish and seafood.

Research and Analyze

1. Research aquaculture operations in the Atlantic Provinces. What types of fish and/or seafood are farmed there? In what quantities? What are some of the advantages and disadvantages of aquaculture operations in these provinces? Have these operations affected the sustainability of nearby aquatic ecosystems? If so, how?

Communicate

2. Make suggestions for how aquaculture operations can be run sustainably.

Harvesting Fish and Seafood Sustainably

There are several methods to harvest wild-caught fish that have low rates of by-catch and minimal impact on the environment. For example, fish that feed near the surface in schools, including herring, anchovies, and mackerel, are caught in purse seines, shown in **Figure 5.21**. A spotter plane finds schools of fish. Then the fishing boat traps them in a large circular net called a purse seine. By-catch rates are low, and since the nets do not touch the bottom, no damage occurs.

Figure 5.21 Purse seine fishing is considered to be a low-impact method of catching fish.

Another low-risk method of fishing is hook and line fishing, which has many fewer hooks per line than longlines. This method is used to catch salmon, Pacific cod, flounder, mackerel, and octopus. Diving for scallops, sea cucumbers, sea urchins, or octopus also has no known by-catch and produces minimal damage to the sea bottom. Swordfish caught using harpoons also has no known by-catch and does not damage the seafloor.

Sustainable Aquaculture

Aquaculture in land-based ponds or warehouses, shown in **Figure 5.22**, can reduce many of the problems associated with aquaculture pens in marine and freshwater ecosystems. This is especially true when raising herbivorous fish, such as catfish, carp, or tilapia, which consume less feed than do carnivorous species.

Figure 5.22 In some enclosed land-based ponds, water from the pond does not mix with natural water sources.

One ecologically balanced system uses four carp species that feed at different levels of the food chain. The grass carp feeds largely on vegetation, while the common carp is a bottom feeder. It feeds on decomposing material that settles on the bottom. Silver carp and bighead carp are filter feeders that eat plankton from the water. Agricultural wastes such as manure, dead worms, and rice straw are used to fertilize the ponds and encourage algal growth. This integrated polyculture system typically boosts fish yields by 50% or more per hectare compared with monoculture systems.

Another system integrates agriculture and aquaculture more closely. In China, for example, certain species of fish are raised in rice paddies. These fish help fight rice pests, such as the golden snail, by consuming them. This system of rice-fish farming increases the yield of rice and provides extra income to farmers when they sell the fish.

Mini-Activity 5-9   What to Choose from the Menu?

Consumers like you can increase the sustainability of fisheries by being aware of which species are harvested sustainably and which are not. SeaChoice is a national program that helps Canadian businesses and consumers make seafood choices that support sustainability. SeaChoice works with the Monterey Bay Aquarium in California to rank the sustainability of both wild-caught and aquaculture fish.

- Examine the brochure provided by your teacher.

- What do the terms *best choice*, *good alternative*, and *avoid* mean?

- What factors are considered when determining the sustainability of each species of fish or seafood? Have you eaten any of the fish in any of the categories?

- Would you make different choices about which fish you will consume now? Why or why not?

Summary

- We currently harvest about 95 million tonnes of wild fish and seafood every year.
- Fish and seafood, including lobsters, shrimp, crabs, clams, and scallops, are the main source of animal protein for about 1.5 billion people in developing countries.
- Fish and seafood are the only wild-caught meat sources still sold commercially on a global scale. Aquaculture is becoming increasingly important as a source of fish and seafood production.
- The United Nations estimates that 70% of the world's marine fisheries are being overexploited or are in danger of being overexploited as the number of fishers increases.

- Rising numbers of boats, along with more efficient technology, allow for the exploitation of a shrinking resource. Boats as big as ocean liners travel thousands of kilometres, using global satellite positioning equipment (GPS), sonar, spotter planes, and other technology to catch large numbers of fish and seafood.
- Fishing using purse seines, hook and line fishing, and diving are considered sustainable fishing methods.
- Aquaculture in land-based ponds or warehouses can reduce many of the problems associated with aquaculture pens in marine and freshwater ecosystems.

Review Questions

1. From which sources do we get our fish and seafood? K/U

2. What does it mean if a fishery is overexploited? K/U

3. Infer how the overexploitation of a fish such as herring could affect the sustainability of the ecosystem of which they are a part. (Hint: Think about the food web the herring are part of.) T/I

4. How has technology influenced the fishing industry? K/U

5. Use a Venn diagram to show the similarities and differences between longline fishing and trawling. K/U C

6. Infer how the clouds of sediment kicked up from trawlers (shown in the photograph below) can affect marine ecosystems. T/I

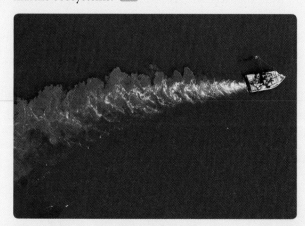

7. How can pots and traps affect both the ecological and economic sustainability of a region? K/U

8. Make a concept map to organize information about how to harvest fish and seafood sustainably. K/U C

9. Make a sketch that shows how the aquaculture system that uses four species of carp is a sustainably balanced system. K/U C

10. The graph below shows the export value in millions of dollars of fish and seafood from different provinces. T/I A

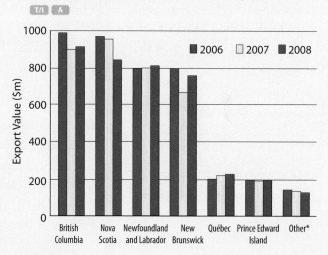

*Ontario, Manitoba, Saskatchewan, Alberta, Northwest Territories, Yukon, Nunavut.
Source: Statistics Canada, International Trade Division.

   a) Which two provinces led exports between 2006 and 2008?

   b) About how much money did Newfoundland and Labrador and New Brunswick earn from exports in 2008, respectively?

   c) Why is it important to all of the provinces and territories in Canada to make efforts to fish sustainability?

11. Why is it important to be an educated consumer? A

Skill Check

Initiating and Planning

Performing and Recording

Analyzing and Interpreting

Communicating

Safety Precautions

- Use appropriate protective equipment such as aprons, goggles, and gloves as well as taking any other safety precautions that are stated in associated Material Safety Data Sheets (MSDS).

- Wash your hands with warm water and soap after completing the Lab.

Materials

- soil samples including clay, loam, potting soil, sand, and silt
- 3 paper plates
- hand lens
- soil pH test kit
- soil nutrient test kit
- triple beam balance or digital scale
- tablespoon or plastic scoop
- microwave oven or conventional oven
- 100 mL graduated cylinder
- water
- paper cup
- marker
- paper towel

Characteristics of Various Soil Samples

Because different species of plants grow in different types of soil, it is important for farmers to know as much as possible about the properties of the soil they are growing crops in. This knowledge helps them monitor soil and maintain or improve its qualities as needed.

Pre-Lab Questions

1. Why is it important for farmers to understand the quality and characteristics of the soil in which they grow crops?

2. What characteristics about your soil sample will you describe and/or measure?

3. What safety precautions will you take during this Lab?

Question

How do the characteristics of different soil types vary?

Procedure

Part A: General Description, pH, and Nutrients in Soil Sample

1. Make a data table like the one shown below

Data Table 1 Soil Characteristics

Soil Sample	General Description	pH	Nitrogen (%)	Phosphorus (%)	Potassium (%)	Moisture (%)	Porosity (%)
Clay							
Loam							
Potting soil							
Sand							
Silt							

2. Obtain a soil sample from your teacher (about 100 grams).

3. Use the hand lens to examine your soil sample. Record your observations in your data table. Describe the general size and shape of the particles. Describe the colour of the soil. Record any other observations about the characteristics of the soil.

4. Use the pH test kit to measure the pH of your soil sample. Record your results.

5. Use the nutrient test kit to measure the percentage of nitrogen, phosphorus, and potassium in your soil sample. Record your results.

Sandy soil

Loam

Part B: Percent of Moisture in Soil Sample

1. Place a paper plate on the balance or scale. Record the mass of the paper plate in grams.

2. Use the tablespoon or scoop to place 25 grams of soil on the paper plate.

3. Place the paper plate with soil into the microwave oven or conventional oven. Follow your teacher's instructions depending on which type of oven you are using.

4. Re-weigh the paper plate and soil sample after it is dry. Record the mass in grams.

5. Subtract the mass of the paper plate you recorded in step 6 from the total mass of the paper plate and dried soil. This is the mass of the dried soil. (Assume that the mass of the paper plate did not change due to any water loss during the drying process.)

6. Calculate the percentage of moisture using Formula 1, below, and record the answer in your data table.

Part C: Percent of Porosity of Soil Sample

1. Measure 100 mL of water into the graduated cylinder, and then pour it into the paper cup.

2. Mark the water line on the outside of the cup.

3. Pour out the water and use a paper towel to thoroughly dry the inside of the cup.

4. Fill the cup with soil to the water line. This 100 mL of soil represents the total volume of soil you are using to calculate the percentage of porosity in your soil sample.

5. Measure 100 mL of water into the graduated cylinder.

6. Slowly pour water into the cup until it reaches the top of the soil. Record the amount of water remaining in the graduated cylinder. Subtract that from 100 mL. The answer represents the volume of pore space in the soil in mL.

7. Calculate the percentage of porosity of your soil sample using Formula 2, below, and record your answer in your data table.

Part D: Compile Data

1. Record your data in the appropriate row of the table your teacher has placed on the board.

2. Once every group has entered their data in the group table, record the results in your own data table.

Analyze and Interpret

1. Which soil sample is the most acidic? Which is the most basic?

2. Which soil sample has the highest percentage of porosity? Which is the least porous?

3. Compare the nutrient content of the different soil samples. Which has the highest amount of nutrients? Which has the lowest?

4. Compare the moisture content on the different soil samples. Which has the highest percentage of moisture? Which has the lowest? How do these results correlate to the sample's porosity?

Conclude and Communicate

5. Summarize the characteristics of your soil sample.

6. Choose one of the other soil samples tested. How did your sample differ from that one?

7. In general, which soil sample do you think would be best for growing agriculture crops? Explain your reasoning.

Formula 1

$$\text{Percent of moisture in sample} = \frac{\text{mass of wet soil (g)} - \text{mass of dry soil (g)}}{\text{mass of dry soil (g)}} \times 100$$

Formula 2

$$\text{Percent of porosity in sample} = \frac{\text{volume of pore space (mL)}}{\text{total volume of soil (mL)}} \times 100$$

Safety Precautions

- Use appropriate protective equipment such as aprons, goggles, and gloves as well as taking any other safety precautions that are stated in associated Material Safety Data Sheets (MSDS).

- Wash your hands with warm water and soap after completing the Lab.

Possible Materials

- compost sample
- paper plates
- soil pH test kit
- soil nutrient test kit
- 100 mL graduated cylinder
- water
- paper cup
- marker
- paper towel

Compost

Characteristics of Compost Samples

Farmers use compost to improve or maintain the quality of soil in several different ways. Compost can improve the drainage of soil, which is important for keeping plant roots healthy. Adding compost to soil improves the physical quality of the soil so that it is easier to till, easier to cultivate crops, and easier to remove weeds. Mixing compost into soil also increases the number of living organisms in soil. In this Lab, you will determine the characteristics of a sample of compost and then answer questions about how your sample might be used.

Pre-Lab Questions

1. Why do farmers use compost?

2. What characteristics about your compost sample will you measure or describe?

3. Why is it important to include safety precautions as part of any laboratory procedure?

Question

How can you determine the characteristics of a compost sample?

Procedure

1. Design a procedure to measure the pH of your compost sample.

2. Design a procedure to measure the percentage of nitrogen, phosphorus, and potassium of your compost sample.

3. Design a procedure to measure the porosity of your compost sample.

4. Include a method by which you will record data in each procedure.

5. Include safety precautions you will take during each procedure.

6. Include a proper method of clean-up after you complete your procedures.

7. Have your teacher approve all of your procedures.

8. Carry out your approved procedures.

Analyze and Interpret

1. Summarize the characteristics of your compost sample.

2. Would your compost be appropriate to add to soil that needs nitrogen? Why or why not?

3. Would your compost help increase the porosity of clay, a soil type that has a very low porosity (assuming the clay and your compost sample were well mixed year after year)? Why or why not?

Conclude and Communicate

4. One possible result of soil that is tilled year after year with nothing added to it is that the soil can become very hard—almost like cement—when dry. When soil reaches this state, it takes extra energy to till the soil. If a tractor is used to till the soil, this means that more fuel has to be used. If compost is added to the soil throughout the year, the soil remains loose and is easier to till. Explain how adding compost to soil can help make farming more sustainable.

Materials

- reference books and/or computer with Internet access
- graph paper

Atlantic Cod Hauls

Year	Haul (tons)
1850	130 000
1860	155 000
1870	150 000
1880	205 000
1890	195 000
1900	200 000
1910	215 000
1920	220 000
1930	205 000
1940	195 000
1950	275 000
1960	500 000
1965	800 000
1970	440 000
1975	195 000
1980	160 000
1985	250 000
1990	140 000
1992	0
1995	0
2000	8 000
2005	8 000

Overfishing and Moratorium of Cod Fishery

The year 2012 was the 20th anniversary of the moratorium (shutdown) of fishing for North Atlantic cod in Canada. The collapse and subsequent shutdown of the cod fishery had many environmental, economic, and social impacts on Canada's Atlantic Provinces. In this Lab, you will analyze the history of the Atlantic cod fishery and research more information about the effects of the moratorium.

Pre-Lab Questions

1. In what year did the cod moratorium begin?
2. Why was the moratorium put in place?

Question

What led to the crash and subsequent shutdown of the Atlantic cod fishery, and how did this affect local ecosystems, economics, and communities?

Procedure

1. Make a line graph of the data in the table on the left.
2. Find out about the history of the Atlantic cod fishery. Use these questions to guide your research:
   - What were people's attitude about the fishery before the crash? Include views from scientists, fishers, and government regulators.
   - How did technology play a role in the crash?
   - What were the economic effects of the moratorium?
   - What were the effects of the moratorium on local communities?
3. Find out about the current status of the moratorium and environmental conditions in the region. Use these questions to guide your research:
   - How much Atlantic cod can be taken with the moratorium in place?
   - There are signs that some species of cod are recovering in other areas of the world. Have populations of Atlantic cod shown signs of recovery? Why or why not? Include statistics as part of your answer.
   - What factors may be playing a role in the recovery of the Atlantic cod?

Analyze and Interpret

1. Use the graph you drew to summarize the history of the Atlantic cod fishery. What happened in 1965? What happened in 1992?
2. In a format approved by your teacher, summarize the results of your research.
3. What factors may be playing a role in the recovery of the Atlantic cod? Specifically, discuss the relationship between the grey seal population and young cod, as well as how other fishing operations may be affecting the recovery of cod. How have capelin and herring affected their recovery? How has climate change affected their recovery?

Conclude and Communicate

4. What things can be learned from the crash of Atlantic cod to help prevent the crash of other fisheries?

Chapter 5   SUMMARY

Section 5.1   Plants and Soil

Soil is an important resource that is a source of nutrients, minerals, and water for plants.

Key Terms

soil
soil profile
topsoil
humus

Key Concepts

- Plants need certain factors and conditions, including light, water, and soil, in order to survive in their environment.
- Soil is a thin covering over the land made up of a mixture of minerals, organic material, living organisms, air, and water that together support the growth of plant life.
- A soil profile is a series of horizontal layers in soil that differ in chemical make-up, physical properties, particle size, and the amount of organic matter they contain.

- Humus is the organic material in soil that results from the decay of plant and animal remains.
- Inhabitants of soil include burrowing animals, such as worms and insects, soil bacteria, fungi, and the roots of plants. Bacteria and fungi are important in the decay and recycling of materials.
- Soil texture is determined by the size of the mineral particles in the soil.
- The porosity of soil refers to the size and number of spaces among particles of soil. Different types of soil have different degrees of porosity.
- Soil pH affects how easily nutrients can be removed from soil, which affects the kinds of plants that will grow.

Section 5.2   Methods of Mechanized Agriculture

Monocultures and the use of synthetic fertilizers and pesticides are part of mechanized agricultural practices.

Key Terms

agriculture
monoculture
pesticide
irrigation

Key Concepts

- Agriculture is the practice of raising plants or livestock for food and other human needs.
- A monoculture is a single crop that is grown on a large area of land.

- Synthetic fertilizers replace soil nutrients removed by plants.
- Farmers use synthetic pesticides to prevent loss of crops and food supplies to pests or infection.
- Worldwide, agriculture accounts for the greatest use of water. Crop irrigation accounts for 70% of water consumption.
- Genetic engineering has allowed scientists to insert specific pieces of DNA into the genetic make-up of organisms. An organism with a genetic make-up (DNA) that has been altered is called a genetically modified organism.

Section 5.3   Sustainable Agriculture

Sustainable agriculture is the production of food to meet the needs of the present without compromising the ability of future generations to meet their needs.

Key Terms

sustainable agriculture
polyculture
companion planting
crop rotation
compost
mulch
integrated pest management (IPM)
organic product

Key Concepts

- Sustainable agriculture is the practice of producing food to meet the needs of the present without compromising the ability of future generations to meet their needs.

- Practices used in sustainable agriculture include polyculture, planting cover crops, using organic fertilizers, using integrated pest management, using green manure, and reducing reliance on fossil fuels.
- Conserving soil and maintaining soil fertility in a sustainable manner are important to all farmers no matter which type of agricultural methods they practise. Soil conservation involves using a variety of ways to reduce soil erosion and restore soil fertility.
- Integrated pest management (IPM) is a system that uses biological organisms, chemical substances, and crop rotation to help keep pest populations under control.
- A specific type of farming that uses sustainable agriculture practices is organic farming. In many countries, including Canada, the United States, France, and Australia, food or other products that are labelled organic meet strict standards defined by the government.

Section 5.4   Livestock Production

Raising livestock can have significant effects on both terrestrial and aquatic ecosystems, so it is important to understand methods of raising livestock sustainably.

Key Terms

desertification   rotational grazing

Key Concepts

- Meat from cows, hogs, chickens, and other animals is an important source of protein, iron, fats, and other nutrients that give us the energy to lead productive lives. Dairy products are also a key protein source.
- About 50% of livestock is raised by grazing on rangelands and enclosed pastures. The other 50% is raised in confined animal feeding operations and feedlots.

- Rangelands, enclosed pastures, and CAFOs have benefits and risks.
- *Overgrazing* occurs when livestock are allowed to eat so much of the grasses and other plants on rangelands that the ecological health of the habitat is damaged.
- Confined animal feeding operations require large inputs of energy and the constant use of antibiotics; they create enormous amounts of waste.
- Methods of raising livestock sustainably include rotational grazing, raising wild species, small pasture operations, raising livestock organically, and using polyculture systems.

Section 5.5   Fisheries and Aquaculture

The harvesting or raising of fish and seafood for human consumption can affect the sustainability of aquatic ecosystems in a variety of ways.

Key Terms

aquaculture   overexploit

Key Concepts

- We currently harvest about 95 million tonnes of wild fish and seafood every year.
- Fish and seafood, including lobsters, shrimp, crabs, clams, and scallops, are the main source of animal protein for about 1.5 billion people in developing countries.
- Fish and seafood are the only wild-caught meat sources still sold commercially on a global scale. Aquaculture is becoming increasingly important as a source of fish and seafood production.

- The United Nations estimates that 70% of the world's marine fisheries are being overexploited or are in danger of being overexploited as the number of fishers increases.
- Rising numbers of boats, along with more efficient technology, allow for the exploitation of a shrinking resource. Boats as big as ocean liners travel thousands of kilometres, using global satellite positioning equipment (GPS), sonar, spotter planes, and other technology to catch large numbers of fish and seafood.
- Fishing using purse seines, hook and line fishing, and diving are considered sustainable fishing methods.
- Aquaculture in land-based ponds or warehouses can reduce many of the problems associated with aquaculture pens in marine and freshwater ecosystems.

Chapter 5 REVIEW

Knowledge and Understanding

Choose the letter of the best answer below.

1. Which condition is *not* required by a plant for its survival?
   a) carbon dioxide for cellular respiration
   b) oxygen for cellular respiration
   c) sufficient space for growth
   d) sunlight for photosynthesis
   e) water intake for photosynthesis

2. Which statement about soil is *false*?
   a) Many plants anchor their roots in soil.
   b) Decomposing plant and animal matter can be found in a soil sample, as well as living organisms.
   c) Most of a plant's roots are found in the topsoil layer.
   d) Humus, organic matter from the decomposition of animal and plant materials, gives soil a black colour.
   e) The proportions of the components that make up soil are always the same, even in different soil types.

3. A property of soil that describes the size and number of spaces found among the soil particles is known as
   a) density
   b) ductility
   c) porosity
   d) malleability
   e) volume

4. Which statement about our food sources is correct?
   a) Agriculture is the practice of raising plants or livestock for food and other human needs.
   b) A monoculture consists of multiple crops growing on a large land area.
   c) Livestock and fisheries do not provide us with protein.
   d) Plants provide food for only a few organisms on Earth.
   e) Different areas in a monoculture on a farm require different amounts of pesticides.

5. Synthetic fertilizers
   a) are made by people using natural, freshly harvested, plant-based resources
   b) commonly contain nitrogen, phosphorus, and potassium
   c) do not affect crop yield
   d) are always safe for use in aquatic ecosystems
   e) are useful for fish, since they increase the oxygen levels in the water

6. An integrated pest management system
   a) uses only living organisms to control pest populations in a safe manner
   b) starts with the application of chemicals to destroy as many pests as possible, and then relies on biological organisms to maintain control of the pests
   c) helps to control pest populations by rotating crops, using biological organisms, and using chemicals
   d) can only be started in an area where adult insects have not mated yet
   e) is only considered to be sustainable if all pests in an area are destroyed so that the farm will make a profit on crop sales during that growing season

7. Livestock that are raised organically
   a) are raised without growth hormones
   b) require low doses of antibiotics in their diet to prevent the occurrence of disease
   c) regularly need to consume foods high in animal by-products to get enough protein
   d) are fed food that contains genetically modified ingredients
   e) are usually fed food that has been grown using either synthetic or natural fertilizers

8. An example of a sustainable method of harvesting wild-caught fish is
   a) dragging heavy nets across the ocean bottom
   b) using wire or wood cages to catch cod
   c) setting up cables with hooks every 2 metres to catch tuna
   d) using a purse seine when a spotter plane finds schools of fish
   e) using sediment clouds to send fish to the surface so that they can be humanely harvested

Answer the questions below.

9. Identify and describe two factors that affect soil pH.

10. Describe why humus is such an important component of soil for agricultural use. Provide three supporting details in your response.

11. Name two living organisms found in the soil, and state how plants benefit from these two organisms.

12. Define the term *mechanized agriculture* and indicate why monocultures are associated with mechanized agriculture.

13. What are two advantages and two disadvantages of growing monocultures?

14. Provide two reasons why surface irrigation can be an inefficient method of delivering water to crops.

15. What is a genetically modified organism? Give one example of a use for this type of organism in agriculture.

16. Sustainable agriculture is a practice that lets us produce enough food to meet our present needs without affecting the food-producing ability of future generations.

   a) Define *polyculture* and briefly explain three ways that it is a sustainable agriculture practice.

   b) List three other examples of sustainable agriculture practices.

17. The "three sisters" are shown in the photo below. Identify the three plants and explain why this is an example of companion planting.

18. Identify three types of organic fertilizers, and give a brief description of each type.

19. What is aquaculture? Give an example of how aquaculture is used in Canada.

20. There is evidence that the world's marine fisheries are being overexploited.

   a) What does *overexploited* mean?

   b) How might the marketing of orange roughy (shown in the photo below) indicate that overexploitation in marine fisheries is likely occurring?

21. Give an example of how the integration of aquaculture and agriculture in a sustainable manner is used in China.

22. Contrast the main method of obtaining wild-caught fish in developing countries with the method used in developed countries.

Thinking and Investigation

23. Ocean dead zones are areas at the bottom of oceans that have such a low oxygen content that organisms that normally live there, such as lobsters, crabs, and various types of fish, usually die. Faster-moving fish might be able to swim away from a dead zone. These dead zones, usually found along coastlines, have several causes, including the presence of fertilizer in run-off. The nutrients in the fertilizer lead to algal blooms. The algae eventually die, sink to the bottom of the ocean, and are decomposed by micro-organisms that need oxygen.

   a) Predict the effects that large-scale dead zones could have on an ocean ecosystem.

   b) Predict the effect that large-scale dead zones could have on commercial fishing operations.

   c) Describe a method of reducing the nitrogen run-off with the spring rains on the east coast of Florida.

   d) How might natural disasters, such as Hurricane Katrina, contribute to ocean dead zones?

24. Windbreaks are tall walls of vegetation that are planted in rows in a variety of locations, such as fields, agricultural areas, and near houses, to help block the wind. The photo below shows windbreaks in an agricultural setting.

   a) Explain how a windbreak helps conserve soil in agricultural areas.

   b) How can windbreaks help protect crops?

   c) Name two possible disadvantages of using windbreaks in agricultural areas.

Chapter 5   REVIEW

25. There are many types of irrigation methods that a farmer can use, depending on the soil conditions, how much water is available, and on the amount of money that is available to purchase and/or maintain a system. Copy and complete the chart below, recommending the best type of irrigation system to use for each condition. Provide at least one reason for each recommendation.

Condition	Suggested Irrigation System	Reason(s) for Your Selection
Sandy soil	Drip irrigation	Sand lets water through readily. Continuous water flow would just seep through the sand. Slower drip will give the plants time to absorb the water.
Clay soil		
Areas with frequent high winds		
Water is in short supply		

26. The photo below shows a farming area that is using a "no-till" approach to sustainable agriculture. No-till refers to the fact that there is no ploughing at the end of the growing season. In the photo, note the area where the soil is covered by the previous year's crop residues. Special equipment is needed to deposit seeds in the ground without disturbing the crop residues.

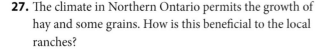

a) Infer two reasons why the no-till approach is a sustainable agriculture practice.

b) What are two possible drawbacks of no-till agriculture?

27. The climate in Northern Ontario permits the growth of hay and some grains. How is this beneficial to the local ranches?

28. The table below summarizes comparisons between wild-caught salmon and salmon raised in aquaculture farms.

a) Why might farm-raised fish have more fat in their bodies?

b) Omega-3 fatty acids are healthy for you. Based on the nutritional information in the table, would you conclude that wild or farm-raised fish are healthier for you to eat?

c) Is the flesh colour of the salmon important to control in fish farms? Explain your answer.

d) Explain why disease is more prevalent in farmed fish than in wild fish.

e) The table gives no indication that the harvesting of wild-caught fish harms the environment. Do you agree or disagree with this the table on this issue? State at least two supporting details.

f) Would you buy wild or farmed fish, based on this table? Keep in mind that many wild fish are overexploited, and many fishing techniques harm the environment. Your response should be at least one paragraph in length.

Wild vs. Farmed Salmon Comparison

	Wild	Farm-raised
Nutrition	Higher ratio of Omega-3 to Omega-6 fatty acids	Lower levels of protein and much fattier
Coloration	Naturally orange or red because of diet	Given pigments to turn colour from natural white
Disease	Low levels of sea lice, disease, and contaminations	High levels of sea lice, disease, and pesticides; given large amounts of antibiotics
Environment	Populations affected by escaped farmed fish	Excess waste and disease harm natural ecosystem
Price	Slightly higher price	Cheaper because already in nets

Communication

29. There is a lot of food waste in grocery stores, as well as in many Canadian homes. We tend to expect our produce to be flawless, and if it is not, we might toss it in the garbage. In a paragraph, describe how this wasteful action affects the environment.

30. Write a short paragraph explaining why you would or would not support each of the following:
   a) an increase in genetically modified foods
   b) an increase in polyculture
   c) an increase in small pasture operations
   d) an increase in industrialized fishing techniques

31. Harpooning is used to harvest large fish such as bluefin tuna.
   a) Is this an example of a sustainable fish harvesting practice? Provide support for your answer.
   b) How do animal-rights activists usually view harpooning? Research this topic if needed, and indicate whether you agree with harpooning animals for food.

32. Make a chart listing three benefits of farmers using synthetic pesticides, as well as three risks of using these pesticides to prevent crop losses. Based on this chart, explain whether you support the use of synthetic pesticides in agriculture.

33. DDT is a pesticide that was readily used in North America until about 40 years ago. DDT is fat-soluble, highly toxic, and does not break down easily. It moves freely in the air, water, and soil. Using this information, sketch a simple terrestrial or aquatic food chain that shows how DDT concentrations might change along the food chain.

34. **Did You Know?** Reread the quotation by Wendell Berry on page 135. What does the quotation mean to you? Write a song, poem, or blog entry, or make a painting or drawing to express your ideas.

Application

35. In 2003, a researcher at the University of Guelph developed the Barcode of Life project. The DNA of different organisms is coded and entered into a database. Species of organisms that cannot be identified due to physical damage or missing parts (such as fish skin, fins, and heads) can then be identified by using a small piece of tissue. Studies have shown that some restaurants in North America are serving cheap fish and labelling them as more expensive fillets, and fish that are overexploited are being sold as fish that are plentiful. Explain how the Barcode of Life project can assist with this marketing problem.

36. Many school cafeterias do not compost their food scraps. Make a list of foods that are commonly discarded in the garbage but could be composted. Plan a compost program for a school, including a brief description of how students could be involved in the composting process. Also provide at least two uses at the school for the resulting nutrient-rich soil.

37. The recycling of animal waste as fertilizer is economical and is generally considered an environmentally sustainable practice. However, care must be taken that manure does not run off into water sources, since it can be contaminated with *E. coli* and other bacteria. In 2000, the well water of Walkerton, Ontario, was contaminated with *E. coli*. Research more information about this event.
   a) Summarize the impact of the event on public health.
   b) What was the source of the contamination?
   c) What other factors contributed to the contaminated water reaching the public?
   d) What steps were taken to avoid something like this happening again?

38. You have likely noticed stickers on various fruits and vegetables, such as the one in the apple photo below.
   Research the meaning of these PLU codes. Use the results of your research to answer the following questions.

   a) What does a four-digit number, such as 4011, mean?
   b) What does a five-digit number mean when the first digit is 8?
   c) What does a five-digit number mean when the first digit is 9?
   d) Indicate whether you would consider purchasing produce whose PLU starts with 4 or 8, and provide support for your response.

Pause and Reflect

How could you incorporate what you have learned in this chapter into your daily actions or choices?

Sustainable Forestry

An increasing number of large cities are converting the dull, grey, lifeless rooftops of many of their buildings to the lush green of living meadows and gardens.

What environmental benefits does a living roof provide to its building, as well as to the city or local community?

How Are You Connected to Forests?

Whether you live within minutes, hours, or days of forests, your life is closely connected to them—perhaps in ways you might not even realize. Use the questions below to help you consider how.

1. Maple syrup is one type of food that comes from forests. What foods do you eat that come from forests?

2. Paper is one type of product that comes from forests. What products do you use that come from forests?

3. Does anyone in your family work in the forest industry or have a job that depends on the forest industry?

4. Do you go hiking, skiing, or snowmobiling in forested areas?

5. What other connections do you have with forests, both nearby and in other countries?

Share your ideas in small groups and with the whole class. Which forest connections surprised you?

In this chapter, you will

• identify ecosystem services of forests
• describe sustainable methods of managing forests
• identify ecosystem services of urban forests
• explain how plants can be integrated into urban development for sustainability

Forestry Management

Canada's Forests

forest an ecosystem in which the dominant plants are trees

A **forest** is an ecosystem in which the dominant plants are trees. Nearly 50% of Canada's land is covered by forests. Forests in Canada provide habitat for about two thirds of the species in Canada and make up 10% of forests worldwide. In particular, the variety of forests in Canada, shown in **Figure 6.1**, play an important role in water filtration, reducing the impact of climate change, and producing both timber and non-timber products.

Figure 6.1 About 400 million hectares of Canada's land is covered by forests.

Temperate Rainforest
The temperate rainforest on the southwest coast of Canada consists of western hemlock, western red cedar, yellow cedar, firs, and spruce trees. Mosses and other small plants grow on larger trees. The forest floor is covered with ferns and wildflowers. Brown bears, grizzly bears, black-tailed deer, frogs, and slugs all live in the temperate rainforest.

Boreal Forest
The main tree species in boreal forests include spruce, birch, pine, larch, poplar, and fir. Canada's boreal forest supports more than 300 species of birds, as well as large populations of wolves, woodland caribou, and grizzly bears.

Acadian Forest

The Acadian forest, found in the Atlantic Provinces, is made up of sugar maple, yellow birch, eastern hemlock, and balsam fir trees. As in other forests, mosses and other fungi, ferns, and wildflowers grow on the forest floor. Deer, foxes, bobcats, and birds all depend on the Acadian forest for habitat.

Map Key

- ■ Boreal Forest
- ■ Deciduous Forest
- ■ Acadian Forest
- ■ Carolinian Forest
- ■ Subalpine Forest
- ■ Columbia Forest
- ■ Montane Forest
- ■ Temperate Rainforest

NON-FOREST
- ■ Tundra
- ■ Grassland

NEWFOUNDLAND AND LABRADOR

Carolinian Forest

The Carolinian forest in southern Ontario makes up only 1% of Canada's landmass. However, it has a higher number of species than any other ecosystem in Canada. Tree species include oak, black walnut, and hickory. Birds, flying squirrels, and snakes are some of the animals that live in the Carolinian forest.

QUÉBEC

PRINCE EDWARD ISLAND

ONTARIO

NOVA SCOTIA

NEW BRUNSWICK

Mini-Activity 6-1 — Canada's Forests

Choose one of the eight types of forests found in Canada. Do research to find the following information:

- its abiotic and biotic characteristics
- examples of ecosystem services the forest provides
- its role in the local, provincial or territorial, and/or federal economy
- examples of past and present controversies or issues involving the forest

Present your findings in a format approved by your teacher.

Basic Structure of Forests and Their Biodiversity

Worldwide, there are many different types of forests, including boreal forests, deciduous forests, temperate rainforests, and tropical rainforests. Although the abiotic and biotic factors that make up each type of forest differ, most forests have the same basic structure. As shown in **Figure 6.2**, forests are made up of three main layers: the canopy, the understory, and the forest floor.

The Canopy

The canopy is made up of the leaves and branches of tall, mature trees. The crowns (tops) of these trees receive most of the sunlight that reaches the forest. The thick mass of leaves and branches form an umbrella-like cover that shades the rest of the forest. The canopy is habitat for animals such as birds and insects, as well as for other kinds of plants. (In tropical rainforests, many species of monkeys live in the canopy and may spend their entire lives there.)

The Understory

The understory is made up of young trees, shrubs, and bushes that are adapted to living in shade. Trees such as dogwoods, berry shrubs, and Canada yew thrive in the understory of many forests. Animals such as insects, snakes, birds, and bats all live in the understory. (In tropical rainforests, larger mammals such as jaguars also live in the understory.)

The Forest Floor

The forest floor is made up of decomposing leaves and trees, animal droppings, and other organic matter. (In tropical rainforests, decomposition occurs very quickly due to the high temperatures and humidity.) The breakdown of all this material is an important part of the nutrient cycle in a forest ecosystem. Nutrients such as nitrogen and phosphorus are released into the soil and are taken up by existing plants and by new seedlings. Ferns, mosses, wildflowers, and fungi all grow on the shaded forest floor. Birds, rodents, and amphibians also live on the forest floor.

Figure 6.2 The basic structure of a deciduous forest such as this one is very similar to the structure of other types of forests.

Inferring *What examples of adaptations would you observe in the animals that live in the different forest layers?*

Ecosystem Services of Forests

Forests provide ecosystem services that include the following:

- reducing soil erosion
- storing carbon
- cycling nutrients such as carbon, nitrogen, and phosphorus
- purifying water
- providing habitats for millions of species on Earth

Forests also provide timber and non-timber resources.

Forests and Timber Resources

Cut trees are a source of wood used as timber. **Timber** is wood that is used for construction and carpentry. Cut trees are also a source of firewood used for heating and cooking. In addition, cut trees are a source of the wood pulp used to make paper products and some building materials, such as pressboard. The forest industry in Canada employs about 230 000 people. In 2011, the forest industry contributed 57 billion dollars to Canada's economy.

timber wood that is used for construction and carpentry

Forests and Non-timber Resources

Non-timber forest resources are biological products that do not come from timber. Table 6.1 shows some examples of non-timber resources that come from forests.

Table 6.1 Non-timber Resources

Food

Forest-based foods, such as mushrooms, wild ginseng, wild leeks, wildflowers, blueberries and other fruits, and nuts, are often collected from different types of forests and sold worldwide. In 2009, Canadian maple sap products brought in over 350 million dollars. Canada produces 85% of the world's maple syrup.

Medicine and Personal Care Products

Medicinal plants and plant extracts are collected from forests. For example, the chemotherapy drug paclitaxel is extracted from Canada yew trees. The essential oils of conifer trees are a popular ingredient in lotions and other personal care products.

Wood-carving and Craft-making Materials

Materials for wood-carving, craft-making, and florist supplies, such as dried greenery, also come from forests. Many First Nations make canoes using bark from birch trees.

Tourism

Forest-related tourism is a multi-billion dollar industry worldwide. People may spend time in forests at home and abroad to hike, backpack, watch birds, ski, snowshoe, or simply enjoy the sights, sounds, and smells. Canada's forests provide almost 400 million hectares of beautiful scenery in which people can connect with nature physically, emotionally, and spiritually.

Silviculture Methods

silviculture a branch of forestry related to the development and management of forests

Silviculture is the practice of developing and managing forests for the timber products they can supply now and in the future. Common silviculture methods include clearcutting, selective cutting, and shelterwood systems. The system used to harvest timber and regenerate trees for a part of a forest depends on several factors. These include the tree species, the ages of the trees, and the conditions of the site. As well, foresters and other people who work in the industry must take into consideration the ecosystem services provided by forests when managing an area of forest.

Some terms that are often used to describe forests and trees include old-growth forest, even-aged forest, uneven-aged forest, and mature trees. An *old-growth forest* is one that has developed for at least 120 years without a severe disturbance such as a fire, windstorm, or logging. An *even-aged forest* is one in which the ages of the trees are within 10 to 20 years of each other. An *uneven-aged forest* is made up of trees with vastly different ages, which results in a complex mix of forest layers. A *mature tree* is one that has grown to reach its greatest economic value for its size and use.

Clearcutting

clearcutting a silviculture method in which most or all of the trees from a chosen area are removed

Clearcutting, shown in **Figure 6.3**, removes most or all of the trees from a chosen area. This method of silviculture is often used to manage an even-aged forest.

Advantages
- mimics a large-scale natural disturbance, such as a fire, flood, windstorm, or disease
- stumps, branches, and some fallen trees are left on the forest floor as habitat and to conserve nutrients in soil
- often safer for forest workers
- efficient and cost-effective, because all harvesting is done at one time

Disadvantages
- habitat loss for species living in all three layers of the forest
- loss of large tracts of carbon-storing trees
- damage to structure and composition of soil
- increased surface-water run-off and soil erosion, especially if done on steep slopes
- takes a long time for trees to regenerate
- reduces the recreational and aesthetic value of a forest

Figure 6.3 Clearcutting

Selective Cutting

Selective cutting, shown in **Figure 6.4**, involves cutting and removing medium-aged or mature trees individually or in small clusters every 10 to 20 years. This method of silviculture is often used in uneven-aged forests.

selective cutting a silviculture method that involves cutting and removing medium-aged or mature trees individually or in small clusters every 10 to 20 years

Advantages

- mimics a small-scale natural disturbance, such as damage caused by trees that are uprooted or knocked down by wind
- preferred method for cutting on steep slopes or in places where permanent tree cover is needed
- retains habitat for species such as certain birds
- more aesthetically appealing than clearcutting
- individual trees of high economic value can be chosen and removed, leaving behind trees that can naturally regenerate the forest

Disadvantages

- costs may be higher than with clearcutting
- risk of damage to remaining trees and plants during harvest
- some animal species may decline after selective cutting occurs

Figure 6.4 Selective cutting

Shelterwood System

The **shelterwood system**, shown in **Figure 6.5**, involves removing trees in a series of cuts over a period of 10 to 30 years. This method of silviculture leaves one third to one half of the mature trees standing.

shelterwood system a silviculture method that involves removing trees in a series of cuts over a period of 10 to 30 years

Advantages

- remaining trees provide seeds for new trees and shelter for seedlings and saplings
- remaining forest, although thinner, is still aesthetically appealing
- encourages regrowth of species that are sensitive to wind and sun, such as spruce and fir

Disadvantages

- young trees may be damaged during harvest of older trees
- more roads must be built into the forest
- planning and harvesting may be more costly than other methods

Figure 6.5 Shelterwood system

Pause and Reflect

1. Identify and briefly describe the three main layers of a forest.

2. What factors may play a role in deciding which silviculture method is used to harvest and regenerate trees?

3. Critical Thinking Why would a forester take environmental, economic, and social factors into account before deciding which silviculture method to use?

The Importance of Sustainable Forestry

Managing forests through sustainable practices is important to current and future generations. As we continue to learn more about the importance of forests and the ecosystem services they provide, governments, which manage a country's publicly owned forests, and private landowners worldwide are turning to sustainable management practices. The following list outlines the goals of sustainable forestry management practices.

- **Protect Biodiversity** Identifying and protecting forest areas that have high biodiversity, such as the Amazon rainforest, the boreal forests of Canada, and old-growth forests worldwide, helps to keep these ecosystems healthy.

- **Harvest Timber Sustainably** Selective cutting and another silviculture method called *strip cutting* are sustainable management practices. Strip cutting, shown in **Figure 6.6**, involves clearcutting a small strip of trees along a natural contour of the land. Using this method, trees grow back within a few years, and then a new strip of trees is cut down nearby.

Figure 6.6 Strip cutting involves clearcutting a strip of trees along a natural contour of the land.

- **Leave Organic Material in Place** After trees are harvested, it is important to leave some organic material behind. Organic material includes dead trees, fallen trees, dead limbs, and the tops of trees; these are usually left lying on the forest floor after a harvest. Leaving this material in place provides habitat for wildlife and allows nutrients to be recycled.

- **Regenerate After Harvesting** Planting tree plantations on deforested land replaces harvested trees. When managed well, tree plantations can produce timber at a fast rate. Tree plantations are managed as even-aged forests that are clear-cut and then replanted on a regular basis. Tree plantations can help reduce cutting in old-growth forests. They also store carbon and provide habitat for wildlife.

- **Certify Sustainably Managed Forests** Timber and other wood products that come from sustainably managed forests can be certified by several different agencies. This enables consumers to actively support sustainable management practices.

- **Value Ecosystem Services** Including the ecosystem services of forests as part of their economic value is being recognized as important. These services are often referred to as *natural capital*. For example, a 2009 report by the Canadian Boreal Initiative estimated that boreal forests and bogs in Canada store about 150 billion tonnes of carbon. This storage decreases the costs associated with carbon being released into the atmosphere by about 580 billion dollars per year. Recreational use of Canada's boreal forests is worth about 4.5 billion dollars per year.

Inquiry Lab 6C, Annual Growth Rings of Trees, on page 196

Society and Sustainability: Third-party Certifications

People who work in the forest industry are not the only ones who can help with maintaining sustainable forests. Consumers can as well. You and other members of society can reduce the use of wood products, which lessens the demand for timber and wood pulp. Recycling paper reduces the number of trees cut, as well as the energy and water needed to process trees into pulp and paper. Buying products made from recycled or repurposed wood products also makes a difference. Another factor to consider when making purchases is whether a wood product is certified as sustainable.

There are several independent agencies that certify wood products as sustainable. These include the Canadian Standards Association Sustainable Forest Management (CSA SFM), the Forest Stewardship Council (FSC), and the Sustainable Forestry Initiative (SFI). By volunteering to participate, companies that make wood products agree to meet the standards set forth by these agencies. In return, the companies may use labels such as those shown in **Figure 6.7**. Labels from all three of these agencies are used throughout Canada, which has the largest area of sustainable forest that is certified by independent agencies.

Figure 6.7 These labels help consumers identify wood products that are sustainable.

Applying *What other products do you know or have you seen that have sustainability-related labels?*

Pause and Reflect

4. Describe one way that sustainable forestry benefits the forest industry.

5. Describe one way that sustainable forestry benefits society.

6. Critical Thinking How can being an informed consumer contribute to sustainability, both for the industries that make products and for you personally?

Mini-Activity 6-2   **Certification Labels**

Examine the certification label provided to your group by your teacher. Use the Internet or other resources provided by your teacher to determine exactly what the certification label means.

- From which agency is the label?
- Does 100% of the product contain materials that met the standards of the certification agency? If not, is a percent designated on the label?

- Does the product contain mixed sources? If so, what does this mean?
- What does chain-of-custody certification mean?

With your group members, discuss the label and answer the questions. As a class, discuss the value of these labels to the consumer and to the sustainability of ecosystems.

Case Study Selective Cutting on Pinkerton Mountain

Pinkerton Mountain is in the Cariboo Mountains in British Columbia. As the name suggests, the Cariboo Mountains are home to mountain caribou (*Rangifer tarandus caribou*), shown in the photo on the right. These animals have one main food source: tree lichen. This hair-like lichen hangs from the branches of evergreen trees that live on the mountain's steep slopes. Commercial logging of these trees can have a huge impact on mountain caribou survival. As a result, silviculture practices must make caribou habitat a top priority. In 1998, scientists at the University of Northern British Columbia set up a long-term study on Pinkerton Mountain. They wanted to find out how different harvesting practices affect caribou habitat.

Results of the Experiment

The Pinkerton Mountain study had a control area and two experimental areas. The control area was a section of unlogged forest. Trees in the experimental areas were removed using two selective cutting techniques that mimic natural forest disturbances on the mountain: group selection and single-tree selection. In the single-tree selection area, only individual trees were removed. Overall, no more than 30% of the trees were harvested.

After 10 years, scientists measured the amount of tree lichen growing at caribou grazing level. They found that the same amount of lichen grew in each area. They also assessed which species of lichen was most common in each area. In the control area, one species of lichen was most common; in the experimental areas, a different species was most common. Since caribou often prefer one species of tree lichen to another, such a change could have an impact on their diet and, thus, on their survival.

In this case, the caribou actually preferred the species that thrived in the experimental areas. Therefore, scientists concluded that changes to lichen growth due to the two selective cutting methods used in the experiment did not affect the mountain caribou.

Research and Analyze

1. This study showed that changes to lichen growth due to selective cutting did not affect the mountain caribou. Describe three other factors that foresters might consider when harvesting trees on Pinkerton Mountain.

2. Suppose you are the owner of a commercial logging company that operates in the Cariboo Mountains. You learn that selective cutting can help protect mountain caribou habitat. Do research to learn more about the advantages and disadvantages of selective cutting for commercial logging. Include information about its environmental, economic, and social impacts. Use a risk-benefit analysis to decide if selective cutting is the best method to harvest trees in the Cariboo Mountains.

Communicate

3. The Pinkerton Mountain study did not investigate clearcutting. Write a short opinion paper of two or three paragraphs to explain how you think clearcutting could affect tree lichen and, therefore, caribou, in the Cariboo Mountains.

Summary

- A forest is an ecosystem in which the dominant plants are trees. Most forests are made up of three main layers: the canopy, the understory, and the forest floor.

- Forests provide many ecosystem services, including being a source of timber and non-timber resources.

- Silviculture is the development and management of forests. Silviculture methods include clearcutting, selective cutting, and shelterwood systems.

- As we continue to learn more about the importance of forests and the ecosystem services they provide, governments and private landowners worldwide are turning to sustainable management practices.

- Companies that make wood products can apply to various independent agencies for permission to apply a label that identifies their products as being managed and manufactured in a sustainable manner.

Review Questions

1. Identify each layer of a forest and describe several ecosystem services provided by that layer. **K/U** **A**

2. Paper and pulp are the fastest-growing sectors of the wood products market, as developing countries such as China and India catch up with the use rates of North America, Europe, and Japan. **K/U** **A**

   a) What effect could this have on the ecosystem services provided by forests?

   b) What are some ways to reduce paper use?

3. Some conservationist groups argue that watershed protection and other ecosystem services of forests are more economically valuable than timber. Some timber companies argue that continued production supports stable jobs and local economies. **T/I** **A**

   a) If you were a judge trying to come up with a compromise between the two groups that balanced ecological and economic sustainability, what questions would you ask both groups? What evidence would help you arrive at a compromise?

   b) How would you gather this evidence? How would you identify bias from any source?

4. List at least four non-timber resources that you use in your daily life. Explain how you might be affected if these resources were no longer available. **A** **C**

5. Copy the table below into your notebook and complete it. **K/U**

	Advantages	Disadvantages
Clearcutting		
Selective cutting		
Shelterwood system		

6. Describe three ways to manage forests more sustainably. **K/U**

7. What does it mean if a wood product is certified as sustainable? **K/U**

8. The illustrations below show models of two silviculture methods that could be used to harvest trees from the same area. Which method do you think would be better to use? Why? **A**

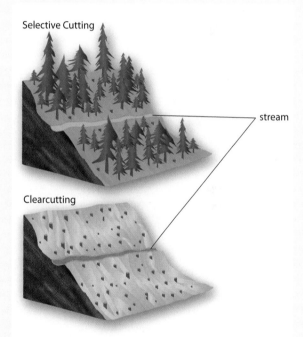

9. Suppose you own 40 acres (about the size of 30 soccer fields) of uneven-aged forest land. The wooded area is a temperate deciduous forest, which includes tree species such as oak, maple, elm, walnut, and beech. Would you harvest trees from your land? Why or why not? If you would, which type of silviculture method would you use? Explain your reasoning. **A** **C**

10. How do you and other consumers exert power and influence each time you choose one product over another when you shop? **T/I** **A**

Urban Forestry

urban forestry the long-term planning, planting, and maintenance of forests, trees, and green spaces in urban environments

Urban forestry is a field of science that deals with the long-term planning, planting, and maintenance of forests, trees, and green spaces in urban environments. The trees and green spaces of urban forestry provide several important ecosystem services. **Figure 6.8** shows three examples. In addition to these services, urban forests

- help to purify city air by removing dust and various polluting gases, including carbon dioxide
- act as buffers to reduce noise pollution
- increase the aesthetic, and often the economic, value of homes and communities
- serve as a source of fresh, locally grown food from individual and community gardens
- help to reduce energy consumption by providing shade and windbreaks

Figure 6.8 There are several important ecosystem services provided by urban forests and green spaces.

Applying *What other ecosystem services do urban forests provide?*

Trees and wooded areas in cities provide habitat and food for a variety of organisms. This blue jay, along with other bird species, lives in wooded areas in urban settings year-round. These spaces can be especially important to migrating birds, which use them as a place to rest and fuel up for the remainder of their flight. Squirrels, deer, and raccoons are just a few of the mammals that live in urban forests.

Urban parks provide space for people to exercise, relax, meditate, and connect with nature. Walking, jogging, or playing organized games in the park are great ways to exercise for free or for very little expense. Cycling on bike paths in parks or through cities is a popular way for people to exercise under the shade provided by trees.

Water from rain and heavy storms easily pools on the concrete and asphalt surfaces of urban areas. This increases the risk of flooding and the movement of pollutants from streets and parking lots into streams, ponds, and other local waterways. Rain gardens are designed to catch surface run-off from storms. They are usually planted in low-lying areas, and they help by increasing the amount of water absorbed into the ground. As well, they provide habitat for wildlife and usually require little maintenance once planted.

Reducing the Heat Island Effect

As urban areas develop, the landscape changes from open land covered with plants to surfaces covered with concrete and asphalt. Land is cleared to make roads, parking lots, and sidewalks. Block-long stretches of concrete buildings often cover much of the downtown part of a city.

Large areas of concrete and asphalt absorb a lot of heat during the day. **Figure 6.9** shows that temperatures in the downtown part of a city can be up to 3°C higher than in surrounding rural areas. This **heat island effect**, as it is called, leads to an increase in energy use for summer cooling. In regions where energy is supplied mainly from fossil fuels, this means an increase in greenhouse gases and a decrease in air quality.

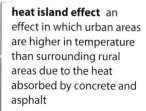

heat island effect an effect in which urban areas are higher in temperature than surrounding rural areas due to the heat absorbed by concrete and asphalt

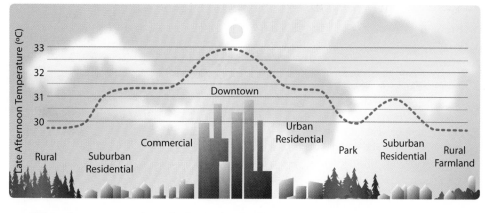

Figure 6.9 The heat island effect leads to increased energy consumption, increased emissions of air pollutants and greenhouse gases, and decreased water quality.

Urban forests can reduce the heat island effect. For example, trees and other plants provide shade. As well, when water evaporates from pores in plant leaves, air is cooled. Notice in **Figure 6.9** that the surface temperature during the day is lower in areas that have trees compared to downtown areas with no plants.

Green roofs such as the one shown on the opening pages of this chapter are also an effective way to help reduce the heat island effect. A green roof is a roof that has a layer of plants growing on its surface. Green roofs may be made up of a simple layer of grass or other ground cover, a flower or vegetable garden, or a complex mixture of plants and trees. Green roofs absorb heat and act as insulators for buildings, reducing the need for cooling in the summer and heating in the winter.

Mini-Activity 6-3   Cool Your School

Many schoolyards are heat islands, due to the high concentration of heat-absorbing materials used for roofs, parking lots, and playgrounds. One study done in Waterloo, Ontario, showed that the average surface temperatures of 15 schoolyards ranged from 48.4°C to 55°C. These temperatures were greater than that of the surrounding land area by as much as 5°C. Survey your schoolyard with the heat island effect in mind.

- Which structures may be contributing to the heat island effect?

- How could you design a procedure to collect data to analyze?

- Based on your results, what changes could be made to reduce the heat island effect, and how could they be implemented?

- What factors could keep the changes you suggested from being made?

- If your school was planned with temperature reduction in mind, identify a schoolyard elsewhere in or near your community that could benefit from your ideas.

Managing Urban Forests Sustainably

Many types of jobs and professions are involved in the planning and maintaining of plants in an urban setting. These include urban foresters, urban forestry technicians, and landscapers, who perform the duties described in **Figure 6.10**.

Inquiry Lab 6A, Which Soil Composition Is Best? on page 194

Figure 6.10 Ways to manage urban forests sustainably include surveying site conditions, planting native plants, and keeping plants healthy.

Inferring *Why do you think it is important for urban foresters to have information about site conditions before deciding what to plant in the area?*

Choosing the Site

Part of the planning stage involves choosing a site with the right conditions to ensure the short- and long-term health of plants. These conditions include
- the type of soil in which the plants will grow
- the amount of sunlight or shade
- the amount of water and drainage
- the amount of pedestrian traffic

native plant a plant that has been growing naturally in an ecosystem without any action, past or present, from humans

Planting Native Plants

Planting the right kinds of plants for the site is also important for urban forestry management. **Native plants** are plants that have been growing naturally in an ecosystem without any action, past or present, from humans. Native plants are better adapted to local site conditions. They are often better suited to resisting drought, disease, and insect pests. As a result, native plants tend to need less water and less maintenance, which saves time, money, and resources. They also increase biodiversity by attracting and supporting bees, butterflies, birds, and other native wildlife species. As well, native plants tend to provide better erosion control and climate control than non-native plants.

Keeping Plants Healthy

An important part of keeping plants healthy in an urban environment includes pruning, watering, fertilizing, and deciding when to remove dying or dead plants. Another important part is protecting plants in wooded areas during construction projects. This helps reduce the impact of changing environmental conditions due to the construction.

Integrated Pest Management

Managing diseases and pests can be a big part of keeping plants healthy in an urban setting. Plants must be monitored for evidence of infestation by insect pests or infection by bacteria or fungi. Areas near and where plants grow must be monitored for the appearance of non-native pests or invasive species. These organisms can cause severe damage and can be costly both economically and aesthetically. Integrated pest management plans are developed, and then must be implemented and evaluated.

The emerald ash borer, shown in **Figure 6.11A**, is a beetle whose larvae destroy the tissues that move water and food in all parts of ash trees. Since 2002, the emerald ash borer has been spreading throughout southwestern Ontario, including the city of Toronto. The beetle is a serious threat to all species of ash trees, which are commonly used to line city streets and are found throughout urban areas. One way to stop the spread of the beetle is to cut down all infested trees within a certain radius. As part of the management plan for the emerald ash borer, more than 80 000 ash trees were cut down in Essex County in 2004.

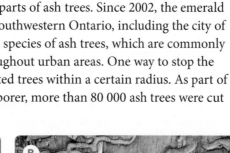
integrated pest management—**see** section 5.3 on page 149

Figure 6.11 (A) The emerald ash borer is a threat to a large area of North America, from Canada to the midwestern and eastern United States. **(B)** Emerald ash borer larva chew through the inside layer of a tree, leaving a characteristic pattern of damage.

In Toronto, the urban forestry management plan initially focussed on using insecticides to eradicate the beetle. However, since 2011, management efforts have been geared toward trying to reduce its impact. These include removing existing ash trees, educating the public, and developing a tree-planting replacement program. Some estimates suggest that the cost to all affected North American cities from the loss of trees, replanting efforts, and general impact of infestation will be in the billions of dollars.

As urban areas grow and expand, urban foresters consider how plants can be integrated into future development. Foresters also identify sites that are suitable for reclaiming as forests and green spaces, and suggest ways to increase their sustainability. Researching and providing long-term plans to municipalities is an important part of urban forestry management.

Thought Lab 6B, Pests and Diseases in Canada's Forests, on page 195

Pause and Reflect

7. Define *urban forestry*.

8. Why is it important to manage urban forests sustainably?

9. Critical Thinking What role could native plants play in integrated pest management in urban forests?

Case Study   National Urban Parks

A national park is an area of scenic, historical, or scientific importance that is set aside and maintained by the federal government. Most national parks are kept in their natural state far from urban areas. For example, Quttinirpaaq National Park in Nunavut includes the most remote, fragile, and northerly lands in North America. Parks Canada describes Fundy National Park as "Atlantic's sanctuary … that encompasses some of the last remaining wilderness in southern New Brunswick … [along] with [the] world's highest tides." Ontario's St. Lawrence Islands National Park is the smallest national park in Canada and is accessible only by boat.

Unlike the national parks that are located outside of and often far from cities, there are many smaller urban parks throughout Canada's cities. Urban parks are usually managed by local governments. However, many countries—including Canada—are now establishing larger national urban parks to conserve important natural and cultural areas.

Rouge National Urban Park

In 2011, the federal government announced that it would create a national urban park in the Rouge Valley in the eastern part of the Greater Toronto Area. The federal budget of 2012 committed 143.7 million dollars over 10 years for the development of the park, with 7.6 million dollars per year after that for the park's continuing operations. The provincial government, non-profit organizations, and community groups also co-operated in the formation of the new park.

The proposed area for Rouge National Urban Park is 5600 hectares, which is the size of nearly 14 000 football fields. The park is also within a short distance (about 100 km) of 20% of Canada's population. The vision for the park is that it will be a "people's park," where visitors can experience the area's unique natural and cultural heritage with no entrance fees.

The park represents diverse natural environments. These include freshwater marsh, rivers, forest, and wetlands. As well, the park will protect the habitats that connect the Oak Ridges Moraine and Lake Ontario. Living in these diverse habitats are 760 plant species, 55 species of fish, 19 species of reptiles and amphibians, 225 bird species, and 27 mammal species. The park will also incorporate the City of Toronto's only working farms and promote sustainable farming techniques.

The Rouge Valley also represents more than 10 000 years of human history, so it will preserve and highlight important cultural heritage sites. These include artifacts and habitation sites from early nomadic hunter-gatherer tribes; the Toronto Carrying Place, an Aboriginal portage trail also used by the fur traders; and Bead Hill, a 17th-century Seneca village. The vision for Rouge National Urban Park is not only to conserve important ecosystems, but also to restore fragile habitats and provide engaging educational experiences for visitors.

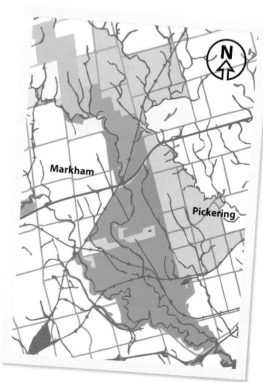

Map of Rouge National Urban Park

Santa Monica Mountains National Recreation Area

Santa Monica Mountains National Recreation Area, in Greater Los Angeles, represents the successful co-operation of a large number of landowners and stakeholder groups to preserve this natural and culturally significant area for future generations. The size of 63 000 football fields, it is the largest national urban park in the United States.

The park was established in 1978 and is made up of many individual parks and nature sanctuaries. These habitats include coastal marine ecosystems and one of the world's best examples of a Mediterranean climate ecosystem. The park is also home to nearly 2000 plant and animal species, including pumas.

Visitors can find relief from the heat of the city in the cooler mountains and enjoy more than 100 km of hiking and walking trails. As well, the park's numerous archeological sites showcase the different human communities that have made the area their home over the past 10 000 years.

Research and Analyze

1. Research what is currently happening with the Rouge National Urban Park. Consider the following questions: Is the project on schedule and on budget? Has the original plan for the park changed? Have there been any controversies during the park's development? If so, how were these issues resolved?

2. Research another possible location for a national urban park in Canada. Outline a short proposal for the new park. Include details such as the site boundaries, important natural and cultural resources that would be included, and which agencies would fund and manage the park.

3. Research another national urban park that is not discussed in this Case Study. Describe how that park was established and any issues that may have arisen as the park was developed.

4. Choose one of the urban parks discussed in this case study or the one you researched in question 3. Make a table that lists at least five ecosystem services the park provides to its city or region. Include specific details for each service you list.

Communicate

5. In pairs or a small group, brainstorm the advantages that a national urban park can provide to the residents of a city. Create a brochure to promote national urban parks to the general public.

Santa Monica Mountains National Recreational Area—Backbone Trail

Case Study Naturalization

Naturalization is an environmentally sustainable technique that is used to create or re-create natural landscapes. Naturalized landscaping uses native plants. This decreases the need for maintenance and creates habitat for wildlife. Naturalization should not be confused with simply abandoning a site and allowing plants to regrow. A successful naturalization project takes research, planning, and management.

Brampton Valleys Re-naturalization

In 2003, the City of Brampton, Ontario, began a project to naturalize 160 hectares of the city's watershed lands along Fletcher's Creek and the Humber River tributaries. The goal was to restore these valleys to their natural state, before they were altered by agriculture and development.

The city committed 8.8 million dollars over 10 years to plant 24 000 trees, 200 000 shrubs, and 100 000 other flowering plants. As shown in the photos, the project has been successful in restoring plant communities, creating fish and wildlife habitats, and stabilizing these important flood-plain lands.

Before naturalization

After naturalization

Glenridge Quarry Naturalization Site

Glenridge Quarry Naturalization Site is located in St. Catharines, in Ontario's Niagara Region. The site was originally a limestone quarry, but was used as a municipal landfill from 1976 to 2001. After the landfill was closed, the city decided to naturalize the site, with the goal of creating sustainable habitats that would also provide recreational and educational opportunities.

The plan for the site used local natural habitats as a guide. It also used ecological principles in its design: low energy consumption, re-use of natural materials, and recycling of building materials. Today, the Site has trails, boardwalks, and picnic areas for visitors to enjoy, and is connected to the Bruce Trail, which is Canada's longest trail system. The naturalization project at Glenridge Quarry has received numerous Canadian and international awards.

Research

1. Research ways that schoolyards can be naturalized and whether there are any organizations that can help. Design a plan for naturalizing a portion of your school's property. Include the native plants you would use in your design.

2. Research an area of land in your community that could be a candidate for naturalization. Create a table to list the benefits and challenges involved in naturalizing your chosen site.

Communicate

3. Write a proposal to present the plan you designed in question 1 to your local school board or to your municipal government. Include a brief description of your plan, the benefits of naturalization, and the advantages the project will bring to the community.

Summary

- Urban forestry deals with the long-term planning, planting, and maintenance of forests, trees, and green spaces in urban environments.

- Urban forests provide ecosystem services such as habitat and food for other organisms, space for people to relax and exercise, reducing the heat island effect, buffering noise, and acting as natural air purifiers.

- Urban forests and green spaces, including green roofs, play an important role in reducing the heat island effect.

- Methods of managing urban forests sustainably include choosing site conditions, planting native plants, and keeping plants healthy.

- Integrated pest management is an important part of managing urban forests sustainably.

- As urban areas grow and expand, urban foresters consider how plants can be integrated into future development and suggest ways to increase their sustainability.

Review Questions

1. Use a graphic organizer of your choice to show the ecosystem services provided by urban forests. **C**

2. Choose one or more of the ecosystem services provided by urban forests and explain its importance to or effect on the environment, the economy, and the local community. **T/I** **C**

3. Which ecosystem service is the urban park in the photo below providing? **A**

4. Explain the heat island effect. **K/U**

5. List three actions that can be taken to reduce the heat island effect. **K/U**

6. Choose one location in your community, such as your own yard, an empty lot, your school, or a group of buildings. Make a list of changes that could be made to increase the sustainability and ecosystem services provided by the area. Use the following questions as a guide. **T/I** **A** **C**

   a) How could native plants be incorporated into the area?

   b) How could run-off be reduced?

   c) Would a community garden be appropriate? Why or why not?

   d) Could a green roof be incorporated? Why or why not?

7. The Asian long-horned beetle (*Anoplophora glabripennis*), shown in the left photo below, is native to northeast Asia. As an invasive species introduced to North America in the 1990s, it poses a serious threat to various urban tree species, including elm, birch, poplar, and maple. The larvae and adult beetles damage trees by feeding on sap, budding leaves, and bark. The photo on the right shows damage to a tree caused by these beetles. Infested trees are weakened and eventually die. The beetle was found in the Greater Toronto Area in 2003, where maple trees make up 50% of the trees that line streets. The City Forester's Office of Toronto counts on alert citizens to report sightings of the beetle or evidence of its infestation on trees. Write a public service announcement to encourage citizens to help reduce the threat of this invasive species. **A** **C**

8. Explain why planting trees around buildings helps reduce costs associated with indoor temperature regulation in both the summer and the winter. **A**

9. Make a concept map that organizes information about the methods used to manage urban forests sustainably. **K/U** **C**

Safety Precautions

- Use appropriate protective equipment such as aprons, goggles, and gloves as well as taking any other safety precautions that are stated in associated Material Safety Data Sheets (MSDS).

- Wash your hands with warm water and soap after each time you come in contact with soil and/or plant material.

Suggested Materials

- reference books and/or computer with Internet access
- materials for soil mixture, which may include compost, manure, vermiculite, potting soil, topsoil, peat moss, powered limestone, coffee grounds, loam, sand, clay, and gravel
- soil pH test kit
- soil nutrient test kit
- soil humus test kit
- plastic pots
- plant seedlings or saplings, such as tomatoes, wheat, jack pine, or cactus
- water
- light source
- fertilizer
- metric ruler

Which Soil Composition Is Best?

The composition and properties of soil affect the health of the plants that grow in it. In this Lab, you will prepare a mixture of soil that is optimal for a plant that you select.

Pre-Lab Questions

1. What variables should you consider as you create your soil mixture?
2. What variables should you consider as you grow your plant?
3. What safety precautions should you take as you carry out your investigation?

Question

How can you prepare a soil mixture for a selected plant species, based on analysis of the criteria for optimal growth for that species?

Procedure

Part A

1. Choose a plant that you would like to try to grow from the selection provided by your teacher.

2. Research the optimal growing conditions for the plant. Consider the following soil properties as you conduct your research: pH, porosity, nutrient balance, and amount of organic material. Research how to recognize if the plant species you selected is not growing in the appropriate soil or getting too much or too little of a certain nutrient.

3. Prepare a procedure to create the proper soil mixture. Consider that you may have to make several mixtures, testing each one and modifying it to create the optimal mixture. Think about how you will use the materials available, including the test kits, to prepare your soil mixture. Think about how you will track data as you test different mixtures and how you will use what you learn from the data to get your mixture closer to the optimal conditions. Include safety procedures as part of your plan. Have your teacher approve your procedure.

4. Make your soil mixture according to your approved procedure.

Part B

1. Once you have created the optimal soil mixture, plant your seedling or sapling.

2. Considering light and temperature conditions, place your plant in an appropriate location to grow. Water your plant as needed.

3. Determine how you will qualitatively and quantitatively monitor the growth of your plant and record your data in a table. Consider the following: How will you know if your plant is getting the proper balance of nutrients? How will you know if the pH of the soil is appropriate? How will you know if the soil has the appropriate porosity?

Analyze and Interpret

1. Briefly summarize how your plant grew.

2. Do you think you created an optimal soil mixture in which your plant could grow? Why or why not? What other variables may have affected growth?

Conclude and Communicate

3. Describe at least two changes you would make to improve your soil mixture, and explain why you would make these changes.

Materials

- reference books and/or computer with Internet access

Gypsy moth

Mountain pine beetle

Apple scab

Pests and Diseases in Canada's Forests

An important part of managing forests sustainably focusses on how to control invasive pests or diseases. For example, the gypsy moth is a non-native species introduced to North America from Europe and Asia. It feeds on the newly budding leaves of trees in the spring. Millions of dollars have been spent trying to control infestations of the gypsy moth. In this Lab, you will research more about an insect pest or disease that is affecting Canada's forests.

Pre-Lab Questions

1. How will you analyze your sources for accuracy, reliability, and bias?

2. How will you organize the information you find as you research?

Question

How do insect pests and/or diseases affect the sustainability of Canada's forests?

Procedure

1. Research more information about an insect pest or disease that is affecting Canada's forests. You may choose from the following list or select an organism you find while researching.

   - Gypsy moth (*Lymantria dispar*)
   - Mountain pine beetle (*Dendroctonus ponderosae*)
   - Asian long-horned beetle (*Anoplophora glabripennis*)
   - Spruce budworm (*Choristoneura fumiferana*)
   - Aspen trunk rot (*Phellinus tremulae*)
   - Apple scab (*Venturia inaequalis*)
   - White pine blister rust (*Cronartium ribicola*)
   - Chestnut blight (*Cryphonectria parasitica*)

2. As you research, analyze your sources for accuracy, reliability, and bias. Determine how you will organize the information you find.

3. Use the following questions to help guide your research.
   - How and when did the insect or disease species arrive in North America?
   - In which areas of Canada is this insect or disease a problem?
   - What is the life cycle of the insect or disease?
   - What types of trees does it infest or infect? How does it damage them?
   - What are the symptoms in infestation or infection?
   - How does the insect or disease affect the sustainability of forest ecosystems in Canada?
   - What methods are used to control the pest or disease? Have they been successful? Why or why not?
   - What impact, if any, could the control methods have on human health?

4. Present the results of your research in a format approved by your teacher.

Analyze and Interpret

1. Why is it important to understand the life cycle of the insect pest or disease?

2. What role does early detection play in managing an insect pest or disease?

Conclude and Communicate

3. Explain how integrated pest management is used to manage the insect or disease you researched.

Skill Check

Initiating and Planning

Performing and Recording

Analyzing and Interpreting

Communicating

Safety Precautions

- Use appropriate protective equipment such as aprons, goggles, and gloves as well as taking any other safety precautions that are stated in associated Material Safety Data Sheets (MSDS).

Materials

- tree samples (cross section or core sample)
- hand lens
- reference books and/or computer with Internet access
- metric ruler

Annual Growth Rings of Trees

In trees, the production of tissue throughout each year produces rings, which are often called growth rings or tree rings. Scientists can analyze these annual growth rings to determine the age and growth rate of a tree, and to make observations and inferences about the environmental conditions in which the tree grew.

In a cross section of a tree that grows in a temperate climate such as southern Canada, an alternating light and dark ring represents one year of growth. The light part of the ring is growth that occurs during the spring and early summer. The dark part marks the end of that season's growth. The age of a tree can be estimated by counting the rings near the base of its trunk.

Environmental conditions, such as rainfall, temperature, and amount of sunlight, influence the width of the rings. For example, in dry years the rings are much thinner, while in years with plenty of rainfall, the rings are much thicker. Other factors that may affect the width of a ring include insect infestation, disease, or a natural disaster such as an ice storm, hurricane, or landslide. Tree rings also show damage from fire, called fire scars.

Pre-Lab Questions

1. What are growth rings?

2. What information can be learned from analyzing the growth rings of trees?

Question

What information can you learn from examining the annual growth rings of a tree?

Procedure

1. Copy the data table shown below into your notebook.

2. Examine the tree sample provided by your teacher.

3. Identify the ring that represents the first year of growth.

4. Count the growth rings and record the age of the tree.

5. Compare the width of each ring to that of the others. In which year of growth was there the most rainfall? In which year of growth was there the least rainfall?

6. Record any evidence that a forest fire occurred during the lifetime of the tree.

Growth Ring Analysis Data Table		
	Data	**Observations**
Age of tree		
Year of growth with most rainfall		
Year of growth with least rainfall		
Evidence of forest fire		
Evidence of pest infestation or disease		
Other patterns or observations		
Diameter (cm)		
Growth rate (cm/year)		

7. Compare the growth rings of your tree sample to the figures provided on these pages, as well as any other figures provided by your teacher. Record any other patterns or observations about your sample.

8. If your tree sample is a cross section, measure the diameter of the trunk. Determine the growth rate of the tree by dividing the diameter of the trunk by its age.

Core sample

Analyze and Interpret

1. Summarize the life history of the tree you observed.

2. Why is it important to know about the growth rate of trees? How might foresters use this knowledge?

3. What are the limitations of the data you collected? What other sources of information could you use to verify the inferences you made?

Conclude and Communicate

4. What can annual growth rings of a tree tell scientists about the environmental conditions in which the tree has grown?

5. How can scientists who are interested in studying current global climate change use annual growth rings to gain information about past climates?

Inquire Further

6. Foresters often bore holes into living trees to take core samples (shown above) so they can observe the growth patterns of the trees. Infer the advantage of doing this over using tree rounds. How might the interpretation of a core sample be different from that of a cross section?

Chapter 6 SUMMARY

Section 6.1 Forestry Management

Forests provide many important ecosystem services, and managing forests through sustainable practices is important to current and future generations economically, ecologically, and socially.

Key Terms
forest
timber
silviculture
clearcutting
selective cutting
shelterwood system

Key Concepts
• A forest is an ecosystem in which the dominant plants are trees. Most forests are made up of three main layers: the canopy, the understory, and the forest floor.

• Forests provide many ecosystem services, including being a source of timber and non-timber resources.

• Silviculture is the development and management of forests. Silviculture methods include clearcutting, selective cutting, and shelterwood systems.

• As we continue to learn more about the importance of forests and the ecosystem services they provide, governments and private landowners worldwide are turning to sustainable management practices.

• Companies that make wood products can apply to various independent agencies for permission to apply a label that identifies their products as being managed and manufactured in a sustainable manner.

Section 6.2 Urban Forestry

Urban forests and green spaces in cities provide important ecosystem services, and as urban areas grow and expand, urban foresters consider how plants can be integrated into future development to increase sustainability.

Key Terms
urban forestry
heat island effect
native plant

Key Concepts
• Urban forestry deals with the long-term planning, planting, and maintenance of forests, trees, and green spaces in urban environments.

• Urban forests provide ecosystem services such as habitat and food for other organisms, space for people to relax and exercise, reducing the heat island effect, buffering noise, and acting as natural air purifiers.

• Urban forests and green spaces, including green roofs, play an important role in reducing the heat island effect.

• Methods of managing urban forests sustainably include choosing site conditions, planting native plants, and keeping plants healthy.

• Integrated pest management is an important part of managing urban forests sustainably.

• As urban areas grow and expand, urban foresters consider how plants can be integrated into future development and suggest ways to increase their sustainability.

Knowledge and Understanding

Choose the letter of the best answer below.

1. Which statement about forests is *false*?
   a) The dominant plants in forests are trees.
   b) The tundra rainforest is one of several types of forests.
   c) The basic structure of most forests consists of the canopy, the understory, and the forest floor.
   d) Different abiotic and biotic factors make up different types of forests.
   e) A forest is an ecosystem.

2. In which part of a forest do shade-loving young trees and shrubs thrive?
   a) the canopy
   b) the forest floor
   c) the understory
   d) the branches
   e) None of these areas support the growth of trees that thrive in the shade.

3. Timber from trees, such as spruce, maple, oak, or pine, might be used
   a) to build frames for houses
   b) by wood crafters to make carvings
   c) as fuel to keep warm or cook food
   d) in a pulp mill to make paper products
   e) All of these are correct.

4. Selective cutting is often used on uneven-aged forests. Which statement about selective cutting is true?
   a) It is a cheaper way to harvest trees than other methods.
   b) Foresters can remove individual trees that are worth a lot of money.
   c) It is often the safest way for workers to harvest trees.
   d) It has the same effect on the forest as a forest fire burning through the area.
   e) Soil erosion will occur if the cut is on a steep slope.

5. The field of science that deals with the long-term planning, planting, and maintenance of forests, trees, and green spaces in urban environments is
   a) urban forestry
   b) selective cutting
   c) shelterwood cutting
   d) silviculture
   e) dendrochronology

6. Which is *not* an ecosystem service provided by forests?
   a) increased soil erosion
   b) carbon storage
   c) cycling of nutrients such as nitrogen and phosphorus
   d) water purification
   e) providing habitats for millions of species on Earth

7. The practice of developing and managing forests for the timber products it can supply now and in the future is
   a) agriculture
   b) ecosystem management
   c) integrated pest management
   d) silviculture
   e) urban forestry

8. Which statement is *false*?
   a) A mature tree does not grow anymore.
   b) An old-growth forest has developed for over at least 120 years.
   c) Trees that are all within 10 to 20 years of age of each other are part of an even-aged forest.
   d) An uneven-aged forest has trees of significantly different ages growing in it.
   e) Clearcutting is usually used to manage an even-aged forest.

Answer the questions below.

9. Decomposing matter in the forest floor is an important part of the forest ecosystem. Explain how this rotting matter benefits new seedlings and existing forest plants.

10. List four non-timber forest resources.

11. Briefly describe two advantages that shelterwood cuttings have over clearcutting methods of tree harvesting.

12. Define *sustainable forestry*.

13. Identify three ecosystem services that are provided by plants in an urban environment.

14. Urban areas tend to have daytime temperatures that are higher than the daytime temperatures recorded in surrounding rural areas.
    a) What is this temperature differential known as?
    b) Explain why it occurs.

15. What is integrated pest management and how might it be used to help maintain healthy trees in an urban area?

16. Identify three site conditions that are usually evaluated by an urban forester before a decision can be made regarding what to plant in the area.

Thinking and Investigation

17. Infer the types of forestry practices that could be used to maintain features of old-growth forests while harvesting trees.

18. A 2012 study published in the academic journal *Science* reported that death rates among trees 100 to 300 years old are increasing compared to past years. The accelerated death rates were found in old trees across many ecosystems in many areas of the world, including forests and urban areas in Brazil, Europe, Canada, and Africa. The scientists who conducted the study compared the death of these large, old trees to the plight of endangered species such as rhinoceroses, whales, and elephants.

a) What ecosystem services are lost when large, old trees like the ones used in the study die?

b) Infer some of the possible causes of the accelerated death rates of these trees.

c) What information would you need to collect to determine whether the causes you listed in part (b) are valid?

19. Without forests, there would not be any non-timber–related products. If you were going to defend this statement, what reasoning would you use to support it? Include at least three facts in your response.

20. Wildfires are usually considered to be very destructive. However, the Ontario Ministry of Natural Resources will conduct a prescribed fire in a forested area as a managed forest practice. The goals of this practice include silvicultural site preparation and maintaining a healthy ecosystem, in which fire is part of the natural processes. For example, this practice could be used to regenerate jack pine populations, as heat from a fire is needed for the cones to open and release the seeds.

a) Describe a simple lab procedure that you could conduct outdoors on the cones of jack pines to determine the effects of fire intensity on the viability of the cones. **Do not conduct this proposed investigation.**

b) Could your results in part (a) be used to predict whether cones exposed to fire and heat in the upper canopy are more viable than those in the lower canopy? Explain the reasons for your response.

c) What would you use as an appropriate control for your proposed investigation?

d) Identify three safety procedures that would have to be followed if this investigation were to be conducted outdoors.

21. Giant sequoias, such as the one shown below, can grow to have diameters over 6 m and height over 100 m. These trees can be grown from seed. The seeds are placed in a refrigerator for about a month to simulate the colder weather, and then left near a warm source for a bit of time to ensure that the seeds germinate. The trees require rich, moist, well-drained soil. They have delicate feeder roots that are needed to obtain water—if these roots are broken, the tree will likely die.

a) Based on your knowledge of urban forestry and the information provided above, do you think that someone could germinate a giant sequoia seed and successfully get it to grow as a seedling in a suburban backyard with dimensions of, say, 25 m by 30 m? Provide support for your response.

b) What other information would an urban forester need to determine whether a giant sequoia could be grown successfully in an urban green space?

Communication

22. Plants in an urban environment provide many ecosystem services, including removing carbon dioxide from the air. In a well-structured paragraph consisting of four or five sentences, describe some other ecosystem services that plants provide.

23. Silviculture can be beneficial to the forest industry as well as the consumer. For example, lower costs might be incurred by the forest industry when a large number of trees are harvested through clearcutting methods. This savings can be passed along to the consumer via lower lumber prices. State one other benefit of silviculture, and give a reason to support your answer.

24. More than 80 000 ash trees were cut down in Essex County in 2004 to stop the spread of the emerald ash borer. Provide a list of the environmental costs associated with this management method.

25. Did You Know? Research more information about Wangari Maathai. What is the Greenbelt Movement? Why was she awarded a Nobel Peace Prize? How does the quote on page 182 emphasize the importance of sustainable forestry? Use the information you find during your research to write a mini-biography of Wangari Maathai.

Application

26. A study by the University College, London, showed that in Toronto, rates of type 2 diabetes were lower in areas that had greater access to parks and other green spaces. Another study completed by a team of Ontario scientists showed that decreased mortality rates in urban areas were associated with access to green spaces. Other studies have shown that access to green spaces in urban areas is also associated with better mental health and less physical illness. Suppose that the development and maintenance of a new urban park would cost taxpayers a lot of money. How could the mayor's office use the facts in this paragraph to justify some of these economic costs?

27. Research more information about the advantages and disadvantages of tree plantations. Include environmental, economic, and social aspects of the issue. Present the results of your research in a table.

28. If you have ever walked on concrete or asphalt in your bare feet, you know that it absorbs a lot of heat from the sunlight that shines on it. These surfaces also transmit heat to the ground under the pavement and release that heat at night, which further increases the heat island effect. Cool pavement, shown in the diagram below, is made with materials that reflect more sunlight so that less heat is absorbed. Cool pavement materials also have pores that let water drain through, which helps to cool the ground under the pavement. How does the use of cool pavement reduce the heat island effect?

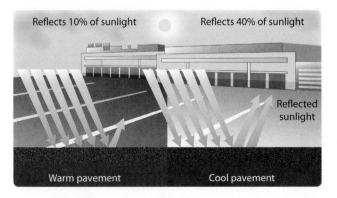

Reflects 10% of sunlight   Reflects 40% of sunlight

Reflected sunlight

Warm pavement   Cool pavement

29. Research more information about how forest thinning can help prevent or reduce the seriousness of a forest fire. Use the following questions to help guide your research.

  a) Does thinning of forests and prescribed burning help prevent catastrophic wildfires? Why or why not?

  b) Is there an ecological advantage to thinning forests and/or using prescribed burns to reduce fuel for fires? Explain your answer.

  c) Are prescribed burns part of Parks Canada's forest management program? Why or why not?

30. Suppose you are buying lumber for a small building project. At the store, you compare prices of different types of lumber. The lumber that is labelled as sustainable is about 20% more expensive than the same lumber that is not labelled sustainable. Which lumber would you buy? Why?

31. Research more information about the long-term effects of clearcutting forests on human health. Present the results of your research in a bulleted list with a brief explanation next to each effect.

32. A typical schoolyard is shown in the photograph below. Provide at least three options that you could present to the school council to change this concrete schoolyard into a green schoolyard. You may use a sketch to help you with your response.

33. Describe three actions that you, as a consumer, can take to encourage businesses to carry wood products that are more sustainable.

> ## Pause and Reflect
>
> How could you incorporate what you have learned in this chapter into your daily actions or choices?

Edmund Zavitz

Edmund Zavitz: Forestry Pioneer

Picture thousands of hectares of farmland turning to desert, floods destroying towns and villages, and cattle grazing on bare rock. No, we are not talking about Russia, Africa, or South America. This was Ontario at the beginning of the 20th century, and Edmund Zavitz was determined to do something about it. Zavitz was born in Ridgeway, Ontario, in 1875. He became passionate about trees during his boyhood as he explored the forests of the Niagara region.

From 1770 until the early 1900s, settlers in Ontario were required to clear their land of all trees. Without trees to act as windbreaks, fertile topsoil was blown away and, in areas such as Norfolk County near Lake Erie and the Oak Ridges Moraine north of Toronto, the sandy soil was exposed and swept up in great sandstorms that buried fences, roads, and buildings. The lack of tree cover to hold moisture in Ontario's watersheds caused disastrous floods in towns such as Guelph and London.

Zavitz travelled the province, often by bicycle, convincing farmers to reforest their land and taking the pictures that he later used to convince the provincial government to take action. "It required a great deal of talking," he said, "But then I always could talk."

After studying forestry at Yale and the University of Michigan, Zavitz's efforts led to the Counties Reforestation Act of 1911. He was appointed Chief Forester for the province of Ontario in 1912. Over the next 37 years, 2 billion trees were planted during his time as Chief Forester, Deputy Minister of the Department of Forests, and Director of the government's Reforestation Branch, giving us the beautiful green country that is ours to protect today.

Edmund Zavitz took this photo as an example of the devastation that can occur when all of the trees are removed from an area of land. It was this type of scene that inspired Zavitz to reforest areas of southern Ontario.

Environmental Science at Work

Focus on Sustainable Agriculture and Forestry

Forestry technician

Farm manager

Arborist

Sustainable
Agriculture and
Forestry

Farmer

Landscape technician

Organic farmer

Organic farmers grow crops using ecological principles that emphasize sustainability and resource conservation, without the use of synthetic pesticides and fertilizers or GMOs. They work for or own certified organic farms of various sizes, from small family farms to larger operations.

Forester

Foresters work to manage forests based on the principles of stewardship and sustainability. They may work for government agencies, as consultants for industry or private landowners, or in forestry research.

Urban forester

Urban foresters care for shrubs and trees in urban green spaces, and they diagnose and treat plant diseases and pests. They work for municipal or provincial governments, utility companies, or corporations with large urban properties, or in environmental consulting.

For Your Consideration

1. What other jobs and careers do you know or can think of that involve the fields of sustainable agriculture and forestry?

2. Research a job or career in agriculture or forestry that interests you. What essential knowledge, skills, and aptitudes are needed? What are the working conditions like? What attracts you to this job or career?

Design a Landscaping Project

Many conditions, such as temperature, rainfall, amount of sunlight, and soil composition, are unique to your local area. Urban foresters and landscapers take these into account when they design, plant, and maintain gardens and natural landscapes.

Question

How can you design a sustainable garden or landscaping project that will thrive in your local area?

Initiate and Plan

1. Decide on the type of garden or landscaping project you would like to design. Ideas include a rooftop garden, rain garden, garden to attract native songbirds or butterflies, community vegetable garden, arboretum (garden devoted to trees), or riparian restoration (replanting native streamside vegetation).

2. Create a list of conditions that describe your local area. Complete any research as directed by your teacher. Include
   • soil properties (both living and non-living)
   • local maximum and minimum temperatures
   • average amount of rainfall and Sun exposure

3. Brainstorm how your design can make use of sustainable practices. For instance, how could use of compost decrease the need for fertilizer that could run off into local streams? Other factors to consider include integrated pest management, companion planting, the use of native plants, soil conservation, and sustainable irrigation practices.

4. Consider health and safety in your plan. Will you need help or specialized training from anyone? What pieces of personal protective equipment must be available? Are there any hazardous materials (WHMIS) or tools required to implement your design?

Perform and Record

5. Incorporate your answers to Initiate and Plan steps 1 to 4 into a garden or natural landscape design.

6. Have your teacher approve your design.

7. If your teacher approves, implement your design.

Analyze and Interpret

1. How do the practices you incorporated into your design support sustainable interactions between your garden or natural landscape and its surrounding ecosystem?

2. If you could redesign your garden or natural landscape, what changes would you make, if any? Explain your reasoning.

Communicate Your Findings

3. Present your design using a format of your choice, keeping your audience in mind.

Assessment Checklist

Review your project when you complete it. Did you ...

☑ **T/I** create a list of conditions that describe your local area?

☑ **A** determine how your design can make use of sustainable practices?

☑ **K/U** create your design, demonstrating your knowledge of sustainable agriculture or forestry?

☑ **A** implement your design, with teacher approval?

☑ **C** choose a format and present your design keeping your audience in mind?

An Issue to Analyze

Investigating Fire in Forestry and Agriculture

Fire can occur naturally or due to human intervention in forested and agricultural ecosystems. The management of fire in these ecosystems is often controversial.

Issue

What role does fire play in forested and agricultural ecosystems? How does it harm or benefit the ecosystem? What consequences are related to fire management practices used in these ecosystems?

Initiate and Plan

1. Use online and other resources to research the occurrence and management of fire in forested and agricultural ecosystems. Based on your research, choose one ecosystem to explore further. Your choice can be general (for example, sugar cane plantations) or specific (for example, the boreal forest in northern Algonquin Park).

2. Before you begin any further research,
   - create a plan to assess the reliability of your resources
   - determine how you will cite the resources you use
   - choose a system to record your research findings
   - have your teacher review your research plan

Perform and Record

3. Research the occurrence and management of fire in your ecosystem. Follow your research plan from step 2.

4. Answer the following questions as part of your research:
   - How would you describe your ecosystem?
   - How does fire occur naturally or due to human intervention in the ecosystem?
   - What role(s) does fire play in the ecosystem? How is fire managed?
   - What are the consequences of fire management in terms of ecosystem sustainability and other factors?
   - Are there any controversies concerning the use or management of fire in the ecosystem? If so, describe them.

Analyze and Interpret

1. Use a cause-and-effect map or other organizer to show how fire affects the sustainability of the ecosystem.

2. Complete a risk-benefit analysis to compare the pros and cons of fire management choices made in the ecosystem.

3. Do you agree with the fire management practices in the ecosystem you researched? Support your response with research findings and logical reasoning.

Communicate Your Findings

4. Share your findings using a format of your choice, keeping your audience in mind. Possibilities include a video documentary, social media website, or blog.

Assessment Checklist

Review your project when you complete it. Did you ...

☑ **T/I** complete background research on the topic?

☑ **K/U** create a research plan and have your teacher review it?

☑ **T/I** research the occurrence and management of fire in an ecosystem that interests you?

☑ **A** prepare a cause-and-effect map and a risk-benefit analysis?

☑ **C** communicate your findings with your audience in mind?

Unit 3 REVIEW

Knowledge and Understanding

Choose the letter of the best answer below.

1. Which of the following activities is *not* an example of a sustainable forestry practice?
   a) planting a variety of trees in a deforested area
   b) protecting forests that have high biodiversity
   c) reducing our use of wood products
   d) removing all fallen trees, stumps, and limbs, as well as all living trees in a large area
   e) selling wood products that have been certified as sustainable

2. Which statement about an urban area where the heat island effect occurs is true?
   a) The application of cool pavement will increase the heat island effect.
   b) Trees should not be planted in urban areas, since water that evaporates from the trees will make the air more humid and hot.
   c) Increased use of air conditioners is likely, which will increase the amount of greenhouse gases emitted.
   d) Urban area temperatures are lower than the surrounding rural area temperatures.
   e) Green roofs are not an effective way to help reduce the heat island effect.

3. A disadvantage of a monoculture is
   a) it has a decreased crop yield
   b) it is as difficult to maintain as a polyculture in the same area
   c) the harvesting of a monoculture is usually more labour-intensive than the harvesting of a polyculture
   d) different irrigation methods must be used in the monoculture
   e) it is more vulnerable to pests that attack that specific crop

4. Apples in Canada that are labelled organic
   a) must have been produced without the use of synthetic fertilizers
   b) must not have been grown using natural fertilizers
   c) have not been exposed to any kind of synthetic or biological pesticides
   d) are still considered organic if they were irradiated, since labels for this process are not required
   e) have likely been grown with low-dose antibiotics to prevent them from bacterial infection

5. An example of a sustainable method of harvesting wild-caught fish with a low by-catch rate and a low impact on the environment is
   a) sediment stirring by trawlers
   b) dragging heavy nets across the ocean bottom
   c) setting pots and traps along the ocean floor
   d) hook and line fishing
   e) longline fishing

Answer the questions below.

6. Plants have many basic needs that must be met in order to ensure their survival in their environments.
   a) Identify two substances that are required for photosynthesis to occur.
   b) Why do plants need oxygen?

7. Many living organisms are needed to enrich the soil in which they live. Identify two such organisms and briefly explain their role in the soil.

8. Synthetic pesticides are beneficial because they prevent crop losses, which in turn increases the food yield. However, the use of these pesticides can have many negative effects on the environment. Identify two disadvantages of using synthetic pesticides and briefly describe how each affects the environment.

9. Forest-related tourism is a multi-billion dollar industry worldwide.
   a) Identify and describe three forest-related tourism activities.
   b) List three ways that forest-related activities can be beneficial to your health.

Thinking and Investigation

10. Suppose plants that were genetically modified to withstand herbicides (pesticides that kill weeds) were growing in a garden and these plants crossbred with nearby weeds.
    a) What kind of a new plant might result?
    b) Identify a negative effect of this type of a cross.

11. Mechanized agriculture has provided many benefits to farmers and to society. For example, instead of manually lifting square hay bales from the field to the truck, large round hay bales can be picked up with a forklift and brought to the barn. Describe two other benefits of mechanized agriculture, as well as two risks or costs that the machinery has caused. Your response can be based on economic, social, or environmental benefits and costs.

Communication

12. In 2001, plans were made to thin out some of the overgrown trees that were blocking scenic areas where tourists take pictures in Yosemite National Park in the United States. Thousands of trees, except for rare or old trees, were to be cut down. Take the position of either a tourist or a state resident. Indicate if you are answering the question as a resident or a tourist, and provide an argument either for or against cutting trees in the overgrown areas. Provide at least three supporting details in your response.

13. More trees are being planted in many urban areas today than in the past, since these trees can reduce the amount of storm water that rushes through the storm sewer system. In turn, this prevents overflow at the waste-water management plant, which is an important part of keeping waste water out of our waterways. Use the diagram to answer the following questions.

Important Ways a Tree Helps with Storm Water Management

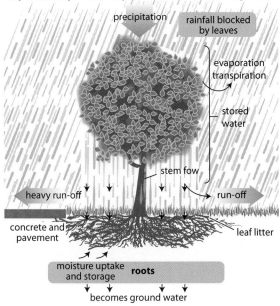

a) Explain two ways in which trees reduce storm water run-off.

b) Sketch a typical urban street in southern Ontario, with a few houses, sidewalks, and storm grates on the edge of the road. On the sketch, show the quantities and locations of the planted trees that would help reduce storm water run-off.

c) Given that this is an urban area, what other modifications could you make as an urban forester to reduce the impact of storm water gathering in low-lying areas?

Application

14. Carbon storage is an important ecosystem service that trees provide. Trees are able to remove some carbon dioxide from the air and use it for photosynthesis. The carbon remains in the wood, and as the tree grows, more carbon is added to the tree. A healthy, well-managed forest can take in more carbon than it releases. Since younger trees grow faster than older trees, there is a greater amount of carbon storage that occurs. If the soil is disturbed in the forest, possibly by heavy equipment, then organic matter in the soil breaks down faster, which releases more carbon dioxide into the air.

a) Do trees give off carbon dioxide, or do they just use it for photosynthesis?

b) How might pest infestations affect the ability of a tree to store carbon? Explain your answer.

c) Why is carbon dioxide released when organic matter breaks down in disrupted soil?

d) What do you think is the best way to maximize the amount of carbon storage in a forest?

15. Suggest at least three ways that you can have a minimal impact on the environment by ensuring that the food you consume has been produced by farmers using sustainable agriculture practices.

16. Genetically modified salmon, which are Atlantic salmon that have rapid growth genes from the Pacific salmon and an eel-like fish, could soon be approved for sale for human consumption. The GMO salmon, shown in the photo below, grow twice as fast as the non-GMO Atlantic salmon.

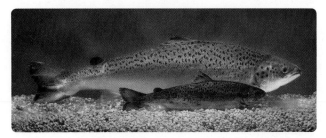

a) Infer three benefits of having the GMO salmon approved for sale in grocery stores and restaurants.

b) One concern surrounding the GMO salmon is that the fish could escape and mix with wild populations. What are some possible ecological consequences of the GMO fish escaping?

c) Currently, there are no laws indicating that the GMO salmon would have to be labelled as such when sold. Is this fair to consumers? Provide support for your response.

UNIT 4 Managing and Reducing Waste

Why do some municipalities ship their garbage to other provinces and even to other countries?

How is reducing waste different from recycling and re-using it? Why is reduction such an important strategy in waste management?

BIG IDEAS

- Well-thought-out waste management plans help to sustain ecosystems, locally and globally.

- By making informed choices, consumers can reduce the amount or alter the nature of the waste they produce.

Overall Expectations

- Analyze economic, political, and environmental considerations affecting waste management strategies.

- Investigate the effectiveness of various waste management practices.

- Demonstrate an understanding of the nature and types of waste and strategies for its management.

Unit Contents

Chapter 7

Solid and Liquid Waste Management

Chapter 8

Managing Hazardous Waste

Topic 1: A Product's Life Cycle

A *product's life cycle* includes all aspects involved in making, distributing, selling, using, and disposing of the product. The impact on sustainability at each stage of a product's life cycle may be questioned and assessed and may affect a consumer's decision to purchase the product.

As shown in **Figure 1**, a product's life cycle begins with the extraction of raw materials needed to make the product. Raw material extraction may have an environmental cost, such as deforestation or the release of pollutants into nearby ecosystems. The raw materials may need to be processed, which can mean the use of energy supplied by fossil fuels and the release of pollutants into the air and/or water.

The manufacture of the product may not be carried out in a socially sustainable manner—the conditions of the workplace and payment of workers may need to be considered before a consumer buys a product. Some products travel a great distance during the distribution process. This can mean the use of fossil fuels to energize the transport.

Once a product is used, disposal can become an issue. Can the product be recycled, re-used, or composted? Will it end up in a landfill, either locally or in another province, or even another country? Does it contain hazardous materials, such as heavy metals or other toxins, which may enter the environment during the product's eventual breakdown? As an educated consumer, you may ask yourself these questions and more before you consider buying a product.

Figure 1 A product's life cycle

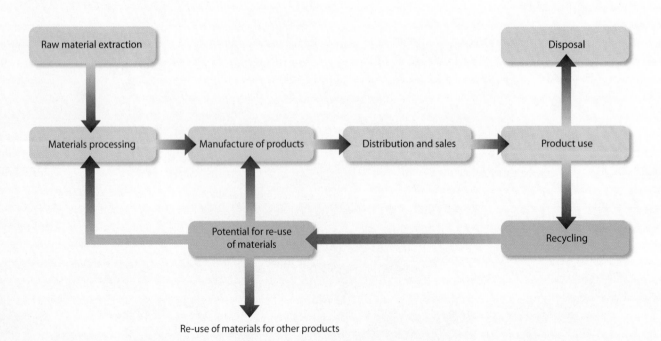

Topic 2: Watersheds

All fresh water on Earth—whether it is surface water such as lakes, rivers, streams, and oceans, or underground water—is part of a watershed. A *watershed* is the area of land that drains into a body of water. Watersheds are also called drainage basins.

When water falls on land, whatever water does not evaporate either filters into water that is under the ground or flows downhill. Water that does not filter into the ground or evaporate but instead flows across Earth's surface is called run-off.

The area of land within a watershed can be small or large. Within large watersheds, there are many smaller ones. For example, every stream is part of a large watershed. A stream in the community where you live flows until it meets other small streams. These streams join larger rivers. Large rivers merge into major waterways.

A watershed includes land as well as water. Each watershed has a variety of habitats. These range from the bodies of water themselves to forests, meadows, and farms. Cities, towns, and villages are also part of watersheds. This means that activities that occur on land or in water can affect the land or water around us. For example, if you pour old milk down the sink or flush a toilet, the wastes go somewhere.

If you live on a farm or in a rural area, the wastes may flow into a septic tank and then into a septic field on your property. There, they soak into the ground. Some of those wastes are taken up by the roots of plants. Some are consumed by soil organisms. Some soak deeper into underground water, where they are carried to other bodies of water, both below ground and above ground.

If you live in a city or in a small urban area, the wastes flow into a system of pipes. These pipes lead to large pipe systems that direct the wastes either to wastewater treatment facilities or directly into bodies of water.

Whether you live in an urban or a rural area, wastes from your watershed eventually feed into a body of water, and they will be carried to a different location. This location might be a place nearby or far away. It might even be in another province or another country. **Figure 2** illustrates why.

Figure 2 Major watersheds (drainage basins) in North America. Notice how and where the various river systems interconnect.

Solid and Liquid Waste Management

Toronto-born artist Aurora Robson created this walk-in exhibit called *The Great Indoors* from 15 000 plastic bottles. Much of her art gives new purpose to recycled plastics, packaging materials, and junk mail.

What are some materials that are thrown in the garbage that could be re-used for other purposes? What are some of those purposes?

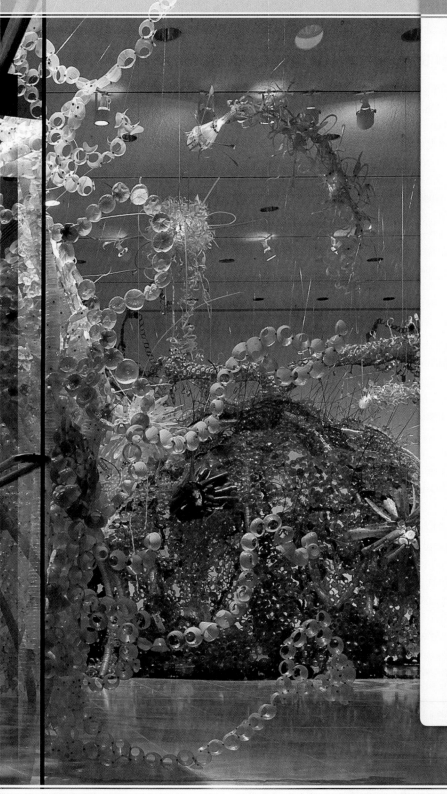

What Do You Know about Recycling?

1. Poll your class to find out how many people know how to recycle the items in the table. Add other items, too.

2. Afterward, convert the numbers to percentages. (Divide the number who know how to recycle each item by the number of students in your class; then multiply by 100.)

3. Repeat this poll with other classes at your school.

4. Discuss reasons for the results you get from these polls.

Item	Percentage Who Know How to Recycle
Cardboard boxes (corrugated)	
Cardboard boxes with tape	
Cloth and other textiles	
Food waste	
Hardcover books	
Leftover paint	
Paperback books	
Paperboard (such as cereal boxes)	
Plastic drink bottles	
Plastic shopping bags	
Stryofoam® food containers	

In this chapter, you will

- identify sources and types of solid and liquid waste
- describe technologies for managing solid and liquid wastes
- assess advantages and disadvantages of various waste management methods

Solid Waste Management

Sources of Solid Waste: Input into the Waste Stream

The term *solid waste* is defined in different ways by different levels of government. These definitions are often highly technical so that producers of large amounts of solid waste understand how and where their wastes are to be disposed. Broadly speaking, however, **solid waste** is any solid or semi-solid material that has been discarded. Each year, Canadians generate 30 million tonnes of solid waste. This much waste is more massive than the combined mass of 250 CN Towers.

The term **waste stream** is a metaphor used to describe solid waste. A real stream of water has one or several sources and travels through other places before reaching its final destination—usually a lake or ocean. In a similar way, the waste stream "flows" from one or several sources through a variety of other places before reaching *its* final destination—usually a landfill or incinerator. Solid waste enters the waste stream from three main sources: agriculture, industry, and municipalities.

solid waste any solid or semi-solid material that is discarded

waste stream the movement of waste from its sources to its final destination

Agricultural Solid Waste

Farms, ranches, feedlots, and many slaughterhouses are the source of agricultural wastes. Agricultural solid waste is made up of animal waste and organic matter that is left over after harvesting and processing crops and animals. Waste that is not used to fertilize crops is often burned, which produces air pollution. Rainwater and irrigation run-off from farms can carry nutrients into surface water and ground water. Livestock waste also contains many micro-organisms, such as *E. coli*, that can harm human health when they enter water systems.

Industrial Solid Waste

Industrial solid waste includes all solid wastes from manufacturing, construction, mining, and other industries. These wastes range from chemicals used in manufacturing to wastes produced during construction or demolition. Mines such as the one shown in **Figure 7.1** produce so much solid waste that mining is sometimes classified as a separate category of solid waste. Most of it consists of mine tailings, which are the particles left over after ores have been processed. Most industrial solid waste is recycled by industry, although it may also be disposed of in private landfills. Some industrial waste can be harmful to human health and the environment and is called hazardous waste.

hazardous waste— **section 8.1 on page 244**

Figure 7.1 The McArthur River mine in northern Saskatchewan is one of the largest uranium mines in the world. Its owner, Cameco, has been working with Jim Hendry, a geoscientist at the University of Saskatchewan, on safe ways to store and dispose of harmful mine wastes.

Municipal Solid Waste

Municipal solid waste (often abbreviated as MSW) includes all garbage—including items that can be recycled and composted—from homes, businesses such as office buildings, and institutions such as schools. (In some cases, MSW also includes wastes from construction and demolition sites.) Canada has the dubious distinction of being among the world leaders in the yearly production of municipal solid waste. In 2008 and 2009, each Canadian produced more than 1030 kg of waste. Most of it—777 kg per person—travelled the waste stream to landfill sites or to incinerators. The remainder was diverted from landfill, which means that it was recycled or re-used in some way. **Figure 7.2** can help you imagine 777 kg of waste.

Municipal solid waste includes a wide variety of items, from apple peels to grass clippings, water bottles, diapers, computers, and motor vehicles. **Figure 7.3** shows the breakdown of types of solid waste that make up the municipal waste stream. Provincial and territorial governments manage MSW. However, responsibility for collecting and treating the waste is often passed on to municipal or regional governments. In Canada, most municipal solid waste is sent to landfills.

Figure 7.2 The maximum mass of a garbage bag for roadside collection is about 20 kg. On average, each Canadian generates enough waste to completely fill about 40 of these bags each year.

paper 36%: newspapers, paper towel, mail, packaging, school and office paper, paper bags

organic waste 29%: food waste, yard waste, diapers, pet waste

metal 10%: cans, electronics, appliances
glass 8%: bottles, jars
plastic 7%: packaging, bags, containers, bottles, toys
other 10%: cat litter, clothing, tires, wood products, furniture

Figure 7.3 Municipal solid waste consists of many different materials.

Applying   Why have paper and organic waste continued to make up the majority of municipal solid waste for decades, even though people are becoming more aware of the value of recycling and composting?

Mini-Activity 7-1   Categories of Solid Waste

There are different ways to classify solid waste. The simplest and most common way is to divide solid waste into two groups: non-hazardous and hazardous. However, some organizations use different schemes to differentiate among the types of solid waste.

1. With a partner or in a small group, devise a classification system that you think each of the following organizations might use:

   • a small family of up to four people

   • a large family of up to nine people

   • a community group or municipality that wants to decrease the amount of waste material going out for curbside collection or being taken to rural transfer stations

2. Explain how you decided on the groupings you came up with in question 1.

3. What might be some advantages and disadvantages of having different classification schemes for solid waste?

Solid Waste Disposal: Output from the Waste Stream

For much of human history, the most common way to dispose of municipal solid waste was to dump it either in the streets or outside the village or city limits. Often, dumping grounds were in wetlands near a river or lake. To reduce the volume of the waste, the items in the dump were often burned. In some parts of the world, this method is still used.

As technologies for managing wastes have developed, there has been greater emphasis placed on the need to safeguard the health of the environment and living things. Today, especially in developed countries, the final destination for most municipal solid waste is a landfill site. Depending on where you live, this site could be in your region, in your province, somewhere else in the country, or even in a different country. Solid waste also may be disposed of at a thermal treatment facility, where this technology exists.

Disposing of Solid Waste in a Landfill

landfill a disposal site for solid waste where the waste is buried between layers of soil, filling in low-lying ground

A **landfill** such as the one in **Figure 7.4** is built into or on top of the ground in a way that isolates waste from ground water, air, and the surrounding soils. Isolation from ground water and soil is made possible by a large, impermeable liner that is buried deep into the ground. The waste is then added and buried over time between layers of soil that are bulldozed on top of it. Alternating layers of soil and waste are eventually compressed. The landfill either is built up into a large mound, or the ground is filled in.

Landfills provide several benefits to the environment. First, they are an immediate solution for solid waste management. In Canada, most solid waste is disposed of in landfills. Without them, this waste would enter ecosystems directly. Some landfills use gases produced by waste to generate electricity. Many landfills use machines called compactors to reduce the volume of waste before it is covered. This reduces the space needed to store waste and stabilizes the landfill. It also reduces offensive smells and makes waste less attractive to insects and rodents.

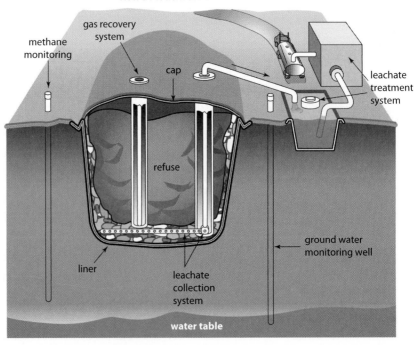

How a Modern Landfill Works

Figure 7.4 This modern landfill has impermeable layers that keep leachate from entering ground water. A collection system carries the leachate to a treatment system. Landfill gas and ground water are also monitored on a continual basis.

Leachate and Landfill Gas

As water seeps down through the contents of the landfill, chemicals from garbage dissolve in it. The liquid that results is called **leachate**. If allowed to leave the landfill untreated, leachate can pollute nearby soil and waterways. In modern landfills, rubber and/or impermeable clay liners are placed at the bottom of the landfill to contain leachate and prevent it from entering the surrounding ecosystem. Drainage systems are installed to monitor and collect leachate before it leaves the landfill. Leachate can also be treated to reduce the harm it can cause. Modern landfills are also built far from waterways to further minimize polluting of ground water.

Landfill gas is created by bacteria as they break down food and other organic material in landfills. The gas is a mixture of mostly methane and carbon dioxide, both of which are potent greenhouse gases. In fact, landfill gas was responsible for 22% of methane emissions in Canada in 2009. Landfill gas is also flammable and explosive.

Technology can help manage landfill gas. Gas capture and combustion (burning) converts methane into carbon dioxide, which is a less potent greenhouse gas than methane. Many landfills also burn landfill gas to generate electricity. In Canadian landfills, nearly 350 000 tonnes of methane are burned yearly; half of this is used for energy. For example, energy from the Landfill Gas Collection and Utilization Project in the Halton Region of Ontario can provide power to more than 1500 homes.

Available Space

Today, about 80% of municipal solid waste from Canada and 55% from the United States goes into landfills. The number of landfills has been declining, however, for two reasons. Many small, poorly run landfills have been closed because they were not meeting regulations. Others have closed because they have reached their capacity. The overall capacity has stayed fairly constant, though, because new landfill sites are much larger than older ones.

An ongoing debate over how to replace lost landfill capacity is developing where population density is high and available land is scarce. Selecting sites for new landfills in places such as Toronto, New York, and Los Angeles is very challenging because of the difficulty in finding sites that are geologically suitable and opposition from the people in the area. Public resistance usually comes from concern over ground water contamination, disease-carrying rats and other organisms, odours, and truck traffic. The acronym NIMBY ("not in my backyard") summarizes the essence of public opposition. Government officials often look for alternatives to landfills to avoid controversy over site selection. Although sites must be chosen for new landfills, politicians are often unwilling to take positions that might alienate their constituents.

leachate liquid from landfills composed of chemicals from garbage

landfill gas the gas produced from decomposition of material in landfills

 InquiryLab 7A, Modelling a Landfill, on page 232

Pause and Reflect

1. Identify the three main sources of solid waste.

2. Describe the main features of a landfill. Why are leachate and landfill gas potential environmental concerns?

3. Critical Thinking The amount of municipal solid waste that Canadians produce each year has been increasing slightly since 2002. Suggest three reasons why this may be the case. Explain your reasoning.

4. Critical Thinking You hear that a new landfill may be built in your community. Would you be supportive of this or against this? Explain your answer. Include at least two statements to support your opinion.

Disposing of Solid Waste in a Thermal Treatment Facility

thermal treatment
processing of solid waste
at high temperatures

With current landfills filling up and limited sites that are suitable for new ones, other ways to manage solid waste are needed. One of these methods is **thermal treatment** of waste. During this treatment, solid waste is processed using high temperatures. As of 2013, thermal treatment facilities for municipal solid waste exist in Ontario, Québec, British Columbia, Alberta, and Prince Edward Island. There are several types of thermal treatment. These are outlined in **Table 7.1**.

Table 7.1 Types of Thermal Treatment		
Type of Thermal Treatment	**Process**	**End Products**
Incineration	• complete combustion (waste is burned at high temperatures in the presence of oxygen) • occurs at temperatures over 850°C	• mainly carbon dioxide and water, as well as solid ash • resulting ash must go to a landfill (some may be hazardous) • a variety of air pollutants may be generated
Pyrolysis	• solid waste is thermally processed in the absence of oxygen • occurs at lower temperatures than combustion, from 350°C to 850°C	• produces solid residue and a synthetic gas called syngas • syngas is a mixture of gases, some of which are air pollutants • some gases in syngas can be condensed to form oils, tars, and waxes
Gasification	• solid waste is thermally processed with limited oxygen, but not enough for total combustion • occurs at temperatures greater than 650°C	• produces syngas and solid residue

One of the main advantages of thermal treatment is that it greatly reduces the volume of waste that would otherwise go into a landfill, generally by 10% to 20%. However, this benefit is undermined by the fact that the thermally treated waste can enter the atmosphere in the form of ash and gases. Some of these are harmful to the environment and human health. Many thermal treatment facilities, including all Canadian facilities, use pollution control technologies to treat or capture this waste before it is released.

air pollution—**see
section 9.1 on
page 280**

Many thermal treatment facilities also produce energy when they burn waste. This includes most Canadian facilities. For instance, the thermal treatment facility on Prince Edward Island helps run Charlottetown's district heating system. Because it generates energy, you may also hear thermal treatment referred to as *energy recovery* or *waste-to-energy* technology. **Figure 7.5** shows how combustion in a solid waste incinerator produces steam. The steam can be used to heat buildings directly. It can also be used to produce electricity.

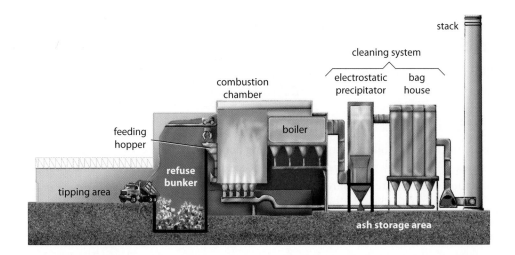

Figure 7.5 This *mass burn* incinerator produces steam in its boiler. The steam is used to generate electricity or heat nearby buildings. This type of incinerator burns everything except very large items.

Exportation

Figure 7.6 shows a barge carrying 2900 tonnes of trash that left a New York City harbour back in 1987. With all the city's major landfill space filled by that time, the plan was to export the waste to North Carolina—about 700 km away. When officials there refused to accept the waste, the barge began a trip in search of a place willing to take it. After unsuccessful stops in Louisiana, Mexico, Belize, and Cuba, and having travelled nearly 10 000 km, the barge returned to New York and the garbage was burned.

Nearly 25 years later, New York still has not solved its garbage problem. In 2011, the city shipped nearly 11 tonnes of waste every day to landfill sites in other states. That is equivalent to throwing away 62 jumbo jets a day.

Transporting solid waste to another region is expensive. (For New York, the cost in 2011 was 300 million dollars.) New York is not the only large city that lacks a local landfill, however. In the 1990s, Toronto's Keele Valley landfill was getting full. The city wanted to send its waste 600 km north by rail to an abandoned mine near Kirkland Lake, Ontario. Angry protests there meant Toronto needed another solution. From 2003 to 2010, the city exported its waste to a landfill in the state of Michigan.

In 2011, Toronto changed its strategy and began keeping its waste closer to home by sending it to the Green Lane Landfill Site 200 km away in southwestern Ontario. Since this landfill is closer to Toronto, transportation-related costs and pollution are reduced. Politically there is a benefit as well. Many people believe that shipping wastes to distant locations is an unacceptable option.

Figure 7.6 The garbage barge, *Mobro*, carrying 2900 tonnes of New York City garbage

---

Pause and Reflect

5. How do the three types of thermal treatment differ?

6. Why are some kinds of thermal treatment sometimes called energy recovery or waste-to-energy technology?

7. Critical Thinking Do you think Canada should export solid waste to other countries? Explain your reasoning.

Recycling and Composting: Diversions from the Waste Stream

Recycling and composting alter the course of waste-stream solids so they no longer end up in disposal sites. Instead, they are diverted to produce other goods and materials.

Diverting Wastes through Recycling

recycling collecting and reprocessing materials so they can be made into new products

Recycling refers to collecting and reprocessing items to make new products. For instance, plastic bottles are used to produce fleece clothing, newspaper is made into insulation, and old tires are shredded to make a new surface for an old road.

The first step in the recycling process is *resource recovery*. Items suitable for recycling are removed from the waste stream. This often begins at the site where the waste is generated: homes, schools, businesses, and industries. Many Canadian communities have blue box programs to collect wastes that are recyclable. There are many benefits to these programs and to recycling in general. However, recycling is not problem-free.

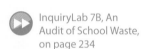

InquiryLab 7B, An Audit of School Waste, on page 234

Advantages of Recycling

- By recycling items that have already been manufactured, fewer raw resources are used to make new products. As well, less energy is needed for production. It takes half as much energy to create a plastic bottle from recycled resources, and production costs are lower. Fewer pollutants are released. For example, 95% less pollution is released when new paper is made from recycled paper.

- Recycling encourages greater responsibility. The act of sorting their garbage makes people and institutions more aware of the solid waste they produce.

- Recycling reduces the cost of municipal waste management. It costs towns and cities much more to process solid waste in a landfill or thermal treatment facility than to recycle it. Profits from selling recycled materials can also offset this cost.

- Recycling decreases the amount of waste flowing through the waste stream. For example, Toronto diverts 70% of its municipal solid waste each year.

Concerns about Recycling

- While recycling makes lighter use of Earth's resources, it still requires water, energy, and other resources. It is neither pollution-free nor energy-free.

- The price of recycled materials is driven by the market. This can affect the demand for and sale of recycled materials.

- Material cross-contamination is an ongoing issue that has not yet been solved. For example, plastic soft-drink bottles are made with a plastic called PET (polyethylene terephthalate). After PET is recycled, it is used to make items such as carpets, packaging, and fleece, which is shown in **Figure 7.7**. PET is sensitive to contamination. If a single bottle made of another plastic finds its way into a load of PET, the load cannot be recycled.

- People must learn how to sort their garbage and recyclables properly.

Figure 7.7 PET can be recycled and used to make new products.

Diverting Wastes through Composting

In **composting**, organisms such as worms, insects, and bacteria decompose organic matter such as food and yard waste. Before the organic materials are added to the compost, they are shredded to make them smaller and increase their surface area. The compost is mixed regularly to provide oxygen-rich conditions. The end result is a nutrient-rich, natural fertilizer.

Composting can be completed in garden composters in private and community gardens. Small composters such as *vermicomposters* can be used in apartments and offices. These rely on worms to break down organic materials, as shown in **Figure 7.8**. Municipal composting facilities produce large amounts of compost every few months. These may include composted wood waste and material from sewage treatment plants. Green box programs in some municipalities pick up household organic waste for composting. Some industries also run their own composting facilities.

composting a process in which organisms such as worms, insects, and bacteria decompose organic matter

Figure 7.8
Vermicomposting takes up little space. Red wiggler worms, such as *Eisenia foetida* and *Lumbricus rubellus*, that live in the bin turn kitchen waste into compost.

Advantages of Composting

- Compost is rich in nutrients that plants need. It also helps soil retain water. Less watering is needed, and there is greater control of sediment run-off.
- Composting decreases the amount of organic waste that enters landfills. This decreases management costs and reduces the amount of methane in landfill gas.
- Composting makes people more aware of the organic waste they generate.

InquiryLab 7C, Vermicomposting, on page 236

Concerns about Composting

- Large municipal and industrial composting facilities suffer from NIMBY. Organic waste may accumulate at these sites and produce a disagreeable odour. Rats and other vermin may be attracted to sites if they are not well maintained. In addition, local residents may be bothered by the sights and sounds of a composting facility.
- Large-scale compost facilities can produce leachate that can enter the surrounding environment if not collected.
- There is some concern that the people working at large-scale composting facilities may be exposed to harmful concentrations of micro-organisms.
- Composting organic waste on a regular basis takes a commitment that some people are unwilling to make.

Pause and Reflect

8. What is resource recovery?

9. Describe two benefits and two concerns associated with composting.

10. Critical Thinking Some municipalities have a green box program that collects household organic waste. Do you think this is a good use of municipal tax dollars? If not, do you think the money should be used to support more plastic recycling? Explain.

Source Reduction: Reducing and Re-using

Wastes are an unavoidable result of using energy and materials to produce goods and services. They are also undesirable, because they often have harmful effects on the health of organisms, the environment, or both. Recycling diverts items from the waste stream, so it prevents them from entering landfills and incinerators, and gives them new or extended lives. Nevertheless, recycled items are still part of the overall waste-stream process. At some time in the future, many recycled items will be thrown away, and so they will re-enter the waste stream on their way to their final destinations once again.

Perhaps the most effective way to deal with wastes is to prevent them in the first place, or at least to limit their impact in some way. **Source reduction** is the practice of designing, manufacturing, purchasing, using, and re-using products in ways that reduce the amount or toxicity of waste created.

- **Source Reduction through Designing** Many products can be designed to use less material without losing their function. By adopting the idea that bigger is not always better, designers can reduce the amount of waste that enters the waste stream at the end of a product's life. This holds true for packaging as well. Discarded packaging makes up about 50% of household solid waste.

- **Source Reduction through Manufacturing** Waste is often generated in the production process itself. Machinery can leak, and spills may occur. Materials can be left unused and discarded as waste if the production process is not planned properly. In industries such as chemical and pharmaceutical manufacturing, fewer reactants can sometimes be used to make the same materials.

- **Source Reduction through Purchasing** When you buy a product, you can make wise choices with the waste stream in mind. Buying in bulk, purchasing items with minimal packaging, bringing your own container—these are all ways that consumers can help manage solid waste. Other helpful practices include buying non-disposable products that are built to last and avoiding products that cannot be repaired, refilled, recharged, or recycled.

- **Source Reduction through Smart Use** Many materials and products are used wastefully and are often thrown out before they are used up. Product sharing and purchasing only what you need, only when you need it, can help reduce this problem. This is especially true with food. Canadians waste nearly 30 billion dollars of food each year.

- **Source Reduction through Re-use** Each item that is re-used can prevent a new item from being manufactured and ending up in the waste stream. On a personal level, this may mean refilling an existing water bottle rather than buying a new one. Another possibility is using an existing bottle for a new purpose, such as building a bird feeder.

source reduction
reducing waste or its ability to cause harm through the design, manufacture, purchase, use, and re-use of a product

Did You Know?
"Use it up, wear it out, make it do, or do without."

— *New England saying*

Mini-Activity 7-2 Something Old Is New Again

Choose a clean, discarded item to make into a new item. For example, a plastic bottle could be made into a sprinkler, a camping shower, a mini-greenhouse, a self-watering seed pot, patio lights, or a toy.

When you present your item, explain how you made it, what materials you re-used, and how it reduces the amount of waste going to the waste stream.

Taking Responsibility for Solid Waste

Everyone—individuals, businesses, industries, and municipalities—plays a role in managing, diverting, and reducing solid waste. But where (and when) does responsibility begin, and where (and when) does it end? For example, imagine that you buy a pair of running shoes, such as those in **Figure 7.9**. After a while, you decide that you need a new pair. Who is responsible for the waste the shoes generate, and to what extent?

Figure 7.9 Eventually, this new pair of running shoes will enter the waste stream.

Analyzing When running shoes (or any other items) are thrown away and enter the waste stream, whose responsibility should they become? Should they become anyone's? Everyone's? What might that mean?

- Are you responsible for the waste? If they are gently used, could you donate them to a program that gives shoes to people in need? If they are worn out, could you donate them to an organization that recycles them into sports surfaces such as tennis courts?

- Is your municipality or region responsible for the waste? Should it bear the cost of picking up the shoes and managing the waste in a landfill or thermal treatment facility, or of shipping it elsewhere?

- Are manufacturers responsible for the products they create? Should you be able to return the shoes to them? Should their responsibility for the product extend to the end of its lifetime?

- What about the industries and farms that produced the materials in the shoes, such as rubber, plastic, and leather? Do they have any responsibility for the final product? Should they?

For example, some municipalities use a pay-per-bag system. This system encourages people to save money by reducing the amount of waste they produce or increasing the amount they recycle and compost. Some nations have passed legislation that extends manufacturer responsibility for the product to recycling or disposal. This type of legislation affects the design, production, and packaging stages as well. Producers have more incentive to reduce wastes in these areas if they are responsible for them through the whole life of their products.

Mini-Activity 7-3 **Whose Responsibility?**

Who is responsible for the waste produced by plastic shopping bags? Many stores are taking more responsibility by creating a plastic bag policy. Stores that sell printers and equipment that use batteries have special bins for people to bring in empty ink and laser cartridges, as well as used batteries. Is this good customer relations, good corporate citizenship, or something else?

Canvas your neighbourhood to learn about the environment-related choices that local stores offer their customers. For each store that offers choices, evaluate the impact of the store's policy on the following stakeholders:

- the environment (consider the short-term and long-term effects)

- consumers

- the store

- your municipality

Choose a policy that you would implement if you owned a small or large retail store. Why would you choose this policy? Refer to both the pros and cons of your choice in your answer.

Case Study Farms Turn Manure into Green Energy

By Nelson Bennett

There are roughly 10 million stabled horses in North America, and each one produces a tonne of waste a month, according to GreenScene Agritek Inc. founder Phil Wilford, who has developed a process for recycling that waste and turning it into energy.

While developing a process for treating chicken manure to address concerns about pathogens during the avian flu outbreak in 2005, Wilford learned that the real market for a manure treatment process was in stables. Racetracks and stables typically use sawdust and shavings in their stalls. Some of the used bedding is used in fields as fertilizer, but its high lignin content means that using too much of it is not good for the soil. Many stable owners simply pile the stuff up and let it decompose. "It was basically being piled on the back 40, hoping it would go away in the middle of the night," Wilford said.

Wilford developed a thermal process that bakes the pathogens out of used bedding. The end product can be sold back to the stables for re-use. Wilford's company has a pilot plant in Ladner [British Columbia] that's now producing recycled horse bedding. But recently, his company received $146 600 in funding under the Canada–B.C. Agri-Innovation Program to develop a secondary product: fuel pellets.

Wilford is now working on a process that would compress the treated bedding into fuel pellets, which greenhouses could then burn for heat and power. Because a processing plant would cost between $1.5 million and $2 million, it only makes economic sense to build it in a region with 1500 to 2000 horses, where stable owners can form a consortium.

Burnaby-based Diacarbon Energy Inc. is also receiving Agri-Innovation Program funding to add farm waste to the stock from which it creates biochar—a kind of clean coal that can be used either as a fuel or for soil enhancement. Diacarbon's pilot plant in Burnaby is already producing biochar from wood waste. With $142 575 from the Agri-Innovation Program, the company aims to develop a blended feedstock that includes chicken manure and waste from mushroom farming.

Animal manure has a high mineral content, so when it's turned into biochar, up to 60% of it becomes ash. That creates a problem of what to do with the ash by-product. Wood waste produces only 2% to 3% ash, said Diacarbon founder and president Jerry Ericsson, so it makes a good blending material.

Currently, farm waste in North America is being composted, landfilled and used as fertilizer. "The problem is we're putting more on the fields than we should be," Ericsson said. "There gets to be problems with water issues. This is—we think—a better way of dealing with that waste stream. What we're doing with this grant is blending wood and other materials to create high-quality fuels."

The U.S. Department of Agriculture estimates that more than 335 million tonnes of manure are produced annually on farms in the U.S.—about one third of its total municipal and industrial waste output. A single dairy farm with 2500 cows produces a volume of waste equivalent to a city with a population of 411 000.

Analyze

1. Often, excess manure is simply piled up and left to sit. Describe some of the negative environmental consequences of this practice. (Hint: Think about nearby water sources, bacteria, and run-off.)

2. Why is biochar a good way of dealing with manure waste? What advantage(s) does it have over other alternatives?

Communicate

3. Do you think the government should be funding projects like those described in the Case Study? Describe your view in a brief paragraph, providing at least two supporting statements for your opinion.

Summary

- Three main sources of solid waste are agriculture, industry, and municipalities.
- Most solid waste in Canada is sent to landfills, which offer an immediate solution to waste disposal. Many landfills are nearing their capacity, and many municipalities are running out of space to develop new ones.
- Thermal treatment of solid waste includes incineration, pyrolysis, and gasification. Although these methods reduce the volume of garbage, they release products that are harmful to the environment.

- Recycling and composting are two ways to divert solid waste from the waste stream. Recycling reprocesses items to make new products, while composting decomposes organic matter.
- The most effective way to reduce solid waste is source reduction. This involves considering how products are designed, manufactured, and used, as well as people making thoughtful choices when purchasing products.

Review Questions

1. In your own words, describe the meaning of the term *waste stream*. Why is it referred to as a stream? **K/U**

2. Agree or disagree with the following statement, and give at least two reasons that support your opinion: "Manure is a natural product and therefore does not harm the environment or human health." **K/U** **C** **T/I**

3. The final destination for most solid municipal waste is a landfill site. **K/U** **C** **A**

   a) Sketch a diagram showing the important features of a landfill.

   b) What is leachate and how is it managed?

   c) Describe how landfills may be used to generate energy for use at the site.

4. Copy the table below into your notebook. Complete it by listing the sources of each type of solid waste and describing what each type is composed of. **K/U** **C**

Sources and Composition of Solid Waste

Type of Solid Waste	Sources	Composed Mostly of
Agricultural		
Industrial		
Municipal		

5. Many municipalities are looking at alternatives to landfill sites. Describe three methods of thermal treatment of solid waste. What is one advantage and one disadvantage of thermally treating solid waste? **K/U** **C**

6. Are you "for" or "against" large cities transporting their solid wastes to other regions, provinces, or countries? Provide two supporting statements for your opinion. **T/I** **C** **A**

7. Many people have grown up learning about "the three Rs"—reduce, re-use, and recycle.

   a) Explain how the three Rs are similar to and different from one another. **K/U** **C**

   b) Which of the three Rs applies to the approach of source reduction of solid waste? Explain your answer. **A** **C**

8. Name 10 items that you threw into the waste stream over the last week. Describe one way that you could have reduced that number of items. **T/I** **A**

9. Imagine that your community is considering collecting household organic waste for composting. Decide whether or not you agree with this approach to reducing solid waste that is generated in your community. Using a format of your choice, such as a 30 second TV or radio announcement, blog, website, or information pamphlet, present your argument for why this is good or bad for your community. **T/I** **C** **A**

10. Many food products are being packaged into single-serving sizes. For example, single servings of crackers are individually packaged and then sold together in boxes of 20 packages, instead of as one larger container. **T/I** **A**

    a) Name another food or drink product that is sold in single-serving sizes.

    b) Describe two advantages that smaller packaging provides to people.

    c) Describe two disadvantages of this type of packaging.

11. Some social commentators refer to Western societies such as those of Canada and the United States as "disposable societies." **C** **A**

    a) Explain what you think is meant by this term.

    b) Do you think that it is a fair description of
    - your own behaviour?
    - the behaviour of the people in your community?
    - the behaviour of people in your country?
    Explain your answer in each case.

Liquid Waste Management

Sources of Liquid Waste

wastewater any waste that occurs in or can be changed to liquid form

Each day, liquids of all kinds flow from sinks, tubs, showers, and toilets into sewer systems or directly into the environment. These liquids include cooking oils, motor oils, and products containing water, such as paints, cleaners, pesticides, and fertilizers. Liquid wastes, with or without water, are typically called **wastewater**.

Agricultural activities are one common source of wastewater. The wastewater from farms can contaminate surface and ground water with run-off from fertilizers and pesticides. Run-off from ranches and animal feedlots can contaminate surface and ground water with nutrients, organic matter, and bacteria. During rainfall and irrigation, soil and other sediments can be washed from open fields into nearby waterways.

Wastewater from industrial sources includes petroleum products, metals, nutrients, and sediments. Acids may also be released as by-products of industrial processes such as leather tanning, metal smelting, and petroleum and chemical production. Many industrial liquid wastes are hazardous to human and ecosystem health.

Wastewater from Municipal Sources

Municipal wastewater comes from buildings such as homes, businesses, and schools, as well as from pavement and other urban surfaces. Wastewater from these sources is often grouped under two headings: storm water and sewage.

storm water wastewater that drains from lawns, driveways, roofs, roads, and other urban surfaces

Storm water includes rain and melting snow that drains from driveways, roofs, lawns, and roads. It also includes liquid waste that is poured down storm sewers. Storm water often contains sediments and organic matter, as well as harmful chemicals.

sewage any materials rinsed down a drain or flushed down a toilet

Sewage includes any materials that are rinsed down a drain or flushed down a toilet. Most sewage is organic matter, which includes flushed urine and feces, food scraps, and water from washing clothes and dishes. Hundreds of other chemicals are also found in municipal sewage. These include

- plasticizers (substances added to plastic products to make them more soft and flexible)
- medicines (either thrown away directly down sinks and toilets or as part of urine and feces)
- pesticides (either down drains or into storm sewers)
- flame retardants (used to make electrical equipment, wires, cables, pipes, car interiors, and furniture)

Many of these chemicals have significant effects on ecosystems. For example, hormones found in some medications and personal care products can change the gender of aquatic organisms, as described in **Figure 7.10**.

Figure 7.10 When estrogens and other hormones from various products enter waterways, they can cause male fish such as this smallmouth bass to produce eggs from female-like sex organs.

Treating Sewage

Liquid wastes cannot be disposed of in a landfill. One reason is that they dramatically increase the amount of leachate generated by the landfill. They also cause landfills to become structurally unstable. Because of the harm it can cause, sewage is treated before it is released into the environment.

The way in which sewage is treated depends mostly on where you live. Rural residents are more likely to have a septic system on their property. Residents of towns and cities have their waste treated in a municipal sewage treatment facility.

ThoughtLab 7D, Greywater: Recycling Household Wastewater, on page 237

Septic Systems

Where land is available and population densities are low, septic systems such as the one shown in **Figure 7.11** are an effective way to treat sewage. In 2009, 14% of Canadian homes used septic systems. In a typical septic system, the following steps occur.

- Wastewater is first drained into a septic tank. Grease and oils rise to the top and solids settle to the bottom, where they are decomposed by bacteria.

- The clarified liquid waste leaves the septic tank and is channelled through a drain field, which includes small pipes with holes in them. These pipes are embedded in gravel just below the surface of the soil.

- Excess water is wicked up through the drain-field gravel and evaporates into the air.

- Periodically, solids in the septic tank are pumped out by a tank truck and taken to a treatment facility for disposal.

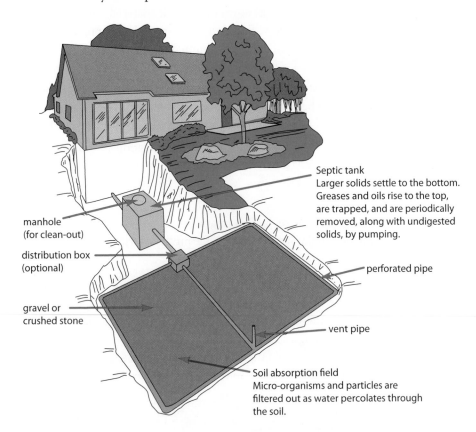

manhole (for clean-out)

distribution box (optional)

gravel or crushed stone

Septic tank
Larger solids settle to the bottom. Greases and oils rise to the top, are trapped, and are periodically removed, along with undigested solids, by pumping.

perforated pipe

vent pipe

Soil absorption field
Micro-organisms and particles are filtered out as water percolates through the soil.

Figure 7.11 Septic systems are often used for the disposal of sewage and wastewater in rural areas. To work properly, the system must have healthy micro-organisms, which digest the organic components of the wastewater.

Inferring *Why should antimicrobial cleaners and chlorine bleach never be poured down the drain in a home with a septic system?*

Municipal Sewage Treatment

Homes and other buildings in towns and cities are linked to municipal sewage treatment facilities through sanitary sewers. Sewage treatment in these facilities generally follows the three main steps described below and shown in **Figure 7.12**.

Step 1: Primary Treatment

Primary sewage treatment physically separates large solids from sewage.

- As raw sewage enters the treatment facility, it passes through a metal grating that removes large debris. A moving screen then filters out smaller objects.

- The sewage moves to a grit tank, where heavier particles such as sand and gravel settle.

- Next, the sewage travels to the primary sedimentation tank. In this tank, about half the suspended, organic solids settle to the bottom as a semi-solid material called *sludge*. Many disease-causing bacteria remain in the sewage at this point. The sewage is not yet safe to release into waterways or ecosystems.

Step 2: Secondary Treatment

Secondary sewage treatment involves the biological breakdown of dissolved organic compounds.

- The sewage from primary treatment flows into either an aeration tank, a trickling filter bed, or a sewage lagoon to be broken down further.

- *Aeration tank digestion:* Sewage pumped into the tank is mixed with aerobic bacteria, which thrive in air/oxygen environments. Air is pumped through the mixture to encourage bacterial activity that breaks down the organic material. The sewage then enters a final settling tank, and settled sludge is removed. This process is called *biological flocculation.* (To flocculate means to clump or separate something from solution.)

- *Trickling filter bed:* If this system is used, sewage drips from perforated pipes or an overhead sprayer through a bed of stones or corrugated plastic sheets. Bacteria and other micro-organisms in the bed decompose organic material as it trickles through.

- *Sewage lagoon:* Where space is available for these outdoor lagoons, exposing sewage to sunlight, algae, and air breaks down dissolved organic compounds. This takes place more slowly than with either of the other two systems, but the energy (and monetary) costs are reduced.

- Finally, the fluid from any of the secondary treatment processes is disinfected with chlorine, UV (ultraviolet) light, or ozone to kill harmful bacteria.

Step 3: Tertiary Treatment

After secondary treatment, sewage is usually free of disease-causing bacteria and organism material. However, it still contains high levels of nitrates, phosphates, and other inorganic substances. These are removed during *tertiary treatment*. In this stage, the sewage may be passed through a natural wetland, which filters out these nutrients, or through an artificial filtering system that does the same thing. Alternatively, chemicals are used to bind to the nutrients so they settle out of solution. This is called *chemical flocculation.*

constructed wetlands—**Case Study, Constructed Wetlands, on page 320**

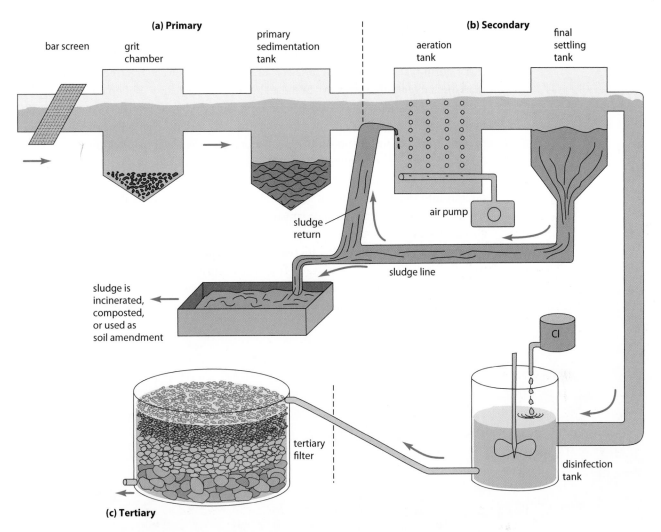

(a) Primary

bar screen

grit chamber

primary sedimentation tank

(b) Secondary

aeration tank

final settling tank

sludge return

air pump

sludge line

sludge is incinerated, composted, or used as soil amendment

CI

tertiary filter

disinfection tank

(c) Tertiary

Figure 7.12 There are three steps to municipal sewage treatment. Primary treatment **(A)** removes solids and suspended sediment. Secondary treatment **(B)** kills harmful micro-organisms and removes most organic material. Tertiary treatment **(C)** removes more inorganic nutrients.

Analyzing *Why is a system like this more suitable for highly populated areas than septic systems?*

Pause and Reflect

11. What is wastewater?

12. What are the two types of systems for treating sewage?

13. Critical Thinking Municipalities spend a great deal of money maintaining and monitoring their sewage treatment plants. Why do you think this is important?

Mini-Activity 7-4   **Exploring Primary Treatment**

Examine the materials and the water sample provided by your teacher. Create a plan to use the materials to remove as many solids as you can from the sample.

Build the filter and run the water sample through your treatment process. When you are finished, review your results and improve your process, if possible.

Treating Storm Water

In many Canadian cities, sanitary sewers are connected to storm sewers. Storm sewers carry storm-water run-off. These sewers are routed to a sewage treatment facility rather than discharged into waterways because the run-off they carry is often contaminated with fertilizers, pesticides, oils, rubber, tars, and other chemicals. This plan works well in dry weather, but heavy storms often overload the system. This can cause raw sewage to back up and harmful run-off to enter waterways directly. To prevent this overflow, many cities are separating their storm drains and sanitary sewers. (See **Figure 7.13**.)

Notice that a separate system of storm drains and sewers diverts storm water directly into waterways without any treatment. As a result, chemicals and sediment from urban surfaces can harm organisms that live in these waterways. Storm water also increases the amount of water that flows through local waterways. This can erode stream banks, and the resulting increase in sediment can harm aquatic organisms. As well, the temperature of storm water from urban surfaces is higher than that of nearby waterways. Warmer storm water that enters these bodies of water reduces the amount of dissolved oxygen in the water. This can kill fish and other aquatic life.

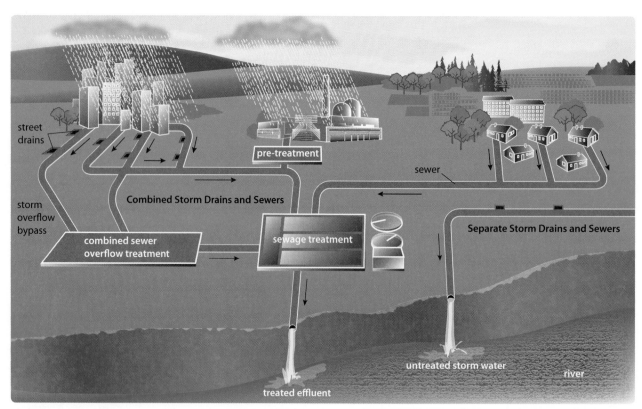

Figure 7.13 A simplified view of how a city could be designed to separate its storm sewers from its sewer systems

Mini-Activity 7-5　　Storm Water Management

Storm water can be managed at its source as well as through the use of infrastructure that takes the impact of storm water into account. Find out about *source control technologies* and *conveyance and end-of-pipe treatment*.

How do these management practices reduce the impact of storm water on the environment? Which cities and towns near where you live use these practices?

Summary

- Agriculture, industries, and municipalities contribute to wastewater.
- Storm water is municipal wastewater that runs off driveways and roads. Sewage is municipal wastewater that contains organic matter and many chemical products that are rinsed down drains and flushed down toilets.
- Sewage must be treated with a septic system or at a municipal sewage treatment facility before it is released into the environment.
- Septic systems use septic tanks, where wastewater is collected and materials separate into layers. Clarified liquids are drained, and solids that are not decomposed are periodically removed from the tank.

- Municipal sewage treatment involves primary treatment, which separates large solids; secondary treatment, which involves the breakdown of dissolved organic materials; and tertiary treatment, which removes nutrients such as phosphates and nitrates.
- In many Canadian cities, storm water is diverted into the sewage system. This allows it to undergo treatment before being released into aquatic environments. However, heavy rain storms can overload the system and cause untreated sewege to enter the environment.

Review Questions

1. What does storm water consist of? K/U

2. Septic systems are used by people who live in rural areas, such as farmlands. K/U T/I C A

   a) What is a septic system?

   b) Draw a flowchart that shows how a septic system works.

   c) Why must people be careful about where septic systems are located on their property?

   d) Why should people with septic systems not add products such as laundry bleach to their drains or toilets?

   e) What are other types of wastewater produced from farms?

3. The table below lists different sources of wastewater and their contributions to the total wastewater that flows in sewer systems of municipalities across Canada. T/I C A

Sources of Wastewater in Municipal Sewers

Source	% of Total
Residential sources	65
Industrial, commercial, and institutional sources	18
Storm water	9
Infiltration of sewer system from surrounding ground water	8

Values are averages based on data collection across Canada by Environment Canada.

   a) What source makes the largest contribution to wastewater that runs through municipal sewers?

   b) How do you think the values in the table would change for communities that rely a great deal on septic systems? Explain your answer.

   c) Many people think that the small amount of wastewater they are responsible for individually does not have much of an impact on the environment. In two or three sentences, develop an argument that counters that opinion using the data in the table.

   d) Draw a pie graph that represents the data in the table. Include the following:
   - a title for your graph
   - labels for each type of source
   - the percentage that each source contributes to the total wastewater in municipal sewer systems

4. Why are liquid wastes not disposed of in landfills? K/U

5. In a short paragraph, describe how sewage wastewater is treated in municipal sewage treatment facilities. K/U C

6. When some people wash their cars, they commonly rinse the soapy wastewater into a nearby storm drain. In 2 or 3 sentences, explain how this practice can affect the environment. A C

7. Storm drains, sewer systems, and local waterways are connected. Develop and draw a label or logo that your community could place on storm sewers in your area to remind people about the impact of wastewater on the environment. C A

8. Describe three disadvantage of having separate storm drains and sewers. K/U

Safety Precautions

- Wash your hands with warm water and soap after completing the Lab and each time you monitor your landfill.

- Do not use meat or dairy products, plate scrapings, or pet waste. Use only yard/garden trimmings and/or fruit and vegetable scraps your teacher directs you to use.

- Wear gloves, an apron, safety goggles, and a mask, and take any other safety precautions that are stated in associated Material Safety Data Sheets (MSDS).

Materials

- plastic container
- plastic shopping bag
- metric ruler
- scissors
- masking tape
- toothpick
- soil
- waste items (cut into pieces about 2.5 cm long) such as plastic, plastic foam (polystyrene), cotton string, cardboard, wooden toothpick, aluminum foil, fruits, vegetables, grass clippings
- camera
- graph paper
- 2 coloured pencils
- straws
- marker
- misting spray bottle
- pH test kit

Modelling a Landfill

Modern landfills are often referred to as "sanitary" landfills, because they have an impermeable liner to isolate decomposing garbage and leachate from surrounding water, soil, and air. In this Lab, you will make a model of a sanitary landfill.

Pre-Lab Questions

1. Describe how modern landfills isolate waste and leachate from ground water.

2. Predict which garbage items will decompose (biodegrade) in your landfill. Rank these items from most to least biodegradable.

3. What safety precautions will you follow during this Lab?

Question

How do the rates of decomposition differ among waste items made of different materials?

Procedure

Part A

1. Line a plastic container with a plastic shopping bag. Suspend the bag so that there is about 1 cm of space between the bottom of the container and the bag. Cut the plastic bag and tape it to the edge of the container.

2. Use a toothpick to poke 15 to 20 holes in the plastic bag. This will mimic gravel and allow leachate to collect inside the plastic container.

3. Place about 2.5 cm of soil on the plastic liner.

4. Obtain a piece of each type of waste item. If possible, take a photo of all of the pieces of waste to use at the end of the Lab for comparison.

5. Place the waste items on the graph paper. Trace each item using a coloured pencil. Label the tracings with the name of the piece of waste.

6. Count the number of squares that each waste item occupies on the graph paper. Record your results in a data table like the one shown below.

Data Table

Waste Item	Number of Squares on Graph Paper Before Added to Landfill	Number of Squares on Graph Paper After Removed from Landfill
Plastic		
Cotton string		
Aluminum foil		
Apple		
Green bean		

7. Add the waste to your model landfill by placing it in the container. Separate the waste pieces so they are not touching one another.

8. Create flags using short lengths of straws and masking tape. Label each flag with the name of a corresponding waste item.

9. Place the labelled flags beside the appropriate waste item to mark their locations in your landfill.

10. Bury all the waste items under about 5 cm of soil.

11. Use a misting spray bottle to add water to your landfill. Add enough water to moisten the soil but not enough to create puddles or mud.

12. Place your landfill in an area where it will not be disturbed.

13. Clean up your work area and wash your hands with warm, soapy water.

14. Every 2 or 3 days, add water to your model landfill using the spray bottle. Only add enough the keep the soil moist.

Part B

1. After 4-6 weeks, unearth each waste item. Carefully place each item on the graph paper over the original tracings. Use a different coloured pencil and re-trace each waste item.

Model landfill

2. Count the number of squares that each waste item now occupies on the graph paper. Record your results in your data table. If you took a photo of your waste items in Part A, compare any visual differences. Record your observations.

3. Carefully remove the plastic bag liner. There should be a small amount of leachate in the container. Test the pH of the leachate by dipping a strip of pH paper into the "landfill" water. Record the pH.

4. Observe the clarity of the water, and record your observations.

Analyze and Interpret

1. Compare the size of each waste item before and after being in your landfill. To do this, compare the number of squares that each item occupied before and after being in your landfill. Prepare a bar graph to illustrate the decomposition of the items.

2. Did any items disappear completely? Which items were smaller? Which remained the same size? Explain your results.

3. Based on your results, would you classify the leachate as corrosive? If yes, was it acidic or basic? What hazards could be caused leachate entered an aquatic ecosystem?

Conclude and Communicate

4. What conclusions can you make about the rate of decomposition of various types of waste in a landfill? Why do some types of waste decompose faster than others?

5. Why is it important to reduce, re-use, or recycle certain types of waste? Explain your reasoning.

Inquire Further

6. The micro-organisms that are responsible for decomposition at landfills are sensitive to temperature and moisture changes. Design an experiment to test how these variables affect the rate of decomposition in a landfill. If approved by your teacher, carry out your plan, and draw a conclusion based on your results.

InquiryLab 7B

Safety Precautions

- Conduct this investigation only under the supervision of your teacher.

- Wear safety goggles, gloves, and a lab apron throughout the Lab.

- **CAUTION:** Do not put your hands into any garbage bags. Your teacher will open the bags and spread the garbage on a tarp. You will assess the waste once it is spread out.

- **CAUTION:** Do not touch any broken glass, needles, or other items that appear unsafe. If you are not certain about whether you should handle an item, ask your teacher for help.

- Wash your hands with warm water and soap when you have completed the Lab.

Possible Materials

- large bags of school garbage, opened and their contents spread out on a tarp by your teacher

- tarp

- tongs or forceps

An Audit of School Waste

Recycling programs have been in effect for a number of years in most communities. However, there is still a great deal of waste that enters the waste stream (and, ultimately, landfills) that could be diverted elsewhere. Many municipalities are asking "How can we increase waste reduction?"

In this Lab, you will assess the types of materials that are being thrown away in the "regular" garbage. Based on your findings, you will propose a plan to reduce the amount of solid waste from your school that enters the waste stream.

Pre-Lab Questions

1. What areas in the school should be monitored to assess what is thrown in the regular garbage?

2. What is your school's policy regarding recycling paper and plastic materials? Does the school collect organic waste for composting?

3. What are some effective ways of increasing awareness of the importance of recycling?

Question

How effective are students and staff at separating the school's solid waste?

Procedure

1. Your teacher will choose one of the following ways to do the waste audit:
   - monitor people in certain areas of the school and keep track of what they throw in the waste
   - analyze the contents of garbage bags from certain areas of the school, which your teacher has spread out on a tarp

2. Working in small groups, plan a procedure for performing the audit. Use the following questions to help you.
   - In what area of the school will you assess the garbage? Note: You cannot monitor any bathroom garbage.
   - To what categories will you assign the different solid wastes?
   - How will you determine the relative amounts of each type of solid waste?
   - What materials are collected for recycling at your school? Symbols like those on the next page appear on many products and are one way to identify the material of which an object is made. Waste management companies use them to indicate materials they do and do not recycle.

3. Have your teacher approve your procedure.

4. Carry out your approved procedure.

5. Record the different types of garbage and the amount of each type of garbage in a table with the headings shown below.

Garbage Audit

Total number of items of garbage:	
Category of Solid Waste	**Amount (number of items)**

6. Once you have completed the audit, dispose of the garbage according to your school's policy for waste management.

Analyze and Interpret

1. What was the total number of items in the regular garbage? What was the number of recyclable items in the garbage that should have been put in recycling bins?

2. Calculate the percentage of waste that could have been diverted to recycling.

3. Approximately what fraction of the garbage could have been composted?

4. How many items needed to be separated in order to dispose of them correctly? For example, some packaging involves both recyclable and non-recyclable materials that can be separated before disposal.

Conclude and Communicate

5. Develop a plan to improve how waste is discarded at your school.

6. Each group should prepare a 5 minute presentation that describes the key points of their plan. Then, as a class, develop one plan for waste reduction that could be presented to the school administration.

Recycling Symbols for Plastics			
Symbol	**Type of Plastic**	**Description**	**Examples**
1 PETE	Polyethylene terephthalate	Rigid; most are clear	Soft drink, juice, and water bottles; peanut butter jars
2 HDPE	High density polyethylene	Somewhat rigid	Milk and water jugs, bleach bottles, shampoo bottles, detergent bottles, household cleaning bottles, butter tubs
3 V	Polyvinyl chloride	Somewhat rigid, glossy	Detergent and cleaning product bottles, peanut butter jars
4 LDPE	Low density polyethylene	Flexible	Plastic bags and grocery bags, dry cleaning bags, flexible film packaging, the plastic rings used to hold 6-packs of cans together
5 PP	Polypropylene	Somewhat rigid	Yogurt tubs, straws, some screw-cap lids
6 PS	Polystyrene	Brittle and glossy	Meat trays, egg cartons, plates, take-out food containers, foam cups, foam packing chips
7 OTHER	Polycarbonate, acrylic, mixed plastics	Variety of appearances	Plastics that do not fit into the other categories. Examples include large water bottles, sunglasses, DVDs, computer cases, and certain food containers

Safety Precautions

- Conduct this Lab only under the supervision of your teacher.

- Use appropriate protective equipment such as apron, safety goggles, and gloves, and take any other safety precautions that are stated in associated Material Safety Data Sheets (MSDS)

- Wash your hands with soap and warm water after handling any materials.

- Dairy, meat products, and fats such as butter and mayonnaise should never be used for composting.

Materials

- plastic storage bin with lid that has holes in it

- enough bedding for the worms to almost fill the bin (can be shredded newspaper, cardboard, potting soil without chemicals added, garden soil, peat moss, fall leaves, straw, or a combination of these)

- tap water, enough to wet the bedding so it is damp

- red wiggler composting worms

- organic food scraps such as vegetable scraps, tea bags, small amounts of cooked pasta, rice, bread, coffee grounds

- work gloves

Vermicomposting

Many organic wastes can be converted to a useful product by composting. Under oxygen-rich conditions, the organic matter decomposes into nutrient-rich material called compost. Compost can be added to gardens to improve plant growth and help retain moisture in the soil.

Worms are one type of organism used in composting to carry out the decomposition. Composting that involves the use of worms is called vermicomposting. The prefix *vermi-* is based on the Latin word *vermis*, which means "worm." Vermicomposting is a practical option for people who live in apartments or small homes with limited space for an outdoor composter.

Pre-Lab Questions

1. What foods and materials are appropriate to compost? What foods and materials should not be used for composting?

2. How can you protect the health and safety of the worms you will be using?

Question

How can a vermicomposter be used to recycle solid organic waste?

Procedure

1. Working in small groups, plan a procedure for using vermicomposting to decompose solid organic matter. Base your procedure on the list of materials that is provided. Consider the following:
   - What materials should be used for composting?
   - When will you monitor the compost and how will you record your observations?
   - How will you make sure there is enough oxygen in the compost environment and enough moisture in the bedding?
   - What safety precautions will you take while working with the composter?

2. Have your teacher approve your procedure.

3. Carry out your approved procedure.

Analyze and Interpret

1. Describe your observations over time. How long did it take for wastes that were added to decompose? Did some decompose faster than others?

2. What problems, if any, occurred with the composter or composting process?

Conclude and Communicate

3. Once you have completed the Lab, describe how you could improve the procedure for vermicomposting.

4. Name two ways that your class can use the compost that is generated.

Skill Check

Initiating and Planning

Performing and Recording

Analyzing and Interpreting

Communicating

Greywater: Recycling Household Wastewater

Greywater is household wastewater except for water from toilets and kitchen sources. It typically makes up 50% to 80% of household wastewater. The main greywater sources are bathrooms (tubs, showers, and sinks) and laundry rooms (washing machines and sinks). Although not yet approved in all jurisdictions in Canada, there is a growing trend to recycle greywater for certain uses.

Pre-Lab Questions

1. Why is water from kitchen sources and toilets not used for recycling greywater?

2. Why is greywater not drinkable?

Question

What are important features of a greywater recycling system?

Procedure

1. Analyze the schematic below, which shows a complete household system that uses greywater from all suitable sources.

2. Make a list of the key features of the system. For example, identify the sources of greywater, what is done with the greywater, and how it is used.

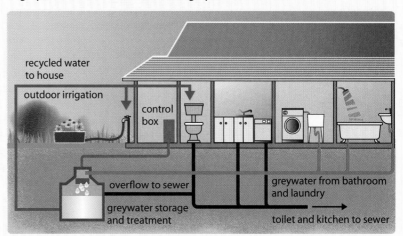

Analyze and Interpret

1. What is the recycled greywater used for?

2. For the system shown, is the recycled greywater used directly or is some type of treatment performed first? Why do you think it is set up this way?

Conclude and Communicate

3. In a chart format, list some benefits and risks of using recycled greywater.

4. Research the status of greywater use in your area. Is it legal? If so, what regulations are in place for its use?

5. Develop a design for a bathroom-specific system that uses greywater from the sink as a source of water to flush the toilet. Include a labelled sketch of your design to explain how it works.

Chapter 7   SUMMARY

Section 7.1   Solid Waste Management

There are different ways to manage solid wastes, including diverting them from the waste stream by composting or recycling. However, reducing the amount we generate in the first place is best.

Key Terms
solid waste
waste stream
landfill
leachate
landfill gas
thermal treatment
recycling
composting
source reduction

Key Concepts
- Three main sources of solid waste are agriculture, industry, and municipalities.

- Most solid waste in Canada is sent to landfills, which offer an immediate solution to waste disposal. Many landfills are nearing their capacity, and many municipalities are running out of space to develop new ones.

- Thermal treatment of solid waste includes incineration, pyrolysis, and gasification. Although these methods reduce the volume of garbage, they release products that are harmful to the environment.

- Recycling and composting are two ways to divert solid waste from the waste stream. Recycling reprocesses items to make new products, while composting decomposes organic matter.

- The most effective way to reduce solid waste is source reduction. This involves considering how products are designed, manufactured, and used, as well as people making thoughtful choices when purchasing products.

Section 7.2   Liquid Waste Management

Managing liquid waste includes treating sewage and storm water. Septic and municipal sewage treatment systems treat sewage. Depending on the region, storm water may or may not be treated before it is released into the environment.

Key Terms
wastewater
storm water
sewage

Key Concepts
- Agriculture, industries, and municipalities contribute to wastewater.

- Storm water is municipal wastewater that runs off driveways and roads. Sewage is municipal wastewater that contains organic matter and many chemical products that are rinsed down drains and flushed down toilets.

- Sewage must be treated with a septic system or at a municipal sewage treatment facility before it is released into the environment.

- Septic systems use septic tanks, where wastewater is collected and materials separate into layers. Clarified liquids are drained, and solids that are not decomposed are periodically removed from the tank.

- Municipal sewage treatment involves primary treatment, which separates large solids; secondary treatment, which involves the breakdown of dissolved organic materials; and tertiary treatment, which removes nutrients such as phosphates and nitrates.

- In many Canadian cities, storm water is diverted into the sewage system. This allows it to undergo treatment before being released into aquatic environments. However, heavy rain storms can overload the system and cause untreated sewage to enter the environment.

Chapter 7 REVIEW

Knowledge and Understanding

Choose the letter of the best answer below.

1. Which of the following statements about solid waste is correct?
   a) Industrial solid waste is generated from manufacturing, construction, and mining.
   b) Most municipal solid waste is diverted from landfills and is recycled or composted.
   c) Agricultural solid waste is generated on farms and ranches and poses minimal threat to the environment.
   d) Most industrial solid waste is sent to landfill and not recycled.
   e) Solid waste from the mining industry is often categorized as agricultural waste.

2. The three main sources of solid waste are
   a) agricultural waste, mine tailings, and industrial waste
   b) municipal waste, agricultural waste, and industrial waste
   c) municipal sewage, agricultural waste, and household hazardous waste
   d) industrial waste, agricultural waste, and hazardous waste
   e) none of the above

3. Which of the following makes up most of the municipal solid waste stream?
   a) paper and plastic waste
   b) plastic and organic waste
   c) glass and metal waste
   d) paper and organic waste
   e) wood products and used clothing

4. Landfills are designed and built to isolate waste from ground water, air, and the surrounding soils. This is accomplished by
   a) burying a large impermeable clay or rubber liner deep in the ground above the water table
   b) using compactors to reduce the volume of waste before it is covered
   c) building landfills close to wetlands, which act as natural filters for the chemicals present in leachate
   d) venting landfill gases, including methane, directly to the atmosphere to prevent the build-up of pressure under the landfill
   e) building a collection system to carry leachate directly to a surface water source to prevent soil contamination

5. Which of the following statements about landfill gas is *incorrect*?
   a) It is a mixture of gases created by bacteria during decomposition of organic material.
   b) It is flammable and explosive.
   c) It is a mixture of gases produced by the incineration of solid waste at thermal treatment facilities.
   d) It is mainly composed of methane gas, which can be converted to carbon dioxide by burning it.
   e) In Canada, some landfills burn landfill gas to generate electricity.

6. Which of the following statements is *incorrect*?
   a) The end products of incineration include carbon dioxide gas, water vapour, and solid ash.
   b) The end products of pyrolysis and gasification are the same: syngas and a solid residue.
   c) Thermally treated waste can enter the atmosphere in the form of ash and gases.
   d) Many facilities use pollution control technologies to capture or treat waste from incinerators.
   e) Steam produced by burning wastes in an incinerator can only be used to heat the incinerator facility.

7. When composting,
   a) large organic matter is often shredded to make it smaller and more compactable, so it can be packaged tightly to avoid oxygen-rich conditions
   b) red wiggler worms can be added to small composters to turn kitchen waste into nutrient-rich, natural fertilizer
   c) there are rarely NIMBY concerns with municipal facilities
   d) leachate is absorbed by the organic material and rarely enters the surrounding environment
   e) items such as wood waste, yard clippings, and materials from sewage treatment plants are removed before the waste enters the compost facility

8. Which of the following substances are *not* commonly found in municipal sewage?
   a) medicines, plasticizers, and flame retardants
   b) pesticides, laundry detergent chemicals, and food scraps
   c) petroleum products, metals, and sediments
   d) urine and feces
   e) bacteria

Chapter 7 REVIEW

Answer the questions below.

9. What is meant by the term *waste stream*? List the three sources of waste that make up the solid waste stream, and give one example of each.

10. Why is the number of landfills in Canada and the United States declining, even though the population in both countries is increasing?

11. Landfills provide several benefits to the environment. List and explain four ways that landfills provide solutions for solid waste management.

12. Describe the construction features of a landfill that allow harmful waste substances to be isolated from air, soil, and water.

13. a) What is thermal treatment of solid waste?

  b) List the three types of thermal treatment, and compare the processes and end products.

  c) What are two advantages of thermal treatment of solid waste?

14. Exporting solid waste from large urban centres is one strategy for managing solid waste. Why is this strategy controversial?

15. What is source reduction? Describe five ways source reduction is possible.

16. Explain the difference between storm water and sewage.

17. What are some strategies for the treatment of storm water in Canadian cities? Why should storm sewers be separate from sanitary sewers?

18. a) List three sources of liquid waste, and provide examples of each.

  b) What are some reasons that the liquid wastes from these sources are considered harmful to human health and to the environment?

19. The ways in which sewage is treated depend mainly on where you live.

  a) Why are septic systems used in rural areas?

  b) Why are septic systems not a good option for urban settings?

  c) Describe the similarities and differences between a septic system and a municipal sewer system.

Thinking and Investigation

20. Waste that comes from essential activities, such as eating, is considered to be unavoidable. Give examples of waste items you produce that you think are avoidable. What are some ways you can decrease your production of this type of waste?

21. Once a landfill is full, it is covered with an impermeable liner and topped with soil. The underlying waste continues to decompose for an undetermined amount of time. The site must be monitored for decades by municipal workers.

  a) What types of hazards would workers monitor the landfill site for?

  b) Make suggestions for what the land could be converted into once the former landfill site is deemed to be safe.

22. Many cities in North America discharge untreated wastewater to local lakes and rivers. What environmental and human health problems could result from this practice? Why do you think untreated wastewater is still discharged this way?

23. The graphs below show water use in a North American district with 15 counties. Many people believe that industry in this region is the major source of water consumption.

  a) Examine the graph titled "District Water Use." Which group is responsible for the greatest water consumption? Suggest a reason why this group consumes so much more water compared to the other groups.

  b) Examine the graph titled "Single Family End Uses." What are the three highest end uses of water?

  c) How is the use of water related to the amount of wastewater produced in a single family home?

  d) The largest end use of water for a single family home is for outdoor use. How could the water used for outdoor purposes eventually end up in a municipal wastewater treatment plant?

  e) Describe at least three ways that single family homes can produce less sewage and wastewater.

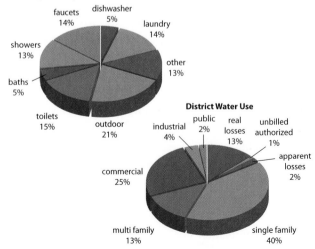

Single Family End Uses

District Water Use

24. Assume that the average Canadian generates 2 kg of unwanted materials each day.

   a) Find out the current population for your province and for Canada.

   b) Calculate the amount of unwanted materials generated per day in your province and in the country.

   c) If 1 kg of trash takes up 0.004 m^3 of space, how tall would a column of the trash produced by your province be if the base of the column has an area of 1 m^2?

Communication

25. Devise a plan to introduce composting to a school cafeteria. Consider the resources you will need, the people who will have to be involved, start up and maintenance costs, as well roadblocks you might encounter. Write a detailed proposal to the school principal for implementing this program.

26. Describe what happens to the water that is flushed down the toilet, beginning from the time of the flush and ending when it is released back into the environment. Describe its path and how it might be treated along the way.

27. Source reduction through designing and manufacturing would greatly reduce the amount of waste that reaches our landfills each year. Choose a product you use regularly that is "over-packaged." Redesign the packaging so that it has less impact on the environment. Include a sketch of your new package design, along with a description of why your packaging design should replace the current one.

28. Biosolids are the nutrient-rich solid residues that remain after the treatment of municipal sewage. They are further treated to reduce the amount of pathogens and to reduce foul odours. Once treated, biosolids can be applied to certain crops as a fertilizer. Do research to find out how and where biosolids are used, the regulations for their use, and the controversies surrounding their use. Present your findings orally or in writing, using a format agreed to by your teacher.

29. You hear a respected environmental scientist say "We have a waste management crisis."

   a) Suggest three possible groups of people that the word "we" could be referring to in this statement.

   b) For each group of people you named in part a), say whether you agree or disagree that there is a waste management crisis. Give reasons to support you opinion in each case.

30. Waste management focuses on the 3 Rs: reducing, recycling, and re-using. Which R is most effective in waste management? Justify your opinion.

Application

31. Tertiary treatment of sewage often involves passing wastewater through a natural wetland, which absorbs and filters out excess nutrients left in the wastewater. This technology is referred to as phytoremediation.

   a) Do research to find out the advantages and disadvantages of phytoremediation. Use at least three reputable sources of information.

   b) Do you support the use of wetlands for phytoremediation as part of a wastewater treatment plan? Give reasons to justify your answer.

32. Archeologists learn about past societies by examining such things as the buildings they lived in and the materials and remains of materials that were left behind. Often only inorganic materials are found, since organic substances usually decompose. Imagine archeologists in the distant future examined your community's garbage. What do you think they would infer about the lifestyle and values of your community?

33. Environmental artists have found creative ways to turn our trash into things of beauty. The violins shown below were made entirely out of material found at a landfill in Asuncion, Paraguay. What other items could be created from the waste in landfills?

34. Did You Know? Reread the quotation on page 222. How easy or challenging would it be for you to live according to this saying? Explain why.

> ### Pause and Reflect
>
> How could you incorporate what you have learned in this chapter into your daily actions or choices?

Managing Hazardous Wastes

Working with hazardous waste often requires special training and protective clothing in order to handle it safely.

What are some hazardous materials that you are exposed to on a daily basis? What makes them harmful?

Design a Hazmat Suit

There are many different professions that involve exposure to hazardous materials. In order to carry out their job safely, people who work with hazardous materials must receive special training. This includes learning how to wear and work in a hazardous material suit, which is often referred to as a hazmat suit.

1. Working in a small group, imagine that your team has been asked to design a hazmat suit to be worn by a firefighter. List the different factors that must be considered.

2. What design features do you need to include?

3. In addition to the use of a hazmat suit, what other controls or considerations can help to ensure the safety of people who work with hazardous wastes?

In this chapter, you will

- identify types of hazardous wastes and their sources, as well as their effects on the environment
- describe the management of hazardous waste
- assess technologies that are used to treat or dispose of hazardous waste

Types of Hazardous Waste

Hazardous Substances and Hazardous Wastes

hazardous substance
a harmful substance that requires special handling

Hazardous substances can cause harm to organisms and the environment, and they require special methods to reduce or eliminate their potential for harm.

Although hazardous substances occur naturally, this chapter will focus on ones that are associated with human activity. In some cases, hazardous substances are used to manufacture products. For example, hydrochloric acid is a hazardous substance that is often used to etch computer circuit boards. Computers may contain components with small amounts of hazardous substances. These include poisonous metals such as mercury and cadmium. Many other common products may be hazardous, such as gasoline, pesticides, household cleaners, and medicines.

Some hazardous substances are converted to non-hazardous substances when a product is used. For example, the burning of gasoline in the presence of plenty of oxygen produces only water and carbon dioxide—two non-hazardous substances. However, if hazardous substances are not used up or changed during their use or manufacture, they will eventually enter the waste stream. At that point, a hazardous substance becomes a hazardous waste. There are many points in the life cycle of a product when its wastes can enter the environment, as shown in **Figure 8.1**.

Figure 8.1 Hazardous wastes are created at many points in the life cycle of a product, from research and development to final disposal.

Inferring Governments set rules and standards to protect people and the environment from hazardous products and wastes. Why is controlling their release still challenging?

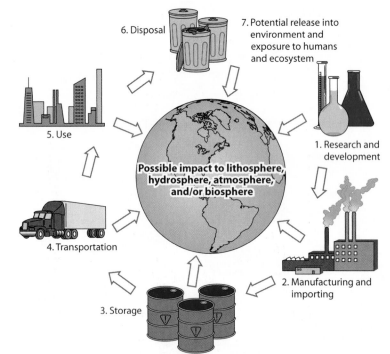

6. Disposal

7. Potential release into environment and exposure to humans and ecosystem

5. Use

1. Research and development

4. Transportation

Possible impact to lithosphere, hydrosphere, atmosphere, and/or biosphere

3. Storage

2. Manufacturing and importing

Properties of Hazardous Materials

hazardous waste
a discarded substance that is or contains a flammable, corrosive, reactive, and/or toxic substance

A **hazardous waste** is any discarded solid, liquid, or gas that is or contains a hazardous substance. **Table 8.1** includes some examples of properties used to identify hazardous substances and wastes. People sometimes use the word *toxic*, which appears in the table, to mean the same thing as *hazardous* when they are talking about hazardous substances and wastes. However, *toxic* and *hazardous* are not synonyms. Toxic substances, such as the one shown in **Figure 8.2**, are a subgroup of hazardous substances that are poisonous to organisms. The broader term *hazardous substance* applies to toxic as well as all other dangerous substances.

Table 8.1 Properties of Hazardous Materials

WHMIS Symbol	Description	Examples
Flammable and Combustible	• materials that present a fire hazard if improperly handled • can ignite and produce smoke and particulate matter • liquids release vapours that are easily ignited at normal working temperatures • solids are unstable and can ignite or explode	• gasoline • paint thinner • kerosene • matches • naphthalene
Corrosive	• materials that corrode (chemically destroy) other materials, including metal and human skin • must be kept in containers made of special materials	• strong acids, such as hydrochloric acid and sulfuric acid • strong bases, such as sodium hydroxide and calcium hydroxide
Reactive	• materials that explode if improperly handled • materials that react with other materials, air, or water to produce harmful gases • materials that destabilize with heat or shock, to produce harmful gases	• gunpowder • nitroglycerine • chlorine bleach mixed with ammonia
Toxic and Infectious 1. 2.　　3.	• materials that can kill or cause serious harm to people and other organisms • these materials are separated into 　1. those that have immediate and very serious effects 　2. those that have other long-term effects 　3. those that are biohazardous	• some medicines • blood products • pesticides • many household polishes and cleaners

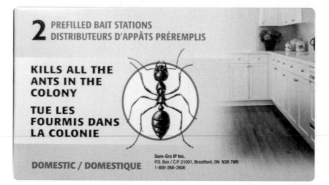

Figure 8.2 Insecticide products like this are commonly available.

Explaining *Why is this substance toxic and hazardous?*

Pause and Reflect

1. Explain the difference between a hazardous substance and a hazardous waste.

2. Distinguish clearly between the terms *toxic* and *hazardous* when they are used to describe certain types of substances.

3. **Critical Thinking** Medicines are used to treat illness. However, some are also classified as toxic hazardous wastes. Explain how this is possible.

Hazardous Wastes from Agriculture

Pesticides are not just deadly to insects and other pests—they are harmful to human beings and other organisms as well. As a result, many pesticides are classified as hazardous substances. Most pesticides are used up when they are applied to crops. Any extra pesticide that is not used becomes waste. This can include leftover solutions and pesticide that has expired and is no longer effective. Pesticide containers and the spray equipment also present a problem, because they often have residues of the pesticides they contained. Pesticide residue can also contaminate other materials, such as soil or sawdust. When this happens, the contaminated material is also considered a hazardous waste.

persistent pesticide
pesticide that does not break down and remains in the environment for a long time

chlorinated hydrocarbon
hazardous chemical used to make products such as pesticides, solvents, and pipes

Some pesticides break down quickly after they are applied, so they become less harmful to the environment. Others do not. Instead, they are said to persist in the environment. One of these **persistent pesticides** is DDT. DDT belongs to a class of hazardous organic compounds called **chlorinated hydrocarbons** (see **Figure 8.3**). DDT was developed to be an especially effective pesticide. However, it has been shown to accumulate in the tissues of humans and other organisms. At high levels, DDT has been linked to nervous system conditions, such as tremors and seizures, lactation problems, and premature births. DDT has also been linked to reproductive problems in some animals and weak eggshells in birds. Consequently, use of this pesticide was banned by most countries during the 1970s and 1980s, and unused stocks became hazardous waste.

Figure 8.3 PVC (polyvinyl chloride) products, such as these pipes, are just one of the many examples of chlorinated hydrocarbons that affect your life. Others include pesticides, solvents, and synthetic rubber products.

The Persistence and Continued Use of DDT

bioaccumulation—**see section 10.1 on page 313**

DDT is easily transported great distances in water and in air, and it persists for decades in the environment. For example even though the use of DDT has been banned in Canada since the early1970s, in 2010 DDT was found in contaminated soil in a former military site in Ivvavik National Park in the Yukon.

Despite the hazards it poses to the environment, DDT is still used in certain parts of the world, where it is considered an essential tool in the battle against mosquitos that carry malaria. This disease causes severe chills and long-lasting fever and is often fatal. It affects millions of people each year. Tools such as mosquito netting are helpful, but they are not always available or affordable. As a result, DDT continues to play a crucial, effective role in controlling this disease.

Hazardous Waste from Manufacturing and Industry

Hazardous substances play a role in most of the items you use or benefit from on a daily basis. Whether these substances become hazardous wastes or not depends on how they are managed during and after their production. **Figure 8.4** lists some of the hazardous wastes that result when common products and materials are made.

The mining, oil and gas, chemical, and pulp and paper industries also use and produce high volumes of hazardous substances. In mining, hazardous chemicals such as cyanide are often used to separate valuable metals from the ore. Naturally occurring hazardous substances may also be released during mining operations. These include **heavy metals** such as mercury, arsenic, and lead. Many of these substances eventually need to be disposed of as hazardous wastes.

The situation is similar in the oil and gas industry. This is partly due to the many operations involved in the industry. Exploring for oil and gas, drilling wells, refining processes, and maintaining equipment and vehicles are just a few of the activities that produce hazardous wastes. Heavy-metal wastes are especially problematic. Managing these wastes is very challenging for provinces such as Alberta, Saskatchewan, and Ontario, which have extensive petroleum industries.

heavy metals highly dense metal elements; includes mercury, arsenic, and lead

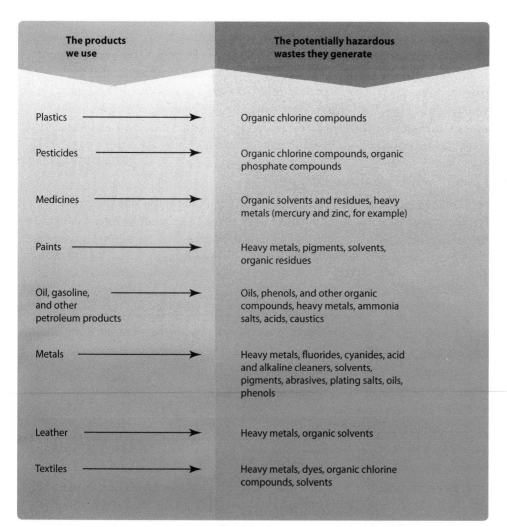

The products we use	The potentially hazardous wastes they generate
Plastics	Organic chlorine compounds
Pesticides	Organic chlorine compounds, organic phosphate compounds
Medicines	Organic solvents and residues, heavy metals (mercury and zinc, for example)
Paints	Heavy metals, pigments, solvents, organic residues
Oil, gasoline, and other petroleum products	Oils, phenols, and other organic compounds, heavy metals, ammonia salts, acids, caustics
Metals	Heavy metals, fluorides, cyanides, acid and alkaline cleaners, solvents, pigments, abrasives, plating salts, oils, phenols
Leather	Heavy metals, organic solvents
Textiles	Heavy metals, dyes, organic chlorine compounds, solvents

Figure 8.4 The manufacture of many common products results in the production of hazardous wastes.

Researching *Work with your classmates to determine how these hazardous wastes affect human health and the environment.*

Hazardous Wastes from Municipal Sources

InquiryLab 8A, An Inventory of Hazardous Materials in Your Home, on page 262

Municipal sources of hazardous wastes are closely linked to the use of consumer products. These products can generate hazardous wastes in two ways: in their production and at the end of their useful life. Many consumer products become hazardous wastes when discarded. These include familiar items such as batteries, antifreeze, paints, compact fluorescent light bulbs, and cleaners.

One significant source of municipal hazardous waste is electronic devices, because they often contain hazardous substances. Electronic devices range from large televisions to small, hand-held devices such as cellphones and portable media players (such as the one shown in **Figure 8.5**). People tend to replace electronic devices frequently, when newer versions become available and because the cost of fixing faulty items is often close to the cost of buying new ones. Unwanted electronic devices that enter the waste stream are often referred to as *e-waste*.

Older electronics present another hazardous waste problem. Many components in them contain **polychlorinated biphenyls (PCBs)**. PCBs are extremely harmful chlorinated hydrocarbons. They are fire-resistant and heat-resistant compounds that conduct little electricity. Due to these properties, they have played a useful role in electronics manufacturing. Like DDT, however, PCBs accumulate in human and animal tissues. Over time, they can harm the immune, nervous, respiratory, reproductive, and urinary systems. PCBs are also thought to cause cancer.

PCBS were banned in Canada in 1977. However, electronics that were produced before the 1980s are still likely sources of PCBs. Any electronic wastes that are suspected to contain PCBs must be handled according to specific government requirements. Like DDT, PCBs continue to be a significant problem in developing nations. In these countries, PCB-containing electronics are still in use and regularly enter the waste stream.

polychlorinated biphenyl (PCB) a specific type of chlorinated hydrocarbon

Figure 8.5 Hazardous wastes can come in small packages. If enough of these wastes are discarded, they can create a large problem.

Landfills and Hazardous Wastes

Some of the solid waste that people put in municipal solid waste landfills is really hazardous waste that has ended up there improperly. These landfills are not designed to handle hazardous wastes. As a result, harmful substances may leach into liquids in the landfill and can sometimes enter surrounding ecosystems.

Mini-Activity 8-1 **Heavy Metals**

Heavy metals have a long history of being used in many products—and just as long a history of poisoning people. For example, lead, arsenic, mercury, and tin were commonly used as medicines from the 1600s to the 1800s. The expression "mad as a hatter" originates from the 18th century, when mercury was used to produce felt for hats. Workers in these factories were continually exposed to mercury, which built up in their bodies over time and often caused dementia due to mercury poisoning.

Today, Health Canada recognizes heavy metals, such as lead, mercury, and arsenic, as being significantly toxic. Because of this, Health Canada has defined allowable levels of these heavy metals as impurities in products.

Products with higher levels are banned from being sold in Canada. Choose one of the following products, which are now banned from sale in Canada:

- lead-containing paint
- mercury thermometers
- lead-containing plumbing in homes
- cosmetics and tattoo inks containing heavy metals

Do research to find out what problems have existed and still exist regarding use and availability of these products. Why were these substances used in the making of these products in the first place? What, if any, alternatives are available?

A main cause of improper disposal is lack of knowledge. In many cases, people do not know that items such as batteries, paints, disposable cellphones, and compact fluorescent light bulbs contain harmful substances that must be taken to a hazardous waste facility (like the one shown in **Figure 8.6**) for proper disposal. As people become better informed, and as they commit themselves to being more attentive, this problem may become less widespread.

Figure 8.6 Most municipalities have locations where people can bring their hazardous waste. The waste is then sorted by technicians.

Analyzing *Why do you think it is important for trained individuals to classify the hazardous waste that is dropped off by the public?*

In developing countries, the disposal of hazardous wastes continues to be a significant environmental problem and health concern. Hazardous wastes are often disposed of in large, open-air sites. Poverty, as well as a lack of understanding of the dangers, often drives people to search such sites for materials to sell.

Medical Facilities and Hazardous Wastes

Medical facilities such as hospitals, clinics, and dental and veterinary practices produce wastes that include blood products, body organs, old medicines, and used swabs, bandages, needles, and syringes. These wastes are hazardous, but they are managed under a special set of regulations. You may have heard these wastes referred to as *biohazards*. Special containers in medical offices, such as the one in **Figure 8.7**, are used to collect these wastes so they can be disposed of properly.

Hospitals and specialized clinics that carry out nuclear medicine tests and treatment, such as cancer therapy, also produce *radioactive wastes*. Radioactive wastes are any wastes that exhibit radioactivity that exceeds a certain level set by government legislation. These wastes are extremely hazardous and are managed in special ways.

Figure 8.7 Biohazard containers like this one are used to collect and store medical wastes until they are disposed of properly.

Pause and Reflect

4. Identify three sources of hazardous wastes.

5. Describe two sources of chlorinated hydrocarbons.

6. Critical Thinking Technological advances occur at such speed that modern electronic devices quickly become outdated and thrown away. Suggest one way that we could reduce this problem.

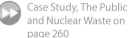

Case Study, The Public and Nuclear Waste on page 260

Effects of Hazardous Substances and Wastes on the Environment

The presence of hazardous substances and wastes can have a variety of effects on the health of organisms and the environment in which they live. These can range from minor temporary effects, such as headaches and nausea, to more serious concerns, such as cancers and birth defects. Factors that determine the nature and severity of health effects include

- toxicity (how harmful the substance or waste is)
- duration of exposure (how long a person is exposed to it)
- amount of exposure (how much of it a person is exposed to)
- route of exposure (whether it is inhaled, ingested, or absorbed through the skin)

Over the past few decades, numerous accidents as well as some deliberate releases of hazardous substances and wastes into the environment have occurred. Many of these have had devastating effects on living organisms. **Figure 8.8** explores the harm such releases have caused around the world over the past few decades.

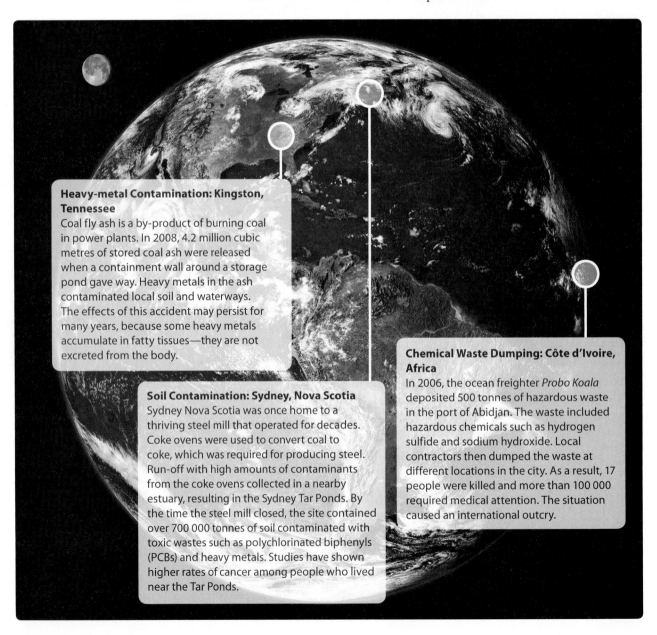

Heavy-metal Contamination: Kingston, Tennessee
Coal fly ash is a by-product of burning coal in power plants. In 2008, 4.2 million cubic metres of stored coal ash were released when a containment wall around a storage pond gave way. Heavy metals in the ash contaminated local soil and waterways. The effects of this accident may persist for many years, because some heavy metals accumulate in fatty tissues—they are not excreted from the body.

Soil Contamination: Sydney, Nova Scotia
Sydney Nova Scotia was once home to a thriving steel mill that operated for decades. Coke ovens were used to convert coal to coke, which was required for producing steel. Run-off with high amounts of contaminants from the coke ovens collected in a nearby estuary, resulting in the Sydney Tar Ponds. By the time the steel mill closed, the site contained over 700 000 tonnes of soil contaminated with toxic wastes such as polychlorinated biphenyls (PCBs) and heavy metals. Studies have shown higher rates of cancer among people who lived near the Tar Ponds.

Chemical Waste Dumping: Côte d'Ivoire, Africa
In 2006, the ocean freighter *Probo Koala* deposited 500 tonnes of hazardous waste in the port of Abidjan. The waste included hazardous chemicals such as hydrogen sulfide and sodium hydroxide. Local contractors then dumped the waste at different locations in the city. As a result, 17 people were killed and more than 100 000 required medical attention. The situation caused an international outcry.

Figure 8.8 Accidental release of hazardous waste into the environment is a global concern.

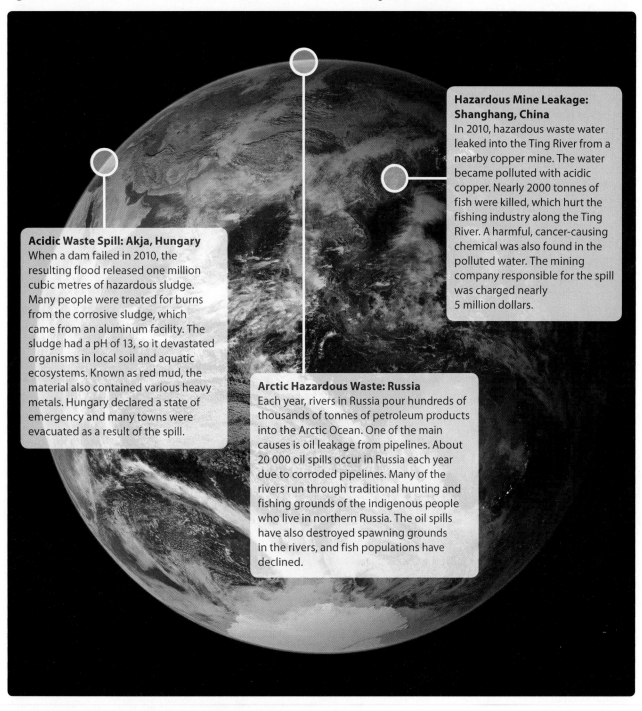

Hazardous Mine Leakage: Shanghang, China
In 2010, hazardous waste water leaked into the Ting River from a nearby copper mine. The water became polluted with acidic copper. Nearly 2000 tonnes of fish were killed, which hurt the fishing industry along the Ting River. A harmful, cancer-causing chemical was also found in the polluted water. The mining company responsible for the spill was charged nearly 5 million dollars.

Acidic Waste Spill: Akja, Hungary
When a dam failed in 2010, the resulting flood released one million cubic metres of hazardous sludge. Many people were treated for burns from the corrosive sludge, which came from an aluminum facility. The sludge had a pH of 13, so it devastated organisms in local soil and aquatic ecosystems. Known as red mud, the material also contained various heavy metals. Hungary declared a state of emergency and many towns were evacuated as a result of the spill.

Arctic Hazardous Waste: Russia
Each year, rivers in Russia pour hundreds of thousands of tonnes of petroleum products into the Arctic Ocean. One of the main causes is oil leakage from pipelines. About 20 000 oil spills occur in Russia each year due to corroded pipelines. Many of the rivers run through traditional hunting and fishing grounds of the indigenous people who live in northern Russia. The oil spills have also destroyed spawning grounds in the rivers, and fish populations have declined.

Mini-Activity 8-2 — Hazardous Wastes in the News

Monitor media sources such as TV, radio, magazines, and websites for news about hazardous wastes. A story might cover an accidental spill or release, or a discovery of contamination from the past. It could also discuss an issue or concern about a specific waste.

• Choose one news story that interests you.

• Make notes of the main points covered in the story.

• Evaluate the short-term and long-term effects of this waste or waste-related event on the environment and human health.

• Present your findings and evaluation to your class in the form of your own news report.

Summary

- Hazardous substances can cause harm to organisms and the environment, and they require special methods to reduce or eliminate their potential for harm.

- A hazardous waste is any substance that is or contains substances that are flammable, corrosive, reactive, or toxic.

- Chlorinated hydrocarbons are a hazardous class of organic compound found in some pesticides, some plastics, and some synthetic rubber.

- Manufacturing and industries such as mining, oil and gas, and pulp and paper use many hazardous substances and produce large amounts of hazardous wastes.

- Consumer products can contribute to municipal hazardous wastes, either due to the way they are made or when they are thrown away. These items include electronic devices, batteries, cleaning products, fabrics, and plastics.

- The effect of hazardous substances and wastes on human health is determined by factors such as length of exposure, how harmful the waste is, how much a person is exposed to, and whether it is inhaled, ingested, or absorbed.

Review Questions

1. State one example of hazardous waste that is produced from each of the following sources. For each example, explain why the waste is hazardous: K/U T/I
 a) industry
 b) agriculture
 c) municipalities

2. What is e-waste? Why is some of it considered hazardous? K/U T/I

3. Describe what the following symbols mean and the properties of material that they represent. K/U

 a)   **b)**

4. Based on your knowledge of WHMIS, draw the symbol for hazardous oxidizing material. C A

5. Many people dispose of hazardous household substances as part of their regular garbage. These hazardous wastes can end up in municipal landfills, where they can contaminate surrounding communities and ecosystems. T/I C A

 a) Describe one way that municipalities can get their citizens to stop putting hazardous waste in regular household garbage.

 b) Design a logo or diagram that informs people of the problems associated with disposing of hazardous waste in regular garbage.

6. Describe one example of the release of hazardous waste into the environment and the harmful effects it had in the area. In your answer, include what the waste was and where the release occurred. K/U

7. Copy the table below into your notebook. Complete it by describing two examples of potentially hazardous waste that each product generates. K/U C

Hazardous Waste Producers

Product	Possible Hazardous Waste from Its Production
Paint	
Leather	
Fabric	
Plastics	

8. Research labs in hospitals and universities generate a great deal of biohazardous waste. Each institution must have government-approved procedures for dealing with these materials. K/U T/I A

 a) What is biohazardous waste?

 b) Why are there such strict regulations for handling and disposing of biohazardous waste?

 c) What impact do you think this has on people such as maintenance and cleaning staff who work in these buildings? What do you think should be done to ensure their protection?

9. In one or two sentences, explain the following statement. "All toxic waste is hazardous, but not all hazardous waste is toxic." T/I C

10. List four factors that determine the effect a hazardous substance may have on a person's health. K/U C

Managing Hazardous Waste

The Pollution-Prevention Hierarchy

Through your own actions, including the choices you make about the products you buy, you contribute to hazardous waste in Canada. In total, Canadians generate more than six million tonnes of hazardous waste each year. Governments at all levels set safety standards to protect workers who must deal with these wastes. Governments at all levels also set regulations and guidelines to oversee the management of these wastes. Government agencies can impose fines or other penalties if a company does not follow those guidelines and regulations.

Environment Canada, like its counterpart in the United States, the EPA (Environmental Protection Agency), promotes a pollution-prevention hierarchy. This strategy is often referred to as P2 (for **p**ollution **p**revention). It emphasizes reducing the amount of hazardous waste produced according to the strategy outlined below and shown in **Figure 8.9**.

1. Reduce the amount of waste at its source.

2. Recycle wastes whenever possible.

3. Treat wastes to reduce their hazard or their volume.

4. As a last resort, dispose of wastes on land or incinerate them.

> ### Did You Know?
> "Why should we tolerate a diet of weak poisons, a home in insipid surroundings, a circle of acquaintances who are not quite our enemies, the noise of motors with just enough relief to prevent insanity? Who would want to live in a world which is just not quite fatal?"
>
> — *Rachel Carson (1907–1964), American scientist, writer, and founder of the modern environmental movement*

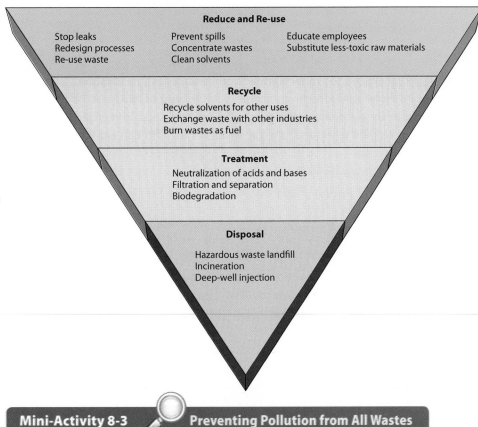

Figure 8.9 The simplest way to deal with hazardous wastes is not to produce them in the first place. The pollution-prevention hierarchy stresses reducing the amount of hazardous waste that is produced.

Mini-Activity 8-3   Preventing Pollution from All Wastes

The strategy shown in **Figure 8.9** refers to hazardous wastes. How well can it be adapted to refer to all wastes, including non-hazardous solid and liquid wastes? Sketch your own version of a pollution-prevention hierarchy that applies to all wastes. Write a brief instruction sheet to explain how it works and how to interpret it.

Figure 8.10 Instead of using harsh cleansers to remove rust stains in the bathroom, you can use a pumice stone with a paste of vinegar and baking soda.

Reducing the Amount of Waste at the Source

Many pollution-prevention changes are simple to perform and cost little to implement. Key among them are activities that result in fewer accidental spills, fewer leaks from pipes and valves, less loss from broken containers, and fewer similar accidents. Sometimes these reductions can be achieved through awareness training for managers and employees.

Sometimes industrial processes can be redesigned so that fewer hazardous wastes are produced. Other times, it may be possible to use a less hazardous substance in a process. These ideas also work well for municipal hazardous waste reduction. For instance, these wastes can be reduced if people only buy the amount of product they expect to use. If this is not possible, sharing leftover hazardous substances with a friend or neighbour is another option. As well, hazardous household products often can be replaced with less harmful substances that do the same job (see **Figure 8.10**).

Re-use

In some industries, hazardous substances used in one process can be cleaned and re-used in the same process, in a different process, or by a completely different industry. For example, waste oils produced by one industry can be used to fuel power plants.

Re-use of hazardous wastes in the home is much more limited. One choice that most people can make, however, is to use rechargeable batteries. All batteries are a source of hazardous waste during their production as well as at the end of their lives. While the start-up cost for rechargeable batteries is higher than for disposable ones, they can be used over and over again. This reduces the price per use over time. The cost to the environment is also reduced—especially if the energy resource that recharges the batteries is a renewable one.

Recycle

 Case Study, New Life for Dead Batteries on page 259

Recycling is best applied to products whose components become hazardous waste when they are thrown away. Some companies recycle paint to make new paint. Electronic waste (e-waste) is another example. Municipal and commercial e-cycling programs help keep these wastes out of the waste stream. Through e-cycling, electronic devices are redistributed to charities for further use. If the electronics cannot be re-used, components can be dismantled and sold for use in other applications. You can also do your own e-cycling by upgrading a device rather than replacing it.

Mini-Activity 8-4   A Personalized Reduction Plan

Create a personalized plan to reduce the amount of hazardous waste that is generated in your home. Try to include each of the three strategies—reduce, re-use, and recycle—in your plan. Discuss your plan with your classmates and teacher, and put it into action, if possible. What kinds of effects do you expect your plan to have on the environment, in the short term as well as in the long term?

Pause and Reflect

7. What has the highest priority in the pollution-prevention hierarchy?

8. How can re-using hazardous substances decrease cost for industries and impact on the environment?

9. Critical Thinking Despite the environmental advantages of using rechargeable batteries, many people still use a lot of disposable batteries. Suggest one reason why, and offer an argument that counters that reason.

Collecting and Treating Hazardous Wastes

Some hazardous wastes cannot be reduced, re-used, or recycled. These wastes must be collected and treated so that they present less of a hazard to the environment and human health.

Collecting Hazardous Wastes

Depending on where you live, household hazardous waste may be collected in several ways in your area. There may be hazardous waste depots, retailer drop-off programs, and municipal curbside collection programs. The Toxic Taxi shown in **Figure 8.11** is one of these services. For people who live rurally or in smaller communities, options can vary greatly in terms of the types of hazardous wastes that can be collected and the times of year when they can be taken to collection depots (see **Figure 8.12**).

Governments fund some of these collection programs through taxes. Industry and manufacturing companies finance others, such as the Orange Drop program of Stewardship Ontario—the same organization that administers Ontario's blue box program. With Orange Drop, companies that make hazardous products collect them when they are discarded as part of their **extended producer responsibility (EPR)**. EPR is the idea that the producer of a product is responsible for all aspects of its production, including its disposal when the useful life of the product is over. EPR encourages manufacturers to design products with waste in mind, since they will ultimately be responsible for those wastes.

Figure 8.11 Toxic Taxis such as this one provide a valuable hazardous waste pickup service.

> **extended producer responsibility (EPR)** producers are responsible for the end-of-life management of their products

Figure 8.12 Many municipalities have scheduled times and locations for people to drop off their hazardous wastes.

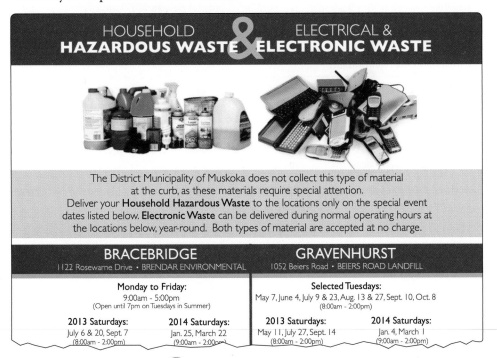

HOUSEHOLD **HAZARDOUS WASTE** & ELECTRICAL & **ELECTRONIC WASTE**

The District Municipality of Muskoka does not collect this type of material at the curb, as these materials require special attention. Deliver your **Household Hazardous Waste** to the locations only on the special event dates listed below. **Electronic Waste** can be delivered during normal operating hours at the locations below, year-round. Both types of material are accepted at no charge.

BRACEBRIDGE	GRAVENHURST
1122 Rosewarne Drive • BRENDAR ENVIRONMENTAL	1052 Beiers Road • BEIERS ROAD LANDFILL
Monday to Friday: 9:00am - 5:00pm (Open until 7pm on Tuesdays in Summer)	**Selected Tuesdays:** May 7, June 4, July 9 & 23, Aug. 13 & 27, Sept. 10, Oct. 8 (8:00am - 2:00pm)
2013 Saturdays: July 6 & 20, Sept. 7 (8:00am - 2:00pm) **2014 Saturdays:** Jan. 25, March 22 (9:00am - 2:00pm)	**2013 Saturdays:** May 11, July 27, Sept. 14 (8:00am - 2:00pm) **2014 Saturdays:** Jan. 4, March 1 (9:00am - 2:00pm)

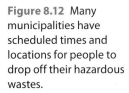

Mini-Activity 8-5   Create a Hazardous Waste Collection Guide

Use a medium of your choice to create a hazardous waste collection education guide. Your guide should explain how hazardous wastes can be properly disposed of in your community. Conduct research as directed by your teacher to complete your guide. Your guide should

- identify and describe the various programs in your area
- explain where, when, and/or how the programs can be accessed
- describe the types of hazardous wastes each program collects

Treating Hazardous Wastes

bioremediation use of organisms to reduce or remove harmful waste

Treating hazardous waste involves converting it to a less hazardous substance and/or reducing its volume before disposal. **Table 8.2** describes a variety of technologies used to treat hazardous wastes. The type of treatment chosen depends on the type of hazardous waste involved. To some extent, the choice also depends on the treatment goal—conversion or volume reduction.

Table 8.2 Types of Hazardous Waste Treatment	
Type of Treatment	**Description**
Chemical	Treatment involves chemically converting hazardous wastes into less hazardous substances. Chemical treatment can also reduce the volume of the waste prior to disposal.
Physical	Physical force and/or mechanical devices are used to isolate hazardous wastes for disposal. For example, filters can remove hazardous substances from liquid wastes. Physical treatment can also decrease the volume of hazardous waste.
Biological	The use of organisms, including micro-organisms, to reduce or remove contaminants from a site is called **bioremediation**. Bioremediation degrades organic wastes so that they are less harmful, and it is used to treat many types of wastes. In some situations, it may be used to treat hazardous wastes. For example, bioremediation may be used to treat agricultural soils and soils around mining sites that have been contaminated with hazardous wastes.
Thermal	Treatment exposes hazardous waste to extreme heat, which makes the substance less hazardous or no longer hazardous. The volume of waste is also greatly reduced.
Immobilization	Treatment fuses wastes at high temperature in glass, ceramics, or cement. These impermeable materials trap the wastes so they can be placed into long-term storage without threatening the environment.
Stripping	A process that separates volatile chemicals from water and collects them for further treatment or disposal.
Precipitation	Also called flocculation, this approach is used mostly for treating sewage. It binds the hazardous components of sewage to chemicals that clump and settle out, making them easy to filter, collect, and dispose.
Carbon absorption	Activated carbon particles bind to hazardous chemicals in waste gases or liquids. Wastes may be removed for disposal. The carbon can be cleaned and re-used.
Phytoremediation	This is a type of bioremediation that uses plants. It often makes use of natural or constructed wetlands to treat sewage before it is released into the environment.

tertiary treatment—
see section 7.2
on page 228

Pause and Reflect

10. What is EPR? Describe how it helps in the management and collection of hazardous wastes in our society.

11. Describe a method of treating hazardous waste that reduces the volume of the hazardous waste and produces waste that is less hazardous.

12. Critical Thinking Thermal treatment can greatly reduce the amount and danger of hazardous wastes. However, most people do not want these facilities in their communities. Should the benefits of thermal treatment be allowed to outweigh people's concerns? Explain your opinion.

Disposal of Hazardous Wastes

Despite the best efforts of governments, industries, communities, and individuals, there will always be some hazardous wastes that need to be disposed of in some manner. There are three main methods used in Canada.

Deep-well Injection

Liquid hazardous wastes are injected into wells to be stored deep below Earth's surface. The wells may be natural geological fissures, or they may be drilled or dug. Modern injection wells have a non-porous casing that can be reinforced with cement to prevent leakage. However, leakage from some older wells remains a concern. Some of these threaten ground water resources. For example, in the past, the ground water in Lambton County, Ontario, and the St. Clair River were contaminated by such leakage.

Non-retrievable Storage: A Hazardous Waste Landfill

As shown in **Figure 8.13**, a hazardous waste landfill is lined with an impermeable plastic liner (1) and several layers of compacted clay (2) to contain leachate. A gravel bed (3) lies between the clay layers to collect leachate. Pipes (4) collect any escaped leachate for removal and treatment. Testing wells (5) and monitoring underdrains (6) check for escaped leachate. Methane may be collected and removed for combustion (7).

Hazardous wastes are placed in the landfill and separated by absorbent soil layers. Techniques such as immobilizing also help decrease the chance of hazardous wastes escaping the landfill. Finally, the landfill is capped with more layers of clay, plastic, and soil (8) when it is full.

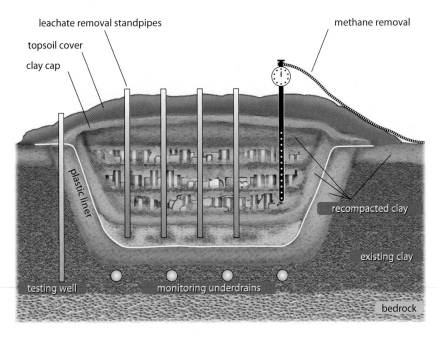

leachate removal standpipes
topsoil cover
clay cap
methane removal
plastic liner
recompacted clay
existing clay
testing well
monitoring underdrains
bedrock

Figure 8.13 A hazardous waste landfill is the most common solution for long-term storage of solid hazardous waste.

Retrievable Storage

Retrievable storage is more expensive than other disposal options. Hazardous wastes are placed in containers in an accessible but secure location, such as a building, cave, or abandoned mine. This storage method is permanent, but it allows for the wastes to be removed if better storage technologies are developed in future. Another benefit of this approach is that the waste can be monitored easily.

Summary

- Environment Canada promotes reducing the amount of hazardous waste by reducing it at the source, recycling it whenever possible, treating it to reduce the volume, and only disposing of it on land or by incineration when there is no alternative.

- Industrial changes that help reduce hazardous waste include minimizing accidents and spills, and properly maintaining equipment used with hazardous substances. Using alternatives to hazardous substances also can result in less hazardous waste.

- Recycling of products that become hazardous waste when they are thrown away can help reduce the amount, especially for electronic waste.

- Municipalities and industries have different programs to help collect hazardous wastes. There are different methods to treat hazardous wastes. Some convert it into non-hazardous waste, and others result in reducing the volume of waste. Any leftover hazardous waste must still be safely and carefully disposed of according to government regulations.

- Deep-well injection, hazardous waste landfills, and retrievable storage facilities are three methods to dispose of hazardous wastes.

Review Questions

1. In a brief paragraph, describe what Environment Canada's pollution-prevention hierarchy is. K/U C

2. Describe one change that you can make that will reduce the amount of hazardous waste you produce. In your answer, be sure to explain why this action will result in less hazardous waste. T/I A

3. In many homes, housecleaning can produce levels of chemicals in indoor air that are more hazardous than the outdoor air of the most polluted cities in the world. However, for hundreds of years people cleaned their homes using common ingredients that were quite safe. Today's ready-made cleaning products only started to be produced after World War II, when different petroleum-based chemicals became widely available. Describe how people can reduce the municipal hazardous waste that is produced from the use of these products, while still maintaining a clean living environment. C A

4. What is one thing that you can do as a consumer to help reduce the amount of hazardous waste that is generated? T/I A

5. How can the "three Rs" apply to the management of hazardous waste? Describe an example for each. K/U T/I A

6. What is the extended producer responsibility? How does it help in the management of hazardous waste? K/U T/I

7. Describe the key features of a hazardous waste landfill. Why must particular attention be paid to collecting leachate and landfill gases? K/U T/I

8. The photo below shows an alfalfa plant, which is commonly used in phytoremediation. K/U

a) Describe what phytoremediation is and what it is used for.
b) Explain why phytoremediation is a type of bioremediation.

9. There are a variety of methods to treat hazardous wastes. The one chosen depends on the type of waste and what the goal of the treatment is. Describe a type of hazardous waste treatment for each of the following. K/U C

a) to reduce the volume of the hazardous waste
b) to convert the hazardous waste to non-hazardous waste
c) to convert and reduce the volume of hazardous waste

10. Imagine that a retrievable storage facility for hazardous waste is being considered in a location close to where you live. Your community will receive money for this storage, which will be used to improve community services. Would you support this type of facility in your community? Explain why or why not. T/I C A C

Case Study New Life for Dead Batteries

Each year, you and other North Americans use about 200 000 tonnes of sealed cell batteries. About 86% of a battery can be recycled. Recovered components such as heavy metals and steel are then re-used for other purposes. Many people, however, do not know that batteries can be recycled. In fact, even though household batteries make up only a small fraction of garbage collected, the toxic substances that many of them contain—mercury, cadmium, and lead— contribute 50% to 70% of all heavy metals in landfills. Currently, only about 5% of batteries are recycled. The situation is changing, though, as municipalities, school and community groups, and retailers work together to promote battery recycling.

Metals can be recycled indefinitely. This not only reduces the need to mine new metals, but also saves energy.

Durham Region Breaks World Record for Battery Collection

Durham Region, just east of Toronto, is the first municipality in Ontario to offer curbside battery recycling to residents—and they started off big! Their first collection week in November 2012 collected almost 24 tonnes of batteries and set a world record for batteries collected in a 24-hour period: more than 5120 kg.

The municipality partnered with waste collection and recycling companies, as well as non-profit groups that provide Ontario consumers with a free and safe way to dispose of household products (including batteries) that are not suitable for the landfill. People place used batteries in special orange bags that go in their blue boxes. This curbside pickup makes battery recycling easier and more convenient than ever before.

Analyze and Conclude

1. Research how batteries are recycled in your community. Suggest ways that battery recycling could be expanded in your region.

2. Take an inventory of the number of single-use (non-rechargeable) batteries in use in your home. Estimate the total number for your class and your school. If your school has a battery recycling program, how many batteries are collected per year?

Communicate

3. Create a brochure about battery recycling that can be part of a school battery collection program for your community.

Case Study   The Public and Nuclear Waste

There are five nuclear power plants with a total of 22 reactors in three provinces: Ontario, Québec, and New Brunswick. These plants provide more than 50% of Ontario's power needs and about 15% of the power needs of all of Canada. While nuclear power plays a vital role in meeting Canada's energy needs, nuclear waste is a cause of public concern. The safe management and storage of nuclear wastes is overseen by government organizations and regulatory bodies.

Short-term Storage

Nuclear wastes are classified based on how much containment and isolation they require to ensure safety. Low-level and intermediate-level wastes tend to be stored and managed at the sites where they are produced. Once their radioactivity has decayed to acceptable levels, they are disposed of by conventional means, such as in landfills.

High-level radioactive waste is used nuclear fuel and waste from producing nuclear power. There are no long-term facilities to store this kind of waste. Instead, it is stored in short-term, on-site facilities. This waste gives off a lot of heat,

so it must be handled in two phases. The wet phase stores used fuel bundles under water in secure, leak-proof pools until they have cooled. This takes 6 to 10 years. Then the bundles are transferred to concrete containers.

Waste produced during the mining and processing of uranium ore contains long-lived radioactive elements. Because so much of this dangerous waste is produced, it is stored and managed near where it is mined and processed.

Long-term Storage

One method of long-term storage is a DGR (deep geological repository). A DGR is a storage facility deep underground (usually more than 300 m below surface level). Ontario Power Generation has proposed such a facility for low-level and intermediate-level waste at the Bruce nuclear plant near Kincardine, Ontario. It would be about 680 m underground in stable rock layers of limestone and shale.

Summary of Current and Future Radioactive Waste Inventories

Waste Category	Waste Inventory to 2010 December	Waste Inventory Projected to End of 2011	Waste Inventory Projected to End of 2050
Nuclear fuel waste	9 079 m^3	9 400 m^3	19 800 m^3
Intermediate-level radioactive waste	32 906 m^3	33 400 m^3	67 000 m^3
Low-level radioactive waste	2338 000 m^3	2 343 000 m^3	2 594 000 m^3
Uranium tailings	214 million tonnes	Not available	Not available
Waste rock	174 million tonnes	Not available	Not available

Source: Low-Level Radioactive Waste Management Office. (2012, March). *Inventory of radioactive waste in Canada*, Table 7.1.

Public Opinion and the Future of Nuclear Power

Case Study, Nuclear Accidents and Public Safety Concerns on page 72

Two main concerns influence public opinion about nuclear power: the danger of nuclear accidents and the risks related with nuclear wastes. In 2010, the Organisation for Economic Co-operation and Development (OECD) conducted a study of international public opinion. Results showed that even though previous nuclear accidents had a negative impact on public opinion about safety, support for nuclear energy had been rising strongly over the past 10 to 20 years in the United States, Western Europe, and Japan. Despite concerns about its waste, nuclear energy was viewed as a dependable energy source that reduced a country's dependence on energy imports and contributed less to climate change.

The accident at Fukushima, Japan, in 2011 changed the way the public views nuclear power. A study of 23 countries one year after the accident showed that attitudes varied by country and were strongly tied to culture and emotions. For example, core support remained strong in the United States and Britain, while Germany and Italy shifted to an anti-nuclear stance. More consistent was the public's perception that the release of information about major nuclear accidents was too slow and contradictory. As well, there was little trust that institutions can prevent accidents or manage them when they occur.

Despite safety concerns, the public still views nuclear power as more dependable than renewable sources such as solar and wind. Overall, public opinion supports stabilizing energy costs, protecting energy supplies, and diversifying energy sources. There is little agreement, though, on how this should be achieved.

Research and Analyze

1. Research the deep geological repository project in Kincardine, Ontario. What is the current state of the project and the public's opinion of it?

2. Choose one of Canada's nuclear power plants. Analyze the way nuclear waste is managed and monitored there.

3. Choose a position for or against nuclear power. Work in a group to prepare for a debate. Consider issues such as safety, costs, waste, reliability, and climate change.

Communicate

4. What role should public opinion play in determining whether nuclear power continues to be part of Canada's energy plan? Write a short (250-word) opinion paper expressing your view.

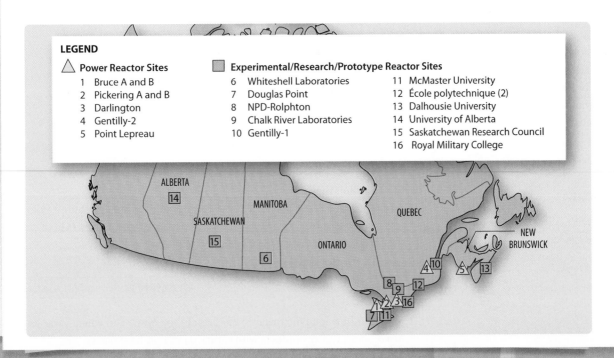

LEGEND

△ **Power Reactor Sites**
1 Bruce A and B
2 Pickering A and B
3 Darlington
4 Gentilly-2
5 Point Lepreau

▢ **Experimental/Research/Prototype Reactor Sites**
6 Whiteshell Laboratories
7 Douglas Point
8 NPD-Rolphton
9 Chalk River Laboratories
10 Gentilly-1

11 McMaster University
12 École polytechnique (2)
13 Dalhousie University
14 University of Alberta
15 Saskatchewan Research Council
16 Royal Military College

Skill Check

Initiating and Planning

Performing and Recording

Analyzing and Interpreting

Communicating

Safety Precautions

- Your teacher must approve your procedure before you carry out the survey.

- Inform parents or guardians about what you are doing. Ensure that one of them accompanies you during your survey, and that they have read your teacher-approved procedure.

- Many products in the home pose hazards to your health, whether or not they are classified as hazardous substances. Be very cautious if you handle or examine products in your home. Use the HHPS warning labels on the next page as a guide.

- While conducting your survey, wear rubber gloves and be careful when handling the products. Do not spill any materials, and do not smell any contents. Do not touch any container that is leaking or broken.

- Wash your hands when you are finished the survey.

Materials

- clipboard with paper
- rubber gloves

A Survey of Hazardous Materials in Your Home

Hazardous substances and wastes are found not only in laboratories or industries, but also in the home. The average home uses 10 kg to 40 kg of hazardous products each year. Some examples are shown below.

Examples of Hazardous Products in the Home

Kitchen Area	Bathroom Area	Garage/Storage Areas
Sink and drain cleaners	Toilet and tile cleaners	Paints and solvents
Oven cleaner	Nail polish and removers	Batteries
Furniture polish	Air fresheners	Automotive products
Disinfectant sprays	Hair dye	Detergent and bleach

These products are required to have labels with warnings about their dangers and how to handle them properly. For consumer products, the system of warnings is called the **H**azardous **H**ousehold **P**roduct **S**ymbols, or HHPS system. These are described on the next page.

Pre-Lab Questions

1. What makes something a hazardous substance?

2. What safety symbols are used for hazardous substances in workplaces and laboratories?

3. Name four locations in your home that may contain hazardous products.

Question

How many types of hazardous materials are in your home?

Procedure

Part A: Make a Plan

1. Make a plan for how you will conduct the survey of your home.

2. Have your teacher approve your plan before performing it.

3. Arrange a time when a parent or guardian can accompany you during your survey.

Part B: Conduct the Survey

1. Prepare a table like the one below to record your findings. Include amounts of hazardous substances. For example, if there are two containers of oven cleaner, record "oven cleaner, 2 containers."

Hazardous Products Identified

Location in Home	Product	Safety Warnings on the Label

2. Find out how long the material has been in the home and when it was last used. Identify any products with expiry dates that have passed.

3. Classify each type of hazardous substance as flammable, corrosive, reactive, and/or toxic.

4. Draw a bar graph that shows the totals of each type of hazardous substance in your home.

Analyze and Interpret

1. Did anything from your survey surprise you? If so, explain why.

2. Did one area of the home contain more hazardous substances than others? If so, which area was it?

3. How much material did you discover that is no longer useful and is therefore now considered hazardous waste?

Conclude and Communicate

4. Compare your survey graph with those of your classmates. Look for and discuss patterns in types and amounts. What, if any, conclusions can you make, based on your discussions?

5. Propose two ways to reduce the amount of hazardous materials in your home.

6. Do research to find two less hazardous alternatives that could replace two hazardous products in your home.

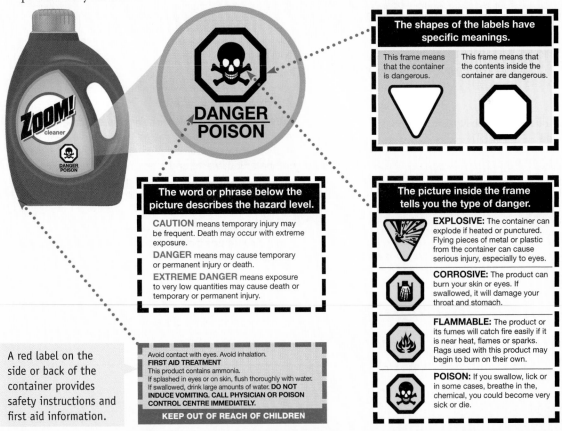

The shapes of the labels have specific meanings.

This frame means that the container is dangerous.

This frame means that the contents inside the container are dangerous.

The word or phrase below the picture describes the hazard level.

CAUTION means temporary injury may be frequent. Death may occur with extreme exposure.

DANGER means may cause temporary or permanent injury or death.

EXTREME DANGER means exposure to very low quantities may cause death or temporary or permanent injury.

The picture inside the frame tells you the type of danger.

EXPLOSIVE: The container can explode if heated or punctured. Flying pieces of metal or plastic from the container can cause serious injury, especially to eyes.

CORROSIVE: The product can burn your skin or eyes. If swallowed, it will damage your throat and stomach.

FLAMMABLE: The product or its fumes will catch fire easily if it is near heat, flames or sparks. Rags used with this product may begin to burn on their own.

POISON: If you swallow, lick or in some cases, breathe in the, chemical, you could become very sick or die.

A red label on the side or back of the container provides safety instructions and first aid information.

Avoid contact with eyes. Avoid inhalation.
FIRST AID TREATMENT
This product contains ammonia.
If splashed in eyes or on skin, flush thoroughly with water.
If swallowed, drink large amounts of water. **DO NOT INDUCE VOMITING. CALL PHYSICIAN OR POISON CONTROL CENTRE IMMEDIATELY.**
KEEP OUT OF REACH OF CHILDREN

The HHPS system uses different shapes and icons, which have certain meanings. A warning symbol shaped like a triangle means the container is dangerous. A symbol shaped like a stop sign, or octagon, means the contents are dangerous. The particular icon that is used indicates the type of danger. The four categories of danger are explosive, corrosive, flammable, and poison.

Chapter 8   SUMMARY

Section 8.1   Types of Hazardous Waste

Hazardous wastes, such as pesticides, heavy metals, and biohazards, are used or produced during the manufacture of many products.

Key Terms
hazardous substance
hazardous waste
persistent pesticide
chlorinated hydrocarbon
heavy metal
polychlorinated biphenyl (PCB)

Key Concepts
- Hazardous substances can cause harm to organisms and the environment, and they require special methods to reduce or eliminate their potential for harm.
- A hazardous waste is any substance that is or contains substances that are flammable, corrosive, reactive, or toxic.

- Chlorinated hydrocarbons are a hazardous class of organic compound in some pesticides, as well as some plastics and synthetic rubber.
- Manufacturing and industries such as mining, oil and gas, and pulp and paper use many hazardous substances and produce large amounts of hazardous wastes.
- Consumer products can contribute to municipal hazardous wastes, either due to the way they are made or when they are thrown away. These items include electronic devices, batteries, cleaning products, fabrics, and plastics.
- Medical and research facilities generate biomedical hazardous wastes. Special procedures are needed to collect and dispose of them.
- The effect of hazardous substances and wastes on human health is determined by factors such as length of exposure, how harmful the waste is, how much a person is exposed to, and whether it is inhaled, ingested, or absorbed.

Section 8.2   Managing Hazardous Waste

Hazardous waste must be handled and disposed of using special methods that are approved by government agencies. Hazardous waste is treated and/or disposed of in highly controlled facilities.

Key Terms
extended producer responsibility (EPR)
bioremediation

Key Concepts
- Environment Canada promotes reducing the amount of hazardous waste by reducing it at the source, recycling it whenever possible, treating it to reduce the volume, and only disposing of it on land or by incineration when there is no alternative.
- Industrial changes that help reduce hazardous waste include minimizing accidents and spills, and properly maintaining equipment used with hazardous substances.

Using alternatives to hazardous substances also can result in less hazardous waste.
- Recycling of products that become hazardous waste when they are thrown away can help reduce the amount, especially for electronic waste.
- Municipalities and industries have different programs to help collect hazardous wastes. There are different methods to treat hazardous wastes. Some convert it into non-hazardous waste, and others result in reducing the volume of waste. Any leftover hazardous waste must still be safely and carefully disposed of according to government regulations.
- Deep-well injection, hazardous waste landfills, and retrievable storage facilities are three methods to dispose of hazardous wastes.

Chapter 8 REVIEW

Knowledge and Understanding

Choose the letter of the best answer below.

1. Hazardous wastes are
   a) used to manufacture products, such as hydrochloric acid that is used to etch computer circuit boards
   b) substances that are created during the life cycle of a product that can cause harm to human health or the environment
   c) regulated by industries through extended producer responsibility initiatives, not by government regulations
   d) always incinerated to convert toxins to non-hazardous substances
   e) only produced from agricultural and industrial sources and processes

2. Which of the following are examples of biohazardous waste?
   a) paint, paint thinner, and furniture polish
   b) nitroglycerine and gunpowder
   c) old medicines, used syringes, and used wound dressings
   d) gasoline and kerosene
   e) heavy metals, acids, and bases

3. In which of the following situations are pesticides *not* considered to be hazardous wastes?
   a) when they are absorbed by other materials, such as soil or sawdust
   b) when they are used up completely during their application on a particular crop
   c) when they are no longer effective at controlling a pest and are disposed of
   d) when they are in small quantities, such as residue left inside a sprayer or other container
   e) when they leach from the soil into ground water sources

4. Heavy metal wastes, such as mercury, arsenic and lead, can result from
   a) mining activities
   b) oil and gas exploration
   c) maintenance of equipment and vehicles
   d) the pulp and paper industry
   e) all of the above

5. Hazardous waste from municipal sources is closely connected to the use of consumer products. Which of the following statements regarding municipal waste is correct?

   a) Older electronic waste items contain PCBs, which are known to harm organisms.
   b) Rechargeable batteries are not hazardous because they are re-used and never end up in a landfill.
   c) Hazardous wastes can come in small packages. Because the quantity of waste is so small, they are not considered to be harmful to the environment.
   d) PCBs from electronic wastes are not considered to be a waste problem in developing nations, because they are not manufactured there.
   e) Household cleaners are rarely considered hazardous because they are packaged in small bottles.

6. What are the four strategies that make up the pollution-prevention hierarchy?
   a) reducing waste at its sources, recycling wastes, reducing volumes of waste by incineration, and disposing liquid hazardous wastes into remote bodies of water
   b) reducing waste at its sources, re-using materials to avoid disposal, sending hazardous waste to landfills, and incinerating wastes
   c) reducing waste at its sources, treating wastes to reduce their hazards, exporting wastes to solve the waste problem, and recycling whenever possible
   d) reducing waste at its sources, recycling wastes when possible, treating wastes to reduce their volume and hazards, and disposing of wastes only as a last resort
   e) none of the above

7. There are many ways to treat hazardous waste to render it less hazardous or reduce its volume. Which of the following statements regarding the types of treatment is *incorrect*?
   a) Stripping separates volatile chemicals from water and collects them for treatment or disposal.
   b) Carbon absorption is a process in which activated carbon binds to hazardous chemicals in gases or liquids so that the wastes can be removed.
   c) Phytoremediation involves the use of micro-organisms, such as bacteria, to degrade organic waste into less hazardous substances.
   d) Immobilization fuses wastes at high temperatures in glass, ceramics, or cement so they can be placed in long-term storage without harming the environment.
   e) Thermal treatment exposes waste to very high temperatures, which reduces the volume of waste and makes it no longer hazardous.

8. Hazardous wastes that cannot be treated are sometimes placed in non-retrievable storage. Which of the following best describes what occurs with this type of disposal?

a) Hazardous wastes are layered between absorbent soil layers above an impermeable plastic liner and layers of compacted clay and gravel.

b) Hazardous wastes are injected into wells to be stored deep below Earth's surface.

c) Hazardous wastes are placed in containers and placed in an accessible but secure location, such as a cave or abandoned mine.

d) Hazardous wastes are placed in a clay-lined pit that has pipes to collect any escaped leachate for removal and treatment.

e) none of the above

Answer the questions below.

9. Define the term *hazardous waste*, and identify three sources.

10. List and describe four main categories of hazardous wastes.

11. DDT is a persistent pesticide.

a) What class of organic compounds does DDT belong to?

b) Why is DDT considered to be persistent?

c) Explain why DDT is still allowed to be used in certain parts of the world, but is banned in most countries.

12. List two common products that are sources of heavy metal wastes. Why are heavy metal wastes problematic?

13. Describe two ways that consumer products generate hazardous wastes.

14. Give two examples of e-waste that contain PCBs, and explain why they are hazardous.

15. Why are hazardous wastes still disposed of in landfills? What can be done to reduce the disposal of these wastes in landfills?

16. List and explain the factors that affect the nature and severity of health effects from hazardous wastes.

17. Environment Canada promotes a pollution-prevention hierarchy, often referred to as P2. What is the purpose of this hierarchy? What are the strategies involved in P2?

18. What are four ways to reduce the amount of municipal hazardous waste at the source?

19. Explain how hazardous substances can be re-used in some industrial processes. Give an example of re-using a hazardous waste in an industrial setting.

Thinking and Investigation

20. The estimated recycling rate for cellphones has decreased, as shown in the graph below. What are possible reasons for this decline? What can be done to improve cellphone recycling rates?

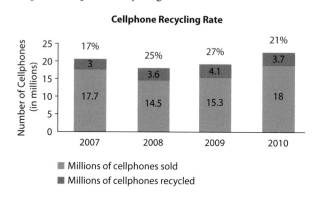

Cellphone Recycling Rate

- ■ Millions of cellphones sold
- ■ Millions of cellphones recycled

21. Identify and describe advantages and disadvantages of extended producer responsibility (EPR) programs.

22. Cleaning products used at home, schools, and other institutions are large contributors of municipal hazardous waste. Most product labels do not provide any information on how to dispose of the product and its waste. In your opinion, who is responsible for changing this? What actions can you take to initiate this change?

23. Why is the study of the life cycle of a consumer product helpful for establishing sustainable waste management strategies? Provide a suitable example to support your answer.

24. What must be considered when selecting a non-retrievable storage site for hazardous wastes?

25. Based on what you know about nuclear power stations, how would you feel if a new reactor was being planned for your community? What are some of the issues that would need to be considered before it is built? Explain any concerns you may have.

26. **Did You Know?** Re-read the quotation from Rachel Carson on page 253. Do you think she is being serious, or is she using hyperbole (deliberate exaggeration) to draw attention to a topic that deserves discussion? How do interpret her intent and her statement itself?

Communication

27. Hazardous wastes sometimes escape accidentally from industries and can cause harm to the environment. Sometimes these accidents are caused by human error, but sometimes they are a result of natural disasters, such as earthquakes or tornadoes. Should industry officials be held responsible for the environmental damage caused by such disasters? Explain your position. Who should be responsible for the costs and labour needed to restore the environment, and why?

28. Brainstorm ways that consumers can influence the decisions made by manufacturers with respect to hazardous wastes. Organize your ideas in a graphic organizer.

29. Think about the products you use at home, school, and at other places in a week. How many of these products could be considered hazardous wastes when they are no longer useful to you? Create a list of all the hazardous waste you are directly and indirectly responsible for creating in a typical day. What are some ways you can reduce the impact of these hazardous wastes on the environment?

30. Phytoremediation is an effective way to treat hazardous wastes before they are released to the environment. The basic principle of phytoremediation is shown below. Using the diagram, develop an explanation that could be used to teach a Grade 7 class about phytoremediation and what plants do in the process.

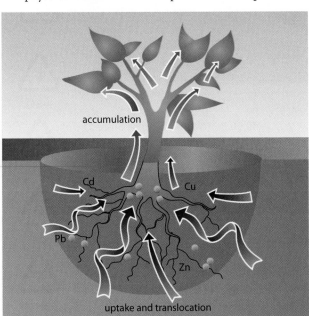

31. While cleaning out the garage or apartment storage locker, you find an old car battery, paint cans, and furniture polish. What hazards are associated with these waste items? Find out what you should do to dispose of them properly in your community. Write a public service announcement for the members of your community so they know the safe way to dispose of household hazardous waste.

Application

32. Choose a traditional product that is used at your home or school. Use the information on the label to do the following.
   a) Identify the hazardous chemicals in the product and any disposal information.
   b) Create a pamphlet that contains information on the safe handling, storage, and proper disposal of the product.
   c) Research a safer alternative to the product. Describe the benefits of using the alternative, as well as any drawbacks.

33. Based on your knowledge of hazardous wastes, do you believe there is a difference between incinerating municipal, medical, and industrial waste? Are you opposed to the construction and operation of an incinerator for one type of waste in your neighborhood, but not other types? Explain.

34. Brownfield sites are abandoned industrial sites that are contaminated with hazardous wastes, but have the potential to be redeveloped once cleaned up.
   a) What are the issues that potential buyers of brownfield sites must consider?
   b) Many brownfield sites in highly desirable neighbourhoods remain undeveloped. Make suggestions for ways the government could change this.

35. Agricultural wastes from pesticides can cause harmful effects on environmental and human health.
   a) When are pesticides considered to be hazardous waste?
   b) Does this mean that pesticides are not hazardous before they become waste? Explain your answer.

Pause and Reflect

How could you incorporate what you have learned in this chapter into your daily actions or choices?

Canadians in Environmental Science

Souad Sharabani: Innovative Chef

One of the greatest stresses on the environment is related to food and the eating habits of people in developed countries such as Canada. One environmental cost of feeding ourselves is the staggering amount of wasted food that ends up in landfills. Almost $28 billion worth of food is wasted each year in Canada. This means about 40% of food produced in Canada becomes waste. Everyone involved in the production, distribution, and purchase of food contributes to this—which means everyone who handles food can help reduce waste.

Food recovery programs are one way to reduce food waste. Many programs involve collecting and redistributing perishable and non-perishable food to feed people in need. People who work in these programs—from the individuals who co-ordinate food pickups, to the drivers who transport the food, and the chefs who prepare it—all play a role in reducing society's waste. One such person is Souad Sharabani.

Souad is no ordinary chef. She leads a team of volunteers who serve lunches to people at the Parkdale Neighbourhood Church in downtown Toronto. In a country where everyone wastes $4 of food a day, Souad manages to feed her clients for less than $1 per person. Preventing food waste means being creative. "We don't have much money, so you have to use whatever you get," Souad says. "One week I was given a 23 kg box of green beans! I had to use them all, so what we couldn't cook for our meal, I cleaned and cut up and put in bags for people to take away with them. For days afterward I saw our people in the neighbourhood, snacking on green beans. I also received a shipment of fennel. I had no idea what to do with it, so I found a recipe for a lovely salad and it was a huge success."

Souad has lived and worked in Asia, Africa, and Latin America. She says that the value placed on food in those countries—where food resources are often scarce for large segments of the population—has been an inspiration in her work. No less inspiring is seeing the positive effect that a healthy meal, creatively prepared and served with respect, can have on someone.

Environmental Science at Work

Focus on Solid, Liquid, and Hazardous Waste Management

Industrial Waste Inspector

Land Reclamation Specialist

Water and Wastewater Laboratory Technologist

Landfill Engineer

Solid, Liquid, and Hazardous Waste Management

Waste Management Specialist

Recycling Co-ordinator

Recycling co-ordinators evaluate and supervise recycling programs for municipalities or large organizations. They may be responsible for specific programs such as curbside and commercial recycling, or for household hazardous waste disposal programs.

Hazardous Waste Management Chemist

In this job, you work as part of a team of scientists who identify pollutants in the air, water, and soil. These chemists also help design techniques that reduce pollution and clean up problems caused by hazardous waste.

Wastewater Collection and Treatment Operator

Collection operators work on sewer systems, while treatment operators work in plants that treat municipal wastewater. Both types of job may involve taking samples for analysis, working with chemicals, and maintaining equipment.

For Your Consideration

1. What other jobs and careers do you know or can you think of that involve managing solid, liquid, and hazardous wastes?

2. Research a job or career in this field that interests you. What essential knowledge, skills, and aptitudes are needed? What are the working conditions like? What attracts you to this job or career?

Investigate a Local Waste Management Practice

In this project, you will investigate a local waste management practice to determine how it helps divert waste from the waste stream.

The blue box program is an example of a national waste management practice. The first community-wide practice in Canada was in 1983 in Kitchener, Ontario.

Question

How is a community waste management practice contributing to reducing the amount of waste going to landfills or incinerators?

Initiate and Plan

1. Choose a waste management practice to investigate. Decide on the type of waste that is being managed, and determine the source of that waste in your community. For example, will you investigate this waste management practice for your community's schools, homes, parks and recreation facilities, or industrial and commercial properties?

2. Decide on the type of information you want to find out. Consider the following points.

   - How much waste has been diverted as a result of the waste management practice?
   - How long has this practice been in effect in the area?
   - How efficient is the program?
   - What type of education or information sessions have been used to inform people about this practice?
   - What have been the financial costs associated with implementing and running the practice?
   - What were the initial goals of the program? Do the individuals who run the program consider it a success?
   - What are their future goals or plans for the program?

3. Identify sources of information to answer your questions. For example, is there a local company in charge of managing this practice that you can contact? Is there a website with information? What information is available through your municipal or district council?

4. Have you teacher approve your choice of waste management practice and location, as well as the sources of information you plan to use.

Perform and Record

5. Collect the information, as outlined in steps 1 to 3.

6. Record all the information you collected and its sources.

7. Make a note of any controversies or problems that have developed due to this waste management practice.

Analyze and Interpret

1. Organize the information you collected using an appropriate graphic organizer.

2. Has the waste management practice successfully diverted waste from the waste stream? Explain your answer, and indicate what factors you used to assess the success.

3. Suggest ways that the waste management practice could be improved.

Communicate Your Findings

4. Communicate your findings in a manner that can be used to inform members of your community.

Assessment Checklist

Review your project when you complete it. Did you ...

- ✓ **K/U** use your knowledge of the types of waste management practices to identify one to study?

- ✓ **T/I** identify appropriate sources of information?

- ✓ **A** carry out your investigative plan, as outlined?

- ✓ **A** determine how the practice has helped to divert waste from the waste stream?

- ✓ **T/I** suggest improvements to the practice?

- ✓ **C** choose an appropriate format for presenting the information, keeping your audience in mind?

An Issue to Analyze

Cradle-to-Grave Analysis of an Electronic Product

Increasingly, governments are passing laws that require businesses and individuals to take into account the environmental costs of electronic products. One way to examine these costs is with a *cradle-to-grave analysis*. This assesses all the environmental impacts that can occur during a product's life cycle, which includes raw material extraction, manufacturing, distribution, product use, and product disposal. In this project, you will conduct a cradle-to-grave analysis of one electronic product.

Issue

How have electronic products had a negative effect on the environment?

Initiate and Plan

1. In a group, choose one electronic product.
2. Before you begin your research,
   - determine what sources of information you will use
   - assign each group member to research part of the life cycle of the product; see step 4 for the different topics
3. Develop a research plan based on steps 1 and 2. Have your teacher approve your plan.

Perform and Record

4. Research each of the following stages of the life cycle of the product:
   - The technology used to design the product and the impact this technology has on the environment.
   - The raw materials used to create the product and the effects of the extraction and manufacturing processes. More than one member of the group may need to research this.
   - How the product is transported and how far it travels to reach the place where it is sold. (Consider transportation to both warehouses and stores.)
   - The approximate distance a consumer travels to purchase the product and the type of transportation used. Analyze the environmental impact of driving a car, riding a bike, taking a bus, or walking.

- How long the product should last. Will it degrade over time or become outdated by new technology? Compare the environmental impact of products that must be replaced frequently with products that have longer life cycles. What factors influence consumers to replace a product before the its useful life is over?
- How the product is discarded. Does it end up in a landfill? Is toxic waste produced from its disposal? Do the federal, provincial, or local governments provide guidelines for disposal or recycling of the product? Do any laws or regulations govern the proper disposal of the product?

5. As a group, combine your research findings to form a complete cradle-to-grave analysis.

Analyze and Interpret

1. Describe the negative effects that the electronic product has on the environment.
2. For each stage of the product's life cycle that you researched, brainstorm some ways in which the ecological impact could be reduced.

Communicate Your Findings

3. Share your findings and analysis with the class, using a format of your choice. Possibilities include a video documentary, social media website, or blog. Keep your audience in mind when designing your presentation.

Assessment Checklist

Review your project when you complete it. Did you …

- ☑ **K/U** create a research plan and have your teacher review it?
- ☑ **T/I** assess the credibility and bias of your research sources?
- ☑ **T/I** research each stage of the product life cycle?
- ☑ **A** collaborate as a group to produce a complete cradle-to-grave analysis?
- ☑ **A** brainstorm ways to reduce the environmental impact of the product?
- ☑ **C** cite your sources using appropriate academic format?
- ☑ **C** communicate your findings using an appropriate format based on your audience?

Unit 4 REVIEW

Knowledge and Understanding

Choose the letter of the best answer below.

1. Which of the following best describes landfills?
   a) Both liquid and solid wastes are disposed of in landfills.
   b) Waste is isolated from ground water, air, and surrounding soil.
   c) The gases they produce are not dangerous.
   d) The leachate is drained into the surrounding soil and water sources.
   e) Only a small percentage of solid waste in Canada goes to landfills.

2. Which of the following statements about recycling is *incorrect*?
   a) It is problem-free and there are only benefits associated with it.
   b) It involves reprocessing items to make new products.
   c) It encourages greater responsibility by citizens.
   d) It decreases the amount of waste in the waste stream.
   e) It relies on initiatives such as the blue box programs

3. Sewage
   a) is a type of storm water.
   b) is composed of only feces and water.
   c) includes any materials flushed down toilets and drains.
   d) is treated by depositing it in landfills.
   e) should be discharged directly into waterways.

4. Which of the following can become hazardous waste?
   a) a computer
   b) motor oil
   c) house paint
   d) drain cleaner
   e) All of these are correct.

5. Which of the following lists three types of hazardous waste treatment?
   a) landfill, bioremediation, and stripping
   b) thermal, recycle, and landfill
   c) recycle, precipitation, and chemical
   d) thermal, chemical, and electrical
   e) chemical, thermal, and physical

Answer the questions below.

6. Solid waste is managed in a variety of ways.
   a) Name three major sources of solid waste. Give an example of a type of solid waste that comes from each source.
   b) Describe the important features of landfills. What are the benefits of landfills?
   c) What does the acronym "NIMBY" mean and how does it apply to public opinion about landfills?
   d) Describe a method of solid waste disposal that is an alternative to landfills.

7. Waste management focuses on "the three Rs." Describe what each of these terms means and how it applies to the management of solid waste. What are the advantages and disadvantages for each of these methods?

8. Liquid waste includes sewage and storm water.
   a) What are sewage and storm water?
   b) Describe the two ways that sewage is treated. Which method is used mainly in rural areas, and which is used in cities?
   c) Sewage systems can back up in homes after a heavy rainfall. What can you infer about the sewer and storm-water systems of cities where this occurs?
   d) What are the disadvantages of diverting storm water directly into local waterways?

9. Describe two properties of hazardous materials.

10. Many common household products can become hazardous wastes.
    a) What are two examples of household hazardous wastes? Describe why they are considered hazardous.
    b) Describe two ways that household hazardous waste collection is encouraged.

Thinking and Investigation

11. Both non-hazardous and hazardous solid wastes are managed according to the three Rs.
    a) How are these three strategies the same for both types of waste?
    b) How are they different, and why do the differences exist?

12. Why is source reduction considered to be the most effective way of managing waste? Describe one way you could apply source reduction to your life in order to reduce the amount of waste you generate.

13. The pie graph below shows the relative amounts, by weight, of types of solid waste generated in the average Canadian household.

Distribution of Types of Wastes from Households

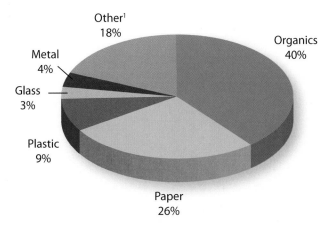

¹Other includes textiles, tires, wood, and animal waste.

a) What type of waste is generated in greatest amount? What items does this include?

b) For each type of solid waste shown, describe one thing people can do to reduce the amount of this waste that enters the waste stream.

c) Describe three methods that divert these wastes from the waste stream.

Communication

14. You are a member of a community where the installation of a nearby landfill is being considered. A town hall meeting is going to be held where citizens can voice their opinions about this issue. Would you support the proposal? Explain your position based on environmental, social, political, and economic considerations. Using a medium of your choice, summarize your opinion. Make sure to include supporting statements.

15. You run a resort that rents out cottages with full greywater systems. Write a notice for guests that explains what greywater is and describes special steps guests should take while staying at the resort.

16. Consider the different solid, liquid, and hazardous wastes that you generate or that are produced as a result of your activities and products you use. Make a table with the following headings: Type of Waste, How the Waste Is Generated, and What I Can Do to Reduce the Amount. Complete the table for ten examples of waste.

17. Draw a diagram or flowchart that summarizes how solid waste is managed. Be sure to indicate the following:
- the waste stream and the final destinations for items in it
- methods used to divert waste from the waste stream
- the role of source reduction

18. Do you agree or disagree with the following statement? Provide at least two reasons that support your answer. "Recycling and composting will solve all of our problems concerning the management of solid waste."

Application

19. Many communities in Canada have banned the use of pesticides for use on homeowners' lawns and gardens. Some people argue that this infringes on their rights as homeowners, and that it harms the local economy by limiting the work of companies that specialize in lawn care. Do you agree or disagree with the banning of pesticides on homeowners' lawns and gardens? Give at least two reasons to support your opinion.

20. Children are more vulnerable to hazardous substances than teens and adults are. What are some reasons for this increased vulnerability? Explain your answer

21. More than 40 million people rely on the Great Lakes, shown below, for drinking water. However, discharge of wastes into the lakes and surrounding areas has reduced the quality of water. Over 360 chemicals have been found in the Great Lakes. These include DDT and heavy metals.

a) Why is DDT a hazardous substance? Why is it called a persistent pesticide?

b) Name two examples of heavy metals. What is a source of heavy-metal waste?

c) In 2011, the Canadian Nuclear Safety Commission wanted to transport 16 steam generators that had been used in nuclear plants through Lake Huron, Lake Erie, and Lake Ontario to the St. Lawrence Seaway. Do research to find out the role that First Nations people played in stopping this and why they did.

UNIT 5 Human Health and the Environment

This satellite image of global air pollution shows suspended particles of dust (red), sea salt (blue), smoke from fires (green), and sulfate particles (white) from volcanoes and from burning fossil fuels.

What distribution patterns do you see? How do air and water pollutants affect human and environmental health?

BIG IDEAS

- Environmental factors can have negative effects on human health.
- It is possible to minimize some of the negative health effects of environmental factors by making informed lifestyle choices and taking other precautions.

Overall Expectations

- Analyze governmental and non-governmental initiatives intended to reduce the impact of environmental factors on human health.
- Investigate environmental factors that can affect human health, and analyze related data.
- Demonstrate an understanding of environmental factors that can affect human health, and explain how their impact can be reduced.

Unit Contents

Chapter 9

Air Quality and Environmental Health

Chapter 10

Water Quality and Environmental Health

UNIT 5 Essential Science Background

Topic 1: Earth's Atmosphere

Figure 1 summarizes information about the structure and composition of the atmosphere. Pressure in the atmosphere decreases with altitude, and this decrease is more rapid at lower altitudes than at higher altitudes. As a result, the majority of the mass of the atmosphere—about 99%—lies within 30 km of Earth's surface. About 90% of the mass of the atmosphere lies within 15 km of the surface, and about 75% lies within 100 km.

The atmosphere is divided into five regions based on temperature changes: the troposphere, stratosphere, mesosphere, thermosphere, and exosphere. In terms of composition, Earth's atmosphere is typically divided into two regions: the homosphere and the heterosphere.

In the homosphere, gases are fairly evenly blended, giving the region a fairly uniform composition. This uniformity results from mixing of the gases by convection—a cyclical movement of air molecules caused by differences in air density. Air near the land is warmer and less dense than the air above it. The cool, dense air above sinks, displacing the warmer, less-dense air below, causing the warm air to rise. Once away from the surface, the warm air cools and its density increases. This cooled air then sinks down to the surface and the convection process repeats itself.

The mixing of gases in the atmosphere by convection does not occur in the heterosphere. Therefore, the gas composition in the heterosphere varies and is limited to a few types of gases. Gas particles are layered by their mass. Nitrogen and oxygen molecules are in the lowest layers, along with ions of oxygen, ozone, and nitrogen. Oxygen atoms are in the next highest layer. Helium and free hydrogen atoms occur in the layers farther above.

Figure 1 The structure and composition of Earth's atmosphere

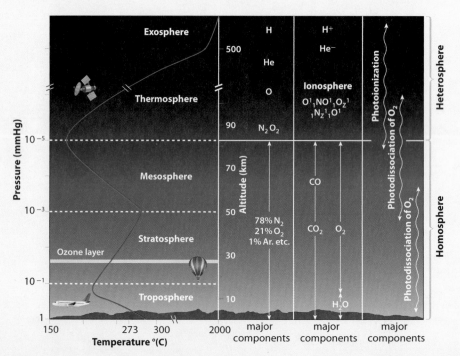

Topic 2: Ground Water

Water that enters soil and is not absorbed by plant roots filters through spaces in the soil and subsurface material until it reaches rock layers through which is can no longer penetrate (impermeable layers). The water that fills the spaces is called *ground water*.

The porous layer that becomes saturated with ground water is called an *aquifer*. An aquifer is an underground layer of gravel, sand, or permeable rock that holds ground water, which can be extracted by wells. Many of the farming operations that feed the world depend on aquifers for their water. There are three basic kinds of aquifers, as shown in **Figure 2**. An unconfined aquifer usually occurs near the surface, where water enters the aquifer from the land above it. The top of the layer saturated with water is called the water table. The lower boundary of the aquifer is a layer of clay or rock that does not let water pass through it. Unconfined aquifers are replenished (recharged) mainly by rain that falls on the ground directly above the aquifer and filters through the layers below.

A confined aquifer is bounded on both the top and bottom by impermeable rock layers. If water can pass in and out of the confining layer, the aquifer is referred to as semiconfined. Both confined and semiconfined aquifers are mainly recharged by rain and surface water that may come from an area many kilometres from where the aquifer is tapped for use. If the recharge area is higher than where is aquifer is tapped, water will flow up the pipe until it reaches the same elevation as the recharge area. Such wells are called artesian wells. If the recharge zone is above the elevation of the top of the well pipe, it is called a flowing artesian well, because water will flow from the pipe.

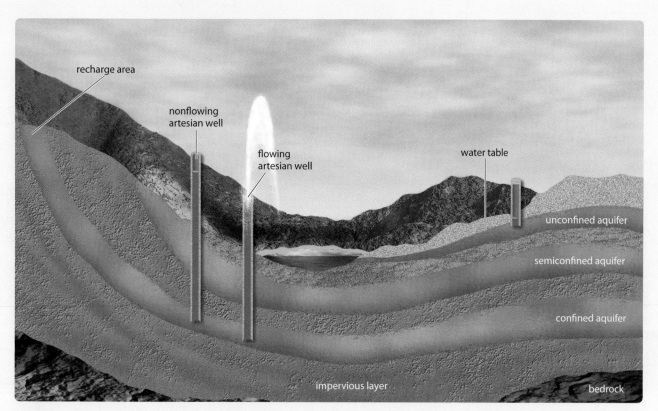

Figure 2 Ground water and aquifers. The various layers of sediment and rock determine the nature of the aquifer and how it can be used.

CHAPTER

9

Air Quality and Environmental Health

Alerts to the dangers of smog are becoming more common in many cities in Canada and around the world.

What kinds of events are likely to trigger a warning like this? Who benefits from the warnings, and how do they benefit?

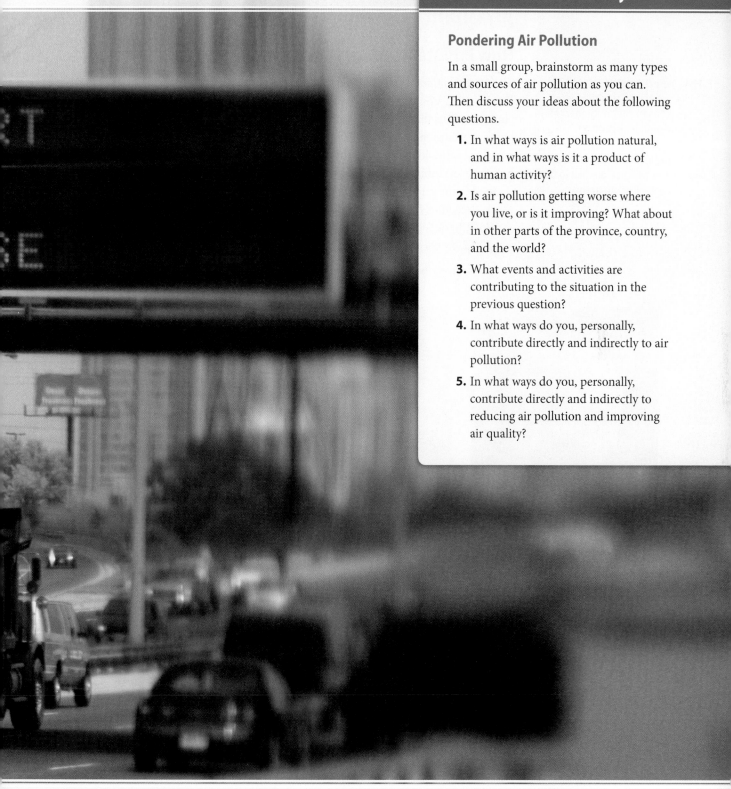

Pondering Air Pollution

In a small group, brainstorm as many types and sources of air pollution as you can. Then discuss your ideas about the following questions.

1. In what ways is air pollution natural, and in what ways is it a product of human activity?

2. Is air pollution getting worse where you live, or is it improving? What about in other parts of the province, country, and the world?

3. What events and activities are contributing to the situation in the previous question?

4. In what ways do you, personally, contribute directly and indirectly to air pollution?

5. In what ways do you, personally, contribute directly and indirectly to reducing air pollution and improving air quality?

In this chapter, you will

- identify primary and secondary air pollutants, their sources, and health effects
- describe how pollutants can enter the body and how people can reduce their exposure to them
- explain how legislation, technology, and personal decisions can improve air quality

Air Quality Outdoors and Indoors

How Pollutants Enter Organisms

contaminant a substance that occurs in concentrations higher than would be expected

pollutant a substance that is or has the potential to be harmful to organisms

The words *contaminant* and *pollutant* are often used to mean the same thing, but they are different. A **contaminant** is a substance in an environment that occurs in concentrations that are higher than what would be expected. A **pollutant**, on the other hand, is a substance in an environment that is harmful or could become harmful to people or other living things. Anything that is a pollutant is therefore a contaminant, but not everything that is a contaminant is a pollutant.

Pollutants can enter the body of an organism in one of three ways.

1. Pollutants can be inhaled (breathed in).

2. Pollutants can be ingested (eaten).

3. Pollutants can be absorbed through the organism's surface layer, such as the skin.

Figure 9.1 describes these three ways in more detail.

Through ingestion, some pollutants that are sprayed on growing crops or their soil can enter the body directly when people eat them. Other pollutants can enter drinking water and become incorporated into the plants and animals that people eat. In some cases, ingested pollutants can irritate or damage the inner lining of the digestive system. In other cases, ingested pollutants can be absorbed by the lining of the small intestine and carried by the blood to other organs, where they can cause harm.

Smoke, dust, gases, and other substances in air enter the body through the nose and mouth when you breathe in. Inhaled substances can directly damage the lungs and the airway tubes that lead to them. Inhaled substances also can enter blood vessels (by diffusion) and be carried by the blood to other organs, where they can cause harm.

The skin is the largest organ in the human body. Many air and water pollutants can enter the body by absorption through the skin. Pollutants also can be absorbed through the eyes, which causes them to be irritated or harmed. If they are absorbed through the eyes, pollutants can be transported to other parts of the body through blood vessels.

Figure 9.1 Pollutants enter a person's body by inhalation, ingestion, and absorption.

Applying *In what ways would a pollutant enter the body of a plant and the body of a single-celled organism such as a bacterium?*

Sources of Pollution

Polluting substances are often classified in terms of their source. *Point source pollutants* come from a source that can be easily identified, because they have a definite source where they enter an ecosystem. Drain pipes and smokestacks from factories are examples of point source pollutants. Since point sources are easy to identify, it is easier to monitor, control, and regulate pollutants that come from them.

Nonpoint source pollutants come from a source that is more difficult to identify and control, because it is scattered or diffuse. Transportation emissions are an example, because they come from many sources on land, in water, and in the air. **Figure 9.2** shows examples of point sources and nonpoint sources of pollutants.

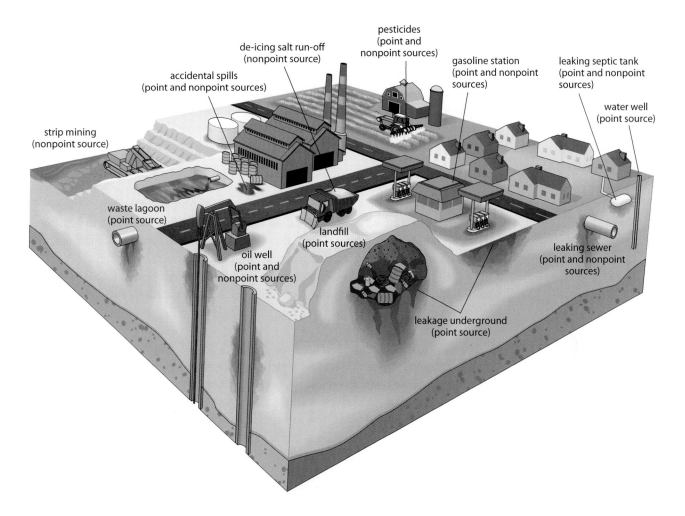

Figure 9.2 Point sources and nonpoint sources of pollutants. Notice that some sources can be both.

Applying *What examples of point source pollutants and nonpoint source pollutants can you name in and around the region where you live?*

Pause and Reflect

1. How are contaminants and pollutants similar to and different from each other?

2. Describe three ways that pollutants can enter a person's body.

3. Critical Thinking Explain why a pollution source can be both point and nonpoint.

Major Air Pollutants

An air pollutant is any gas or particle in the atmosphere that can cause harm or damage to organisms and to the environment. Air pollutants can come from natural sources as well as from human activities. Natural sources of air pollutants include

- ash and gases from volcanoes and other land formations

- gases given off by decaying plants

- gases (belches and flatulence) expelled from animals

anthropogenic made (produced) by humans

primary air pollutants substances released directly into the atmosphere in amounts that pose a health threat

secondary air pollutants substances that result from primary air pollutants interacting with one another and with other substances

Pollutants that result from human activities are known as **anthropogenic** sources. This term comes from two ancient Greek words. *Anthropo* refers to humans, and *genic* means makes or produces. So *anthropogenic* means made by humans.

Air pollutants from anthropogenic sources are classified as primary or secondary. **Primary air pollutants** are substances released directly into the atmosphere in amounts that pose a threat to health. There are five primary air pollutants, outlined in **Table 9.1**. As part of their air quality monitoring systems, Environment Canada and the provinces and territories refer to these air pollutants as *criteria air contaminants* (CACs).

Primary air pollutants can interact with one another and with gases and particles that are naturally present in the atmosphere. The substances that result from these interactions are called **secondary air pollutants**. Smog is the most familiar example.

How Smog Forms

air quality health index (AQHI)—**see section 9.2 on page 290**

On sunny days in the summer, you might notice a yellow-brown haze above or near densely populated areas. This haze is called smog—or *photochemical smog*, to be more precise. It forms when pollutants from the exhaust pipes of cars and trucks and the smokestacks of industrial plants react chemically in the presence of sunlight. (See **Figure 9.3**.) This reaction takes place up to about 1 km above the ground. One of the main compounds that forms in the reaction is ground level ozone. Ground level ozone can irritate the eyes, nose, throat, and lungs of people and other animals. It also has harmful effects on plants.

Note: Do not confuse ground level ozone, which occurs close to Earth's surface, with the ozone layer, which is nearly 30 km above Earth's surface. The ozone in the ozone layer is beneficial to life on Earth, because it filters out harmful ultraviolet (UV) radiation. Ground level ozone, on the other hand, can cause harm to life.

Figure 9.3 Main steps in the formation of photochemical smog

Source	Necessary Conditions	Reactions Take Place in Atmosphere		Products
Mostly motor vehicles	volatile organic compounds (VOC) present	VOC + O* or O_3 → highly reactive compounds + NO_2 →		A variety of irritant and toxic nitrogen compounds
Mostly motor vehicles	nitrogen monoxide (NO) present	NO + very reactive compounds → NO_2		
From motor vehicles and industrial smokestacks	nitrogen dioxide (NO_2) present	NO_2 → NO + O* (atomic oxygen); O* + O_2 → O_3 (ozone) ; sunlight		Ozone
Sun	sunlight			
Sun (summer temperatures)	heat	Reactions take place more rapidly at higher temperatures.		

Table 9.1 Sources and Effects of the Primary Air Pollutants

Air Pollutant	Description	Sources	Effects on Human Health
Carbon monoxide (CO) Solvents, waste disposal, etc. Metals — Other Non-road engines Transportation	• colourless and odourless gas formed when wood, gasoline, and other fuels are burned	• burning fossil fuels for transportation and generating electricity • burning wood for heat and clearing land	• reduces amount of oxygen that blood can carry to cells • high levels in urban areas can cause headaches, fatigue, and blurred vision • prolonged and/or high exposure is deadly
Nitrogen dioxide (NO_2) and other nitrogen oxides (NO_x) Non-road Industry Transportation Other Power plants	• reddish-brown gas that readily reacts with other compounds • responsible for photochemical smog	• motor vehicle exhaust • burning fossil fuels and biomass • forest fires	• irritates eyes and lungs • increases chance of respiratory infections • causes chronic bronchitis
Particulate matter (PM) Transportation Agriculture Other Construction	• solid particles or liquid droplets (collectively called aerosols) suspended in air • range in size from what can be seen with the eye to those requiring powerful microscopes	• burning plant material • agriculture • travel on roads • industrial processes • construction • mining • volcanoes • sea spray	• small particles can enter the nose and mouth and damage tissue of lungs and tubes leading to them • worsens asthma, bronchitis, and other respiratory conditions • some increase chance of cancer
Sulfur dioxide (SO_2) and other sulfur oxides (SO_x) Other Industry Power plants	• colourless gas with a sharp odour • reacts with water vapour in the atmosphere to form sulfur-containing acids	• burning sulfur-rich coal in power plants • burning biomass • industrial gases • volcanic activity	• sulfur-containing acids can burn lung tissue if inhaled • inhalation symptoms include chest pain, sore throat, headache, nausea
Volatile organic compounds (VOCs) Other Transportation Solvents Waste disposal Non-road	• wide range of very reactive, toxic compounds that vaporize readily and exist as gases in air • many undergo reactions in the presence of sunlight to form ground level ozone	• burning of fossil fuels and wood • refineries and other industrial plants • burning tobacco • paints, glues, fabrics, building materials, copy machines/ printers, personal care products	• irritate eyes, nose, and throat • can cause headaches, nausea, dizziness • higher concentrations can damage internal organs such as the lungs, liver, and brain • some increase chance of cancer

Mini-Activity 9-1 Make a Particulate Collector

Cut a hole about 6 cm across in each of several index cards. Cover each hole with clear packing tape. Then make a small hole in each card and put a string through it so the cards can be suspended. These are your particulate collectors.

Predict places at your school where pollution will be greatest, and hang collectors there for a few days. After, examine them with a hand lens, and use a flashlight to observe how much light shines through. Design a system to rank the pollution in the locations you chose.

Types of Indoor Air Pollution

Many outdoor air pollutants also act as indoor pollutants when trapped or generated indoors. These include carbon monoxide, nitrogen dioxide, VOCs, and particulate materials. Indoor levels of these pollutants often can build up to become much greater than outdoor levels. Other air pollutants are unique to the indoor environment or only act as pollutants when they are trapped indoors. **Figure 9.4** describes several indoor air pollutants, along with their sources and their effects on human health.

Figure 9.4 Examples of air pollutants that commonly affect the indoor environment

Second-hand Smoke
This is smoke that is exhaled by people who smoke tobacco products or that comes from the burning of those products. This smoke contains over 4000 chemicals. Many of these are toxic, including carbon monoxide, formaldehyde, ammonia, and arsenic. More than 30 of the chemicals in second-hand smoke are known to cause cancer. It also causes headaches, sore throat, and nausea, as well as an increase in respiratory infections and irritations. It may also increase the risk of heart disease.

Volatile Organic Compounds (VOCs)
All VOCs are gases at room temperature. Formaldehyde, acetone, and benzene are examples of VOCs, but there are thousands of others. Many synthetic household items, such as furniture, carpets, adhesive and cleaning products, and paint, release these compounds. Building materials such as insulation and polyvinyl chloride (PVC) pipes also emit VOCs. VOCs can cause headaches, dizziness, and eye and respiratory irritation. Nerve, kidney, and liver damage have been linked to long-term exposure to these pollutants. People with chemical sensitivities are especially vulnerable to VOCs.

Radon
Radon is a natural indoor air pollutant. The element uranium is sometimes found underground in rocks and soil. This element breaks down over time and releases radon gas. Radon frequently leaks into homes and buildings. Pores or cracks in concrete and brick, loose-fitting floor joints, and well water are common points of entry. Once inside, the gas can build up to high concentrations that are hundreds or thousands of times the average outdoor level. People exposed to this amount can have a lung cancer risk equal to a person who smokes several packs of cigarettes a day.

Asbestos

Asbestos is a naturally occurring mineral. It can be spun into lightweight fibres that are fireproof and have good insulating properties. As a result, it is often used in insulation. It is also used in walls, ceilings, and floors to provide soundproofing. When asbestos is disturbed, it easily enters the air. Once inhaled, it can damage the lungs and cause cancer. Because of the health hazards, its use has been phased out over past decades in Canada. Today, new sealing techniques that prevent the escape of asbestos fibres have made the material safe to use again. However, old asbestos in homes and buildings remains a danger. People who are renovating old buildings must wear protective gear when disturbing asbestos.

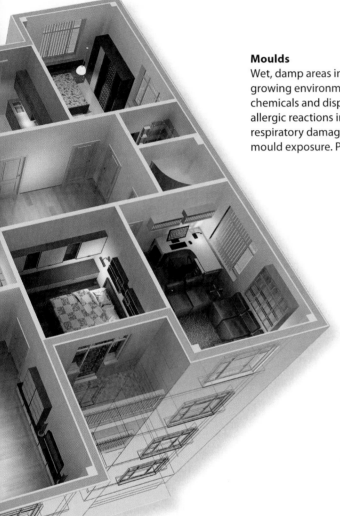

Moulds

Wet, damp areas in buildings and moisture trapped between walls provide an excellent growing environment for microscopic fungi called moulds. They release various chemicals and disperse reproductive spores through the air. These spores cause allergic reactions in many people. Inhaling mould spores and chemicals can also cause respiratory damage. Seniors, infants, and pregnant women are especially vulnerable to mould exposure. People with chronic health problems are also at higher risk.

Bacteria

Many bacteria are transmitted through the air. For example, *Legionella* bacteria normally live in moist soil or water. Indoors, these bacteria can be dispersed through the air in tiny water droplets by cooling towers that spray water into circulating air. Air conditioners can spread the bacteria-rich droplets even farther. When the water droplets are inhaled, the bacteria cause Legionnaires' disease. Symptoms include fever, cough, chills, and pneumonia. Legionnaires' disease is rare. It occurs most frequently in regions where air conditioning keeps buildings cool during hot summers. Most outbreaks of the disease occur in large buildings such as hospitals or hotels. Seniors, smokers, and people with poor immune systems are at greater risk.

Mini-Activity 9-2   Design a Clean Air Garden

Growing a variety of certain plants in your home or school, under the right conditions, can reduce exposure to indoor air pollution. Long-term experiments done by NASA show that VOCs especially can be reduced with certain house plants. Some of the most effective plants from this study are bamboo palm, rubber plant, ficus, Boston fern, and peace lily.

Criteria used to assess these plants included how well they removed chemical vapours, ease of growth and maintenance, and insect-resistance. Use the plants listed here or do your own research to design a small indoor garden to reduce indoor air pollutants. Create a flowchart of steps to build the garden and keep it healthy. With your teacher's permission, build and maintain your garden.

Sick Building Syndrome

sick building
syndrome the adverse
health effects due to time
spent in a building

In Canada, health issues related to indoor air quality are often labelled as **sick building syndrome** (SBS). SBS can affect people who live or work in a building and who experience adverse health effects that appear linked to time spent there. Because many buildings are tightly sealed, air pollutants can become trapped inside. Over time, concentrations can build up to levels that affect human health. People who live or work in such buildings may experience effects such as those listed in **Table 9.2.**

Table 9.2 Pollutants and Health Effects Associated with Sick Building Syndrome

Examples of Pollutants That Cause Sick Building Syndrome		Examples of Health Effects	
Chemical sources such as • carbon monoxide • toluene • benzene	• tobacco smoke • formaldehyde • some air fresheners	• respiratory problems • headaches • fatigue and lethargy	• memory lapses • blurred vision
Allergens such as • dust • pollen • mould • mildew	• tobacco smoke • wood smoke • pet dander • insect feces	• eye, nose, and throat irritation • congestion • sneezing	• coughing • asthma flare-ups
Infectious agents such as • bacteria (for example, *Legionella* and *pneumococcus*) • viruses (for example, influenza and cold virus) • spores from moulds and other fungi		• upper respiratory infections • throat and ear infections • bronchitis	• pneumonia • sinusitis

Most of the pollutants that cause SBS are released within the buildings themselves. For instance, high levels of VOCs are often found in buildings linked to SBS, especially in those that are new or have been remodelled. Over 3000 building products, including particleboard and insulation, release formaldehyde. Other products, such as carpets and upholstery, are treated with chemicals that resist bacteria, mould, and stains. Many of these products emit VOCs. In older buildings, mould often plays a key role in SBS.

Outdoor air pollution can also contribute to SBS when it enters a building and becomes concentrated over time. For example, motor vehicle exhaust can enter buildings through windows or air intake vents. Pollutants, such as carbon monoxide, nitrogen dioxide, and VOCs, can quickly build up to high levels that harm human health.

Sick building syndrome has sparked education and awareness campaigns in various workplaces. Educating people who live and work in these buildings, as well as people who own and maintain them, is important. It is the first step in preventing SBS and solving indoor air quality problems where they already exist.

Mini-Activity 9-3 Education Campaign

Use the information about indoor air pollutants and sick building syndrome to create an information campaign to educate people about SBS. At a minimum, your campaign must answer the following questions:

• What are the main causes of sick building syndrome?
• Which people are especially vulnerable to the health effects of indoor air pollution, and why?

• How can green building techniques and other methods help reduce the effects of sick building syndrome?

You may use any medium to complete your campaign. Ideas include creating a pamphlet, video documentary, podcast, or computer presentation.

Case Study   Ozone Destruction and Health

The ozone layer is a region in the lower stratosphere, about 20 to 30 km above Earth. Here, ozone is destroyed as it absorbs UV radiation. As a result, the ozone layer naturally prevents UV radiation from reaching organisms on Earth. While some UV does reach the surface, specific wavelengths known as UV-B are largely screened out. UV-B is harmful to most living things and is the main cause of sunburn. It can also damage eyes and cause skin cancers—in humans as well as in other animals such as sheep, cows, and surfacing ocean mammals such as orcas.

Some substances that humans release into the atmosphere destroy ozone. Those that contain chlorine (Cl) and bromine (Br) are two of the most destructive. Human-made compounds called *halocarbons* increase atmospheric concentrations of Cl and Br. Halocarbons rise to the stratosphere, where UV radiation breaks them down to yield Cl or Br atoms. Because these atoms persist for a long time in the stratosphere, each atom can break down over 100 000 ozone molecules.

In the late 1970s, scientists began to observe a decrease in stratospheric ozone levels of about 4% per decade. As the density of the ozone layer decreases, more UV-B radiation gets through. Above the poles, especially Antarctica, there is so much seasonal decline that an ozone hole is often created. The first hole was discovered over Antarctica in 1985. Holes have also occurred over the Arctic.

Government Initiatives to Combat Ozone Depletion

In 1978, CFCs in aerosol sprays were banned in Canada, the United States, and Scandinavia. As research revealed the severity of the damage to the ozone layer, many other countries acted to limit CFCs. As a result, the atmospheric concentration of chlorine has been declining since it peaked in 1994. A 2005 study showed that, on average, global ozone depletion has stabilized.

Threats to the ozone layer still exist, however. CFCs and other ozone-depleting substances continue to be used in countries where bans are not yet in place. These substances are sometimes smuggled into Canada and the United States. In addition, many of the substances invented to replace CFCs are greenhouse gases and contribute to climate change.

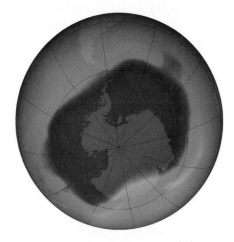

The largest ozone hole, with an average size of about 27 million km², was recorded over Antarctica in September 2006.

Research and Analyze

1. Research Canada's regulations for CFCs and other ozone-depleting substances under the Canadian Environmental Protection Act (CEPA). How are regulations enforced and how are alternative technologies encouraged? Assess the success these regulations have had in reducing damage to the ozone layer.

2. Since CFCs have been banned in North America, some people consider the ozone hole to be "old news." Why is this view not supported by the scientific community? (Hint: Find out how long it would take the ozone hole over Antarctica to be restored if all CFCs were immediately banned.)

3. Find out more about the dangerous effects of UV radiation to plants, microscopic organisms, humans, and other animals.

Communicate

4. Create a plan to protect yourself against UV rays. How could you limit your exposure and minimize risk when you are exposed? Communicate your plan to the class.

Case Study Carried in Air: Noise Pollution

Whether or not a sound is pleasant is a matter of personal taste, of course. Generally, people who find certain sounds unpleasant and unwanted tend to call such sounds noise. Noise, however, can be more than just unpleasant. Exposure to noise can cause physical and psychological harm. The loudness of a sound is measured in units called decibels (db). The frequency or pitch of a sound is also a factor in determining its degree of harm. The most common scale used for high-pitched sounds is the A scale, whose units are written as dbA. This scale describes the amount of pressure that sound energy exerts on the eardrum. Hearing loss begins with prolonged exposure (eight hours or more per day) to 80 to 90 dbA levels of sound pressure. Sound pressure becomes painful at about 140 dbA, and it can kill at 180 dbA.

In addition to hearing loss, noise pollution is linked to a variety of other ailments ranging from nervous tension headaches to psychological problems. Research also has shown that noise may cause blood vessels to constrict (which reduces blood flow through the body), disturbs unborn children, and sometimes causes seizures in people who have epilepsy. Many cities and towns try to protect their citizens by passing laws aimed at regulating noise levels.

Intensity of Noise (Sound)

Source of Sound	Intensity in Decibels (db)
Jet aircraft at takeoff	145
Hydraulic press	130
Jet airplane (160 metres overhead)	120
Unmuffled motorcycle	110
Subway train	100
Farm tractor	98
Gasoline lawn mower	96
Food blender	93
Heavy city traffic	90
Vacuum cleaner	85
Normal speech	60

Research and Analyze

1. Gordon Hempton is an audio ecologist. He travels the world in search of places that are completely untouched by the sounds of technology. Such places are surprisingly few, but one of them is found here in Canada: Grasslands National Park in Saskatchewan. Find out more about Gordon Hempton, the work he does, and his views on noise, sound, and silence.

Communicate

2. Cases can be made for and against considering noise pollution to be a form of air pollution. Make an argument either in favour of or against classifying noise pollution as air pollution. Justify your opinion.

3. Do you think noise pollution should be governed in the same way as other types of pollution? Hold a debate about this question. To prepare, you might want or need to do some research about the laws (federal, provincial, and municipal) that govern the control of noise.

Summary

- Pollutants can enter the body through inhalation, ingestion, or absorption through the skin.
- A point source pollutant has a definite source, while a nonpoint source of pollution is more difficult to identify because it results from many sources.
- Primary air pollutants are released directly into the air and include carbon monoxide, nitric oxides, sulfur oxides, particulate matter, and volatile organic compounds (VOCs).

- Secondary air pollutants result from interactions of primary air pollutants. Photochemical smog is an example of a secondary air pollutant.
- Some outdoor air pollutants can also act as indoor pollutants. There are also indoor-specific air pollutants that can be hazardous when their levels build up. Examples include radon, moulds, and asbestos.
- Sick building syndrome refers to health issues that are due to the poor air quality inside a building.

Review Questions

1. In your own words, describe what a pollutant is. Give an example and a non-example. K/U C

2. Pollutants can enter the body of an organism in different ways.
   a) Describe three ways they can enter the body.
   b) Give an example of a common activity or process for each way a pollutant can enter the body. K/U

3. Explain why all pollutants are contaminants, but not all contaminants are pollutants. T/I C

4. The photo below shows a source of pollution.
   a) Identify it as a point source, nonpoint source, or both. Explain your answer.
   b) Describe an example of the other two sources of pollution. T/I A

5. Copy the table below into your notebook and complete it based on the headings provided. K/U

Pollutants

Primary Air Pollutant	Sources	Effects It Has on Human Health
Nitrogen oxides (NO$_x$)		
Sulfur oxides (SO$_x$)		
Particulate matter (PM)		

6. Give an example of a primary air pollutant not listed in the above table. K/U

7. What is photochemical smog? Why is it called a secondary air pollutant? K/U T/I

8. It is recommended that people have carbon monoxide detectors, such as the one shown below, in their homes. Why do you think these are important to have? Provide at least two reasons. C A

9. Pollutants can occur indoors.
   a) Name and describe three examples of indoor air pollutants.
   b) For each of your examples in part (a), describe the effects the pollutant has on human health.
   c) How are indoor air pollutants related to sick building syndrome?
   d) Suggest one way that indoor air pollution can be reduced in your home. K/U T/I A

10. The Blue Ridge Mountains in Georgia get their name from the blue-coloured gases given off by plants that grow there. In the Northwest Territories, the Smoking Hills are rich in rocks made up of carbon and sulfur compounds that can spontaneously ignite, releasing the smoke that gives them their name.
    a) What other natural sources of air pollution are there?
    b) Should people be as concerned about these natural sources as they are about anthropogenic sources? Give reasons to justify your opinion. T/I C A

Improving Air Quality

Government and Air Quality

InquiryLab 9A, Trends in Air Pollutants in the Past 25 Years, on page 298

The effects of poor air quality on human health and the environment are well documented and widespread. Higher levels of air pollutants, as well as long-term exposure to them may have a detrimental effect on health. People who are sick, infants, children, and seniors are most vulnerable to pollution-related illness and disease. Many health problems are linked to poor air quality. Examples of these health problems include asthma, allergies, lung disease, and heart disease. In addition, there is growing concern about the possible links between air pollutants and environmental events such as extreme weather. These events pose a risk to many of the same people—seniors especially—who are vulnerable to illness and disease brought on by poor air quality.

Air Quality Health Index (AQHI) a system to indicate air quality and its associated health risks

To help people understand and make decisions related to the health effects of air quality, Environment Canada, with the provinces and territories, has devised the **Air Quality Health Index (AQHI)**. The AQHI is based on the health risks linked to common air pollutants, including ozone and particulate matter. **Figure 9.5** shows that the AQHI uses numbers from 1 to 10+ and a rating scale of "low" to "very high" to indicate the air quality and its associated health risks. Each AQHI report for a selected location indicates who may be at risk on any given day or time. The reports also give health messages and a short-term forecast to help people who are planning outdoor activities. These reports are available through a variety of media outlets, including the Internet, radio, TV, and newspapers.

Figure 9.5 The AQHI is based on the known health risks that are associated with common air pollutants that include ground level ozone, particulate matter, and nitrogen dioxide.

Applying What other air quality health index does your province use? How does it work, and how does it compare with the AQHI?

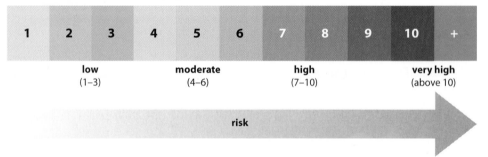

In Canada, the management of air quality is mainly the responsibility of the provincial and territorial governments, and it is overseen by the federal government. The main federal environmental law is the Canadian Environmental Protection Act, 1999, which supports this role. The law addresses the government's responsibility to work with the provinces and territories on ways to prevent pollution and to create national environmental standards. One such standard focusses on particulate matter and ground level ozone. It commits governments to meet defined pollution reduction targets. Governments have passed regulations and set up programs to reduce harmful gas emissions from all sectors of the economy. Governments also encourage individuals, businesses, and industries to join voluntary programs to achieve this goal.

Pause and Reflect

4. What is the AQHI, and how does it help people make decisions related to the health effects of air quality?

5. Critical Thinking Why are the very young and the very old especially vulnerable to pollution-related illness and disease?

Technology, Legislation, and Emissions Reduction

ThoughtLab 9B, Canada and Air Quality, on page 301

Laws enacted by governments can require industries to limit emissions of air pollutants from their manufacturing plants and from their products, such as motor vehicles and engines. Industries also play an important role through the invention of technologies and techniques to capture pollutants when they are created, before they enter the air. Described on the next three pages are examples of how emissions have been reduced from three major sources of air pollution: motor vehicles, particulate matter, and power plants.

Motor Vehicle Emissions

Motor vehicles release more carbon monoxide, nitrogen dioxide, and VOCs than any other activity or industry. Government legislation plays an important role in keeping these emissions in check. Provincial programs such as Drive Clean in Ontario and Air Care in British Columbia set emission standards that motor vehicles must meet in order to be legally driven on the road. Regulating VOC and nitrogen dioxide emissions plays an important role in reducing smog in large Canadian cities. This is especially important, because car travel within cities has increased over the past few decades.

Many engineering changes in cars and trucks have reduced the amount of VOCs that escapes from the gas tank and crankcase. Modifications to the pumps at gas stations and to the filler pipes of cars have also helped. Better fuel efficiency and specially blended fuels that produce less carbon monoxide and unburned compounds have also played a role. Catalytic converters, shown in **Figure 9.6**, help reduce emissions. These devices use materials called catalysts that convert potential pollutants into harmless substances before they enter the air.

Figure 9.6 Catalytic converters work best when the catalysts are warmed up, which takes about 90 seconds after a car starts.

Inferring How effective is the converter during the 90 second warm-up? Why?

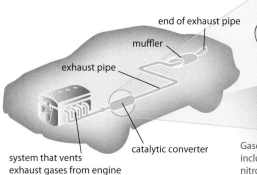

Gases from tail pipe: water vapour (H_2O), carbon dioxide (CO_2), and nitrogen (N_2)

heat shield

thin layer of precious metals, including platinum and palladium

end of exhaust pipe

muffler

exhaust pipe

system that vents exhaust gases from engine

catalytic converter

Gases into catalytic converter include carbon monoxide (CO), nitrogen oxides (NO_x), and hydrogen (H_2).

Mini-Activity 9-4   Effectiveness of Regulations

Governments use a variety of methods to reduce motor vehicle emissions. Consider these three approaches:

- Drivers are denied a privilege as an incentive for them to reduce motor vehicle emissions. Programs such as Drive Clean and Air Care require that cars pass an emission standards test to remain road legal.

- Citizens are rewarded for reducing motor vehicle emissions. People who carpool can take the less congested HOV (high occupancy vehicle) lanes to get to and from work or school.

- People pay a fee when they make choices that increase motor vehicle emissions. For example, taxes are applied on motor vehicle fuel purchases across Canada.

In small groups, discuss which method you think is most effective in reducing motor vehicle emissions and why. Consider both short-term and long-term effectiveness. Record your ideas and share your opinions with your class.

Particulate Matter Emissions

In 2012, federal, provincial, and territorial governments joined forces to set new air quality guidelines. They created the Canadian Ambient Air Quality Standards for particulates and most outdoor air pollutants. These guidelines are becoming stricter over time to meet air quality improvement goals. Industrial emission requirements help make sure that emissions from Canadian industry meet targets in line with these goals.

Industries, meanwhile, have been working for many decades to reduce the amount of particulate matter that results from their activities. Various devices can be used to trap particles so they do not escape from smokestacks. These include the following:

- Industrial filters trap fine particulates and remove them from the air. The filters look and function a lot like the bags found in vacuum cleaners, but on a much larger scale. Up to 15 m long and 3 m wide, the bags are made of cotton, asbestos, cellulose, or glass fibres. These porous materials allow air to pass through while particulates are trapped.

- **Electrostatic precipitators**, shown in **Figure 9.7**, are more expensive to operate than filters, but they are extremely effective. As a result, these devices are the most commonly used fine particulate controls. Particulates pick up an electrostatic charge as they rise in a smokestack. This is very similar to how clothes pick up a static charge in a dryer. The charged particulates are attracted to a plate that carries an opposite charge. Electrostatic precipitators stop up to 99% of particulates from entering the air. The tonnes of collected material may be treated as solid waste or hazardous solid waste and disposed of accordingly in appropriate landfills.

electrostatic precipitator a device that uses electric charge to remove particulates from industrial emissions

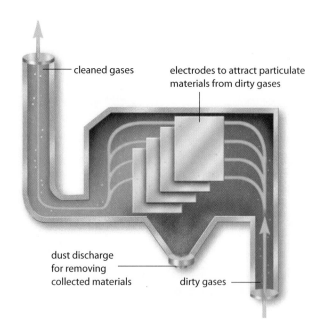

Figure 9.7 Electrostatic precipitators use static charge to remove fine particulates from industrial emissions.

Applying *What other technologies can you think of that use static charge to reduce pollution?*

cleaned gases

electrodes to attract particulate materials from dirty gases

dust discharge for removing collected materials

dirty gases

 **Mini-Activity 9-5**  **Model an Electrostatic Precipitator**

Note: Your teacher may choose to do this activity as a demo with volunteers to assist.

Charged plates in an electrostatic precipitator attract particulates the same way a statically charged piece of clothing collects lint. Brainstorm how you could use common objects to create a model of an electrostatic precipitator. Make sure any electrical equipment you use is CSA-approved or approved by the electrical safety authority (ESA). After you check your plan with your teacher, build your model.

Power Plant Emissions

The main pollutants associated with electrical power plants are particulates, sulfur dioxide, and nitrogen oxides. Most particulate emissions are controlled with filters and precipitators. The control of sulfur dioxide requires more fundamental changes to the way electricity is produced. In the United States, the Environmental Protection Agency has approached the problem by setting limits and allowing electric utilities to decide which options are the best for them.

In Canada, provinces such as Ontario and Nova Scotia have committed to phasing out all coal-burning power plants by setting target dates for increasing the use of other energy sources for generating electricity. These include wind, natural gas, and nuclear energy. Each of these carries its own set of advantages and disadvantages.

Electric utilities in North America that still rely on coal have used several strategies for reducing their emissions. Switching from a high-sulfur to a low-sulfur coal reduces the amount of sulfur dioxide released into the atmosphere by as much as 66%. Chemical or physical treatment of coal to remove sulfur before it is burned can remove nearly 40% of the sulfur. Using **scrubbers** is a third alternative (see **Figure 9.8**). The technology for these controls is costly to install, maintain, and operate.

Nitrogen oxides are produced wherever high-temperature combustion occurs with air. Control of the release of nitrogen oxides often involves removing the compounds from the stack with chemical processes or using catalysts to encourage the breakdown of nitrogen oxides into nitrogen and oxygen. These processes are complex and expensive. However, emissions of particulates, sulfur dioxide, and nitrogen oxides from coal-burning power plants have been reduced greatly. In some cases, this has resulted in increased costs to consumers, since the cost of improving emission control technology is passed along as a rate increase to the consumer.

nonrenewable and renewable energy— see Section 3.1 on page 58

scrubber a device that moves gases through a liquid spray that reacts with pollutant chemical compounds to remove them from industrial emissions

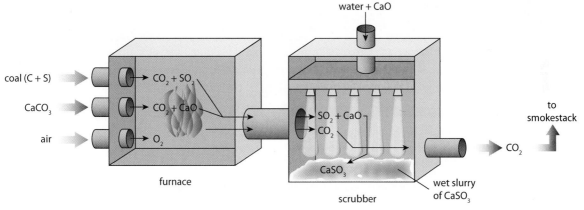

Figure 9.8 Scrubbers are effective for removing oxides (such as nitrogen oxides and sulfur oxides). Shown here, the compound calcium oxide (CaO) reacts with sulfur dioxide to form a solid compound, calcium sulfite ($CaSO_3$), which can be washed away with water. This solid must be disposed of, however—often in landfills.

Pause and Reflect

6. Explain how government regulations help reduce industrial emissions.

7. Describe one way that technology can reduce sulfur dioxide emissions from coal-burning power plants.

8. Critical Thinking The materials collected by industrial filters and electrostatic precipitators must be buried in landfills as solid waste. What new problems might this cause?

Improving Air Quality through Personal Actions

The choices you make and the actions you take each day build on one another over time. A lifetime spent making choices that promote clean air can have a major impact on both indoor and outdoor air quality. **Figure 9.9** suggests just a few of the ways that you can reduce your personal emissions and exposure to air pollution.

Get involved in activities that promote clean air. Plant a tree or join a community garden. Walk, bike, bus, or carpool to reduce your personal emissions.

Write a letter to the editor of your local paper or an elected official in support of air quality improvement initiatives where you live.

Improve ventilation in your home and school by opening doors and windows. This allows indoor air pollutants to escape and keeps mould-friendly moisture from building up.

VOC levels are typically five times higher indoors than outdoors. Purchase low-VOC products whenever possible to reduce these pollutants. Low-emission paints, cleaning products, and adhesives are widely sold.

Figure 9.9 Actions you take each day can reduce your personal emissions. They can also reduce your exposure to pre-existing air pollution.

Applying *Brainstorm at least five other ways that you can reduce your personal emissions of and exposure to air pollution.*

Mini-Activity 9-6   Make a Filtration Mask

Imagine that you are a bicycle courier in a large Canadian city. Wearing an air pollution filtration face mask can help reduce your exposure to air pollutants as you cycle along busy streets. Design and create your own face mask filter with the materials provided by your teacher. (Do not share masks with others.) Your face mask must have the following characteristics:

- covers your nose and mouth
- allows you to breathe normally
- prevents air pollution from entering your respiratory system as best as possible
- is durable and comfortable enough to wear all day
- works when wet (on a rainy day)

Air Quality and Grassroots Initiatives

Another way to help improve air quality is to join a *grassroots initiative* that reduces air pollution. A grassroots initiative like those shown in **Figure 9.10** involves people working together at a local level to bring about positive change. A few types of initiatives are described below. Notice that many of these help to improve air quality indirectly. Some do this by promoting activities that you may not automatically link to air quality.

Walking and Cycling
Many grassroots initiatives promote walking and cycling. Some groups provide cycling skills workshops and repairs, refurbish used bikes, or provide free bikes for use in urban centres. Others advocate safe cycling, installing more bike lanes, and constructing pedestrian paths. Families in the same neighbourhood organize "walking school buses" that reduce the need to drive children to school.

Car Sharing
Car sharing is popular in Europe, as this example from Germany shows. Many communities in Canada have car sharing programs that were started by grassroots organizations. Members join the program and borrow cars only when they need them. The rest of the time, they walk, cycle, or use alternative transportation.

Focus on Plants
Many grassroots initiatives focus on growing, maintaining, and protecting plants. Plants play an important role in improving air quality, especially in urban centres. Some groups work to protect urban forests and woodlots. Others plant trees to promote clean air, or attend political events to advocate for more trees in urban settings. Even groups that maintain community gardens help improve local air quality.

Figure 9.10 Three examples of grassroots initiatives.

Applying *What other examples of grassroots initiatives have you heard of, and what have they accomplished?*

 Mini-Activity 9-7   **Who Should Have a Say?**

Your city council is considering a proposal to install more bike lanes on a busy city street that has many small businesses. The plan requires the loss of either one lane of traffic for cars or one lane for parking spaces. A grassroots group is being organized to make sure that citizens can raise any concerns they have. The group will be made up of the stakeholders who will be affected if the bike lane is built.

• As a class, come up with six or more community members you think have a stake in this issue.

• Your group will be assigned one of these stakeholders. Consider the position these stakeholders will take on installing more bike lanes, based on their personal and professional concerns.

• Share your stakeholder's opinion with your class.

• After hearing each opinion, decide whether you think this process was an effective way to make a decision that will affect the air quality of a community.

Case Study Beijing's Polluted Air

On January 15, 2013, a British media outlet, *The Guardian*, published the graphic reproduced here. At the time, the city of Beijing was experiencing several days of severe air pollution due to photochemical smog.

WHAT IS IN BEIJING'S AIR?

The two different US and China **Air Quality Indexes** (AQI) both measure, and then rate, **particulate matter.** They are indentical scales above 200, and only go up to 500.
[Hazard level provided is from the US index]

755 ● **Jan. 12:** Beijing's AQI (recorded by the US embassy), soared to **755**, with levels 30 times those recommended by WHO for a 24-hour period.

BEYOND THE SCALE

500+

301-500 **HAZARDOUS**
In the US, 300+ triggers emergency conditions. The entire population is more likely to be affected.

201-300 **VERY UNHEALTHY**

151-200 **UNHEALTHY**

101-150 **UNHEALTHY FOR SENSITIVE GROUPS**

51-100
US cities normally have AQI ratings well below 100.

0-50

PARTICULATE MATTER

Tiny particles less than **2.5 micrometers** (PM2.5) in diameter and smaller—1/30th of a human hair—can penetrate lungs and lead to increased risk of heart and lung cancer deaths. PM includes acids, metals, and allergens. In December 2012, Chinese officials vowed to cut PM2.5 5% each year in 13 major areas through 2015, according to *China Daily.*

NITROGEN OXIDES

Nitrogen oxides (NOx) form when fuel is burned at high temps, such as in car engines and electric utilities. Beijing is ringed by coal-fired power plants. Dangers include emphysema, bronchitis and acid rain. In the NO family is **nitrous oxide** (N2O), a nasty **greenhouse gas.**

CARBON MONOXIDE

Carbon monoxide (CO) is a colorless, odorless gas emitted from combustion. At extremely high levels, CO can cause death. Exposure to CO can reduce the oxygen-carrying capacity of the blood. As with other pollutants, many blame China's reliance on coal and explosive growth in car ownership.

SULFUR DIOXIDE

Sulfur dioxide (SO2) is produced by burning fossil fuels or during mining. Dangers include increased asthma symptoms. Studies also show increased visits to hospitals for respiratory illnesses, particularly amongst children and the elderly. Beijing hospitals are already showing a spike, according to *Xinhua.*

OZONE

Ozone is the main part of smog, formed when NOx and **volatile organic compounds** (VOCs) react in the presence of heat and sunlight. Dangers include chest pain, coughing, throat irritation, and congestion. It can worsen bronchitis, emphysema, and asthma. Repeated exposure may permanently scar lung tissue.

Sources: *China Daily, Xinhua,* EPA
Produced by: James West

CLIMATE DESK

Communicate

1. How are the U.S. and China Air Quality Indexes different from Canada's AQHI?

2. Why is air quality a problem in Beijing?

3. How does this graphic add to your understanding of air quality issues?

Summary

- Environment Canada and provincial and territorial agencies developed the Air Quality Health Index (AQHI). The AQHI uses a rating scale that indicates the quality of air in a given region.

- In Canada, air quality is managed by provincial and territorial agencies and regulated by the federal government. Laws have been enacted to help reduce emissions of air pollutants by major sources such as motor vehicles, particulate matter, and power plants.

- Motor vehicle emissions have been reduced through programs that require testing of vehicles, improvements in the quality of fuel for the vehicles, and the development of technologies, such as catalytic converters.

- Industries have worked to reduce emissions of particulate matter by using technologies such as filter traps and electrostatic precipitators.

- Reducing coal-burning power plant emissions has involved developing methods to reduce emissions of particulate matter, nitrogen oxides, and sulfur dioxide.

- There are many simple acts that people can do to reduce their exposure to poor quality air, as well as grassroots initiative groups that work to help improve air quality.

Review Questions

1. What is the Air Quality Health Index? Why is reporting it for a given region important for the people who live there? K/U T/I A

2. In a short paragraph, describe the roles that provincial and federal governments play in managing air quality. Do you think these actions have had a positive effect? Provide at least two reasons that support your opinion. C T/I

3. Describe how a catalytic converter helps to reduce emissions of air pollutants from motor vehicles. T/I A

4. Name and describe two devices that industry uses to help reduce the amount of particulate matter that is released into the air. K/U

5. Design a one-page information bulletin that discusses the contribution of coal-burning power plants to poor air quality in Canada. Your bulletin should include at least
   - a figure or table
   - information on the air pollutants released by these power plants and the harm to human health they cause
   - measures the power plants have taken to help reduce emissions
   - one or two suggested alternatives for generating power that are possible in Canada K/U C A

6. Describe three things you did in the last week that helped reduce your exposure to air pollution. K/U A

7. How do urban forests contribute to improving air quality? A

8. Sudbury's Inco Superstack is the tallest smokestack in North America and the second tallest in the world. The nickel smelter's 380 m chimney disperses pollutants, including fine particulates, high into the atmosphere. Identify one positive and one negative aspect of dispersing pollution in this manner. C A

9. Many grassroots initiatives have been developed to address air pollution in our society.

   a) Describe a grassroots initiative that helps reduce air pollution.

   b) Do you think this is an effective way of helping reduce air pollution? Support your opinion with at least one reason. K/U C T/I

10. A significant factor that contributes to the success of many grassroots initiatives is to get people to change habits or behaviours that directly or indirectly support air pollution. Suggest one way that could be effective in getting people to change a habit that contributes to air pollution. Explain why you think that approach would be effective. A T/I

Skill Check

Initiating and Planning

Performing and Recording

Analyzing and Interpreting

Communicating

Materials

- computer graphing program or graph paper
- Table 1 on page 299 and Table 2 on page 300

Trends in Air Pollutants in the Past 25 Years

Industries must report their air emissions to Environment Canada's National Pollutant Release Inventory (NPRI). Summaries of these data are published yearly to inform and to help people assess how effective regulations are.

Pre-Lab Questions

1. Name five primary air pollutants.

2. How many kilograms are equal to one tonne (1 t)?

3. When is it appropriate to round off values you plan to graph?

Question

How have emissions of certain air pollutants changed since 1985?

Procedure

1. Graph the data in **Table 1** to show changes in SO_x and CO emissions between 1985 and 2010 from three sources.

2. Discuss how to graph the data. Use the Analyze and Interpret questions to know what type of analysis you are to do, and use these questions:
- What variable will you plot on the x-axis and the y-axis? How will you label them?
- How many different line graphs do you need to draw?
- How can using different colours help you compare the data?

3. Graph the data.

4. **Table 2** shows emissions from individual non-industrial SO_x sources. Decide how to graph the data. Then graph them.

Analyze and Interpret

1. Use these questions to summarize the trends in SO_x and CO emissions.
- From 1985 to 2010, did emissions from industrial sources increase or decrease? What about emissions from non-industrial and transportation sources?
- What is the overall percentage of the increase or decrease in emissions for each pollutant for the time period? (For each pollutant, divide the emission value from 2010 by the emission value from 1985, and then multiply by 100%. For example, for SO_x emissions from industrial sources,
$$\frac{889\ 492}{2\ 670\ 956} \times 100\% = 33.3024\%$$
Therefore, the value in 2010 is approximately 33% of the value in 1985; there has been a 67% decrease in emissions).
- Has there been a greater change in emissions levels for either of the air pollutants? How do you know?
- What other trends can you describe?

2. Summarize the trends in the graph for the data from **Table 2**.

Conclude and Communicate

3. Based on your knowledge of the sources of sulfur dioxide, what can you infer about the sources based on the trends that the data show?

4. What do the data in **Table 2** indicate about the value of looking at contributions of individual sectors? How can this improve guidelines for emissions and properly target the greatest contributors of emissions?

5. Why must managing and reducing air pollution be a global undertaking? Hint: Think about Canada and industries and populations that are close to our borders.

Table 1 Air Pollutant Emissions in Tonnes from 1985 and 2010					
	Industrial Sources*		Non-industrial Sources**		Transportation Sources***
Year	SO$_X$	CO	SO$_X$	CO	SO$_X$	CO
1985	2 670 956	1 868 167	879 113	756 651	172 873	12 546 102
1986	2 324 568	1 836 215	835 034	590 932	165 485	12 667 647
1987	2 594 273	1 904 930	848 253	754 891	167 576	12 506 450
1988	2 589 927	1 924 621	960 738	767 859	183 586	12 560 160
1989	2 435 483	1 909 142	736 651	590 886	190 546	12 419 619
1990	2 269 455	1 970 671	746 446	744 702	181 828	11 741 517
1991	2 415 987	2 013 777	754 507	738 850	162 530	11 469 668
1992	2 189 546	1 978 572	728 166	704 119	165 974	11 512 782
1993	1 658 281	1 988 868	597 837	692 765	173 506	11 129 199
1994	1 629 918	1 995 913	592 311	699 306	168 180	10 849 150
1995	1 763 115	1 985 876	566 465	680 468	150 472	10 273 540
1996	1 659 663	1 953 086	589 290	673 980	138 813	9 758 388
1997	1 628 735	1 935 772	577 180	535 863	143 663	9 458 444
1998	1 603 909	1 869 648	586 933	533 487	125 618	8 859 726
1999	1 576 521	1 888 913	593 817	668 591	124 142	8 843 674
2000	1 524 723	1 637 306	669 097	676 115	122 446	8 868 144
2001	1 571 906	1 623 006	665 309	694 947	123 023	8 508 379
2002	1 534 442	1 566 840	653 243	782 980	114 798	8 277 583
2003	1 456 912	1 705 587	688 308	767 126	112 128	7 998 140
2004	1 466 876	1 402 611	634 993	741 185	109 413	7 381 280
2005	1 418 443	1 556 510	576 690	776 508	110 409	7 096 585
2006	1 353 695	1 465 883	506 308	779 635	107 537	6 964 414
2007	1 259 014	1 488 994	539 975	775 947	101 530	6 852 157
2008	1 160 440	1 483 994	476 966	764 246	93 285	6 746 884
2009	948 343	1 425 090	433 052	764 570	95 355	6 605 699
2010	889 492	1 420 652	382 857	773 084	94 737	6 514 674

* Examples of industrial sources include aluminum industry, asbestos industry, cement and concrete industry, chemicals industry, smelting and refining industry, pulp and paper industry, and petroleum industry.
** Examples of non-industrial sources include commercial fuel combustion, residential fuel combustion, and power utilities.
*** Examples of transportation sources include air travel, railways, light-duty vehicles, heavy-duty vehicles, and marine vessels.

Continued on next page >

Table 2 Non-industrial SO$_x$ Emissions in Tonnes from 1985 and 2010				
Year	Commercial Fuel Combustion	Electric Power Generation	Residential Fuel Combustion	Residential Fuel Wood Combustion
1985	23 048	817 728	36 936	1401
1986	20 486	776 813	36 676	1410
1987	15 472	801 670	29 701	1410
1988	17 263	911 965	30 082	1428
1989	16 154	690 921	28 531	1045
1990	17 353	696 491	31 216	1386
1991	16 024	708 704	28 413	1366
1992	16 131	681 770	28 966	1299
1993	15 706	551 323	29 445	1363
1994	15 135	549 045	26 756	1375
1995	8 329	539 409	17 393	1334
1996	10 620	550 790	26 565	1315
1997	11 637	539 529	24 986	1028
1998	12 805	552 356	20 743	1029
1999	12 476	558 956	21 084	1310
2000	14 792	639 777	13 213	1315
2001	22 150	627 608	14 169	1382
2002	20 830	619 906	10 961	1546
2003	46 410	629 443	10 988	1467
2004	43 202	579 371	10 988	1432
2005	46 282	518 028	10 884	1496
2006	35 459	459 886	9 446	1516
2007	35 267	493 072	10 137	1499
2008	35 081	430 612	9 791	1483
2009	39 383	384 897	7 307	1466
2010	38 738	335 363	7 307	1449

Materials

- reference books and/or computer with Internet access

Canada and Air Quality

According to the Canadian Medical Association, more than 20 000 people die prematurely each year due to air pollution. By 2031, the estimated costs associated with the health effects of air pollution will be more than 250 billion dollars. Yet, on a global scale, Canada's air quality ranks much higher than that of other countries. According to the United Nations, urban air pollution is linked to up to 1 million premature deaths each year. In this Lab, you will consider Canada's role in improving global air quality.

Pre-Lab Questions

1. How can you assess Canada's efforts to improve air quality?

2. How will you analyze information sources for accuracy, reliability, and bias?

Question

What is Canada doing to improve air quality, and should we be doing more?

Procedure

1. Research Canada's involvement in the following protocols or agreements that are aimed at improving global air quality.
   - Montreal protocol
   - Kyoto protocol
   - Climate and Clean Air Coalition
   - Convention on Long-range Transboundary Air Pollution
   - Canada–United States Air Quality Agreement

2. As you research, analyze your sources for accuracy, reliability, and bias. Determine how you will organize the information you find.

3. Use the following questions to help guide your research.
   - What are the aims of each protocol or agreement?
   - What is Canada's involvement in them?
   - How has Canada's agreement or involvement changed environmental policies within our country? What has Canada agreed to do as part of each agreement?
   - Are the agreements enforced in some way, or is compliance voluntary for the countries who have agreed to support it?
   - If Canada is not involved in an agreement, determine why not.

4. Present your findings in a format approved by your teacher.

Analyze and Interpret

1. Develop a grading system to assess Canada's role in improving global air quality. Explain the reasoning behind your grading system.

Conclude and Communicate

2. Explain the grade you gave Canada. If you think we can improve, suggest one improvement and explain why you think it would help.

3. Why is international co-ordination important for dealing with environmental issues such as air pollution?

4. Describe one aspect of international environmental agreements that you think is needed in order for them to be successful. What do you think are some barriers to implementing these international agreements?

Chapter 9 SUMMARY

Section 9.1 Air Quality Outdoors and Indoors

Air pollutants can be released into the air from point sources and nonpoint sources of pollution. Primary air pollutants can interact to produce secondary air pollutants, such as smog. Poor air quality can occur indoors and outdoors.

Key Terms
contaminant
pollutant
anthropogenic
primary air pollutants
secondary air pollutants
sick building syndrome

Key Concepts
- Pollutants can enter the body through inhalation, ingestion, or absorption through the skin.

- A point source pollutant has a definite source, while a nonpoint source of pollution is more difficult to identify because it results from many sources.

- Primary air pollutants are released directly into the air and include carbon monoxide, nitric oxides, sulfur oxides, particulate matter, and volatile organic compounds (VOCs).

- Secondary air pollutants result from interactions of primary air pollutants. Photochemical smog is an example of a secondary air pollutant.

- Some outdoor air pollutants can also act as indoor pollutants. There are also indoor-specific air pollutants, such as radon, moulds, and asbestos, that can be hazardous when the levels build up.

- Sick building syndrome refers to health issues that are due to the poor air quality inside a building.

Section 9.2 Improving Air Quality

Air quality is monitored and reported through programs administered at all levels of government. Government regulations have placed limits on emissions of air pollutants. This has resulted in industries developing new technologies to lower emissions of pollutants, as well as making products that pollute less. Individuals and groups of like-minded people working together also contribute to improving air quality.

Key Terms
Air Quality Health Index (AQHI)
electrostatic precipitator
scrubber

Key Concepts
- Environment Canada and provincial and territorial agencies developed the Air Quality Health Index (AQHI). The AQHI uses a rating scale that indicates the quality of air in a given region.

- In Canada, air quality is managed by provincial and territorial agencies and regulated by the federal government. Laws have been enacted to help reduce emissions of air pollutants by major sources such as motor vehicles, particulate matter, and power plants.

- Motor vehicle emissions have been reduced through programs that require testing of vehicles, improvements in the quality of fuel for the vehicles, and the development of technologies, such as catalytic converters.

- Industries have worked to reduce emissions of particulate matter by using technologies such as filter traps and electrostatic precipitators.

- Reducing coal-burning power plant emissions has involved developing methods to reduce emissions of particulate matter, nitrogen oxides, and sulfur dioxide.

- There are many simple acts that people can do to reduce their exposure to poor quality air, as well as grassroots initiative groups that work to help improve air quality.

Chapter 9   REVIEW

Knowledge and Understanding

Choose the letter of the best answer below.

1. Which is the best example of a contaminant?
   a) smoke that is inhaled in a forest fire
   b) DDT that was sprayed on fruit that was not washed before being eaten
   c) asbestos in a home renovation project that was removed without using respiratory system protection
   d) a few grains of sand that stick to an ice cream cone dropped on the ground
   e) carbon monoxide emitted by a propane heater operated in a home without ventilation

2. The Air Quality Health Index (AQHI)
   a) is a joint venture between the Canadian and American federal environment agencies
   b) includes a rating scale of "low" to "very high" to indicate air quality and its associated risks
   c) is administered by the provinces and territories, but not by the federal government
   d) measures the health effects of only secondary air pollutants
   e) is part of a grassroots initiative to safeguard health

3. Which one of the following statements is *incorrect*?
   a) Anthropogenic pollutants result from human activities.
   b) A point source pollutant is easily identified, since it enters an ecosystem from a definite source.
   c) All air pollutants result from human activities.
   d) VOCs (volatile organic compounds) can cause headaches or dizziness.
   e) Many chemicals in second-hand smoke are known to cause cancer.

4. When some fuels are burned, they can release a colourless, odourless gas that can be deadly with prolonged exposure. This gas attaches to red blood cells, which reduces the amount of oxygen that is available to cells. This released gas is likely
   a) carbon dioxide
   b) nitrogen monoxide
   c) carbon monoxide
   d) sulfur monoxide
   e) calcium oxide

5. Why is smog a secondary air pollutant?
   a) It is released directly into the atmosphere.
   b) It results when pollutants that are in quantities that threaten human health interact with one another and with naturally occurring particles in the air.
   c) It is released from the decay of plants.
   d) It causes harm to life at ground level.
   e) It breaks down over time and produces radon.

6. Which pollutant is most likely to cause sick building syndrome (SBS)?
   a) insect feces on the picnic table
   b) wood smoke from the campfire
   c) pollen that is released in the spring
   d) tobacco smoke from a smoking area near a mall parking lot
   e) urea formaldehyde foam insulation in an older home

7. Which of the following technologies is used to reduce motor vehicle emissions of pollutants?
   a) catalytic converters
   b) scrubbers
   c) electrostatic precipitators
   d) carbon monoxide detectors
   e) None of the above are used.

8. An electrostatic precipitator is used
   a) in a car to reduce the amount of VOCs that escape from the gas tank
   b) to convert potential pollutants from vehicle emissions into harmless substances
   c) to trap particulates from the air inside a motor vehicle
   d) in a smokestack to attract charged particulate matter to a plate
   e) to alert residents about the air quality on a particular day

Answer the questions below.

9. a) What is the difference between natural sources of pollution and anthropogenic sources of pollution?
   b) Identify one example of each type of pollution from part (a).

10. Name four sources of sulfur dioxide and other sulfur oxide emissions.

11. a) What is second-hand smoke?
    b) Identify three toxic chemicals found in second-hand smoke.
    c) Name two health hazards to humans associated with exposure to second-hand smoke.

Chapter 9 REVIEW

12. Give one example for each of the three ways that pollutants can enter the human body.

13. Briefly describe how photochemical smog forms, and state three effects that this smog could have on humans and other animals.

14. a) What are VOCs (volatile organic compounds), and how are they dangerous to human health?
   b) List four common household items that are likely to contain VOCs.

15. Air quality management is mainly a provincial and territorial responsibility, but the federal government oversees this management.
   a) What is the name of the main federal environmental law that supports air quality management?
   b) Identify at least three ways that air quality management is addressed by this law.

Thinking and Investigation

16. Shown below are the Air Quality Index (AQI) ratings for the Ontario location of Windsor West from Monday, June 25 to Tuesday, July 3, 2012.

Air Quality Index (AQI) Categories

AQI	Colour
0-15 Very Good	
16-34 Good	
32-49 Moderate	
50-99 Poor	
100+ Very Poor	

			AQI		
Jul-05-12	4:00 pm EDT	30			Ozone O₃
Jul-04-12	4:00 pm EDT	36			Ozone O₃
Jul-03-12	4:00 pm EDT	40			Ozone O₃
Jul-02-12	4:00 pm EDT	57			Ozone O₃
Jul-05-12	4:00 pm EDT	40			Ozone O₃
Jun-30-12	4:00 pm EDT	42			Ozone O₃
Jun-29-12	4:00 pm EDT	37			Ozone O₃
Jun-28-12	4:00 pm EDT	68			Ozone O₃
Jun-27-12	4:00 pm EDT	38			Ozone O₃
Jun-26-12	4:00 pm EDT	29			Ozone O₃
Jun-25-12	4:00 pm EDT	21			Ozone O₃

a) Which air pollutant is monitored in this chart?
b) On what days would people with asthma not have to take precautions when outdoors?
c) Suggest at least two possible reasons for the different AQI values on June 25 and July 2, 2012.

d) Would you expect the AQI reading for June 28 to improve later in the night? Give reasons for your answer.
e) Predict what the AQI data might look like over a few days in January in Windsor West. Explain your prediction.
f) Does the chart provide you with enough information to make valid predictions? Provide support for your decision.

17. Industries must meet air quality guidelines that are set by federal, provincial, and territorial agencies. Identify and briefly describe two devices that industries use to trap particles in order to reduce particulate matter emission levels.

18. Read the following information, and answer the questions that follow.
"SESAA is the Southeast Saskatchewan Airshed Association, incorporated in October 2005. We are a non-profit organization of public, industry, government, and non-government members. Our goal is to collect credible, scientifically defensible air quality data for the southeast Saskatchewan region, and to make this data freely available to all stakeholders. Our objective is to bring together stakeholders from all backgrounds to identify local air quality issues and to develop innovative solutions for managing these issues. Diverse stakeholder representation recognizes concerns specific to the region, and encourages solutions that are tailored to address local needs."
(Source: http://sesaa.ca/)

a) A watershed is a region of land that is drained by a river. Pollution that affects a watershed can therefore affect the water quality of a river. Based on this explanation, as well as the description of SESAA, propose a definition for the term *airshed*.
b) Do you think SESAA is an example of a grassroots initiative? If so, explain why. If not, explain why you do not think so, and give an example of something that you think is a grassroots initiative.

19. Collect some local store flyers or investigate store products on the Internet that show the types of low VOC products that are available for renovating a bedroom.
a) Briefly describe a renovation project, and produce a list of at least three low VOC products that could be used in the project.
b) Compare these low VOC products to similar products that are not low VOCs. Is there any reason why someone would choose not to use low VOC products? Provide support for your response.

20. Sulfur dioxide emissions in the Sudbury area have been declining over the years due to various pollution control initiatives. Vale, the company that owns the iron smelting plant in Sudbury, will invest a lot of money in the next few years with the Atmospheric Emissions Reduction Project (AERP). This will reduce the sulfur dioxide emissions in the Sudbury area to levels that are below government-regulated limits. This project will also reduce dust and metal emissions.
   a) How will this AERP contribute to the sustainability of the environment? Provide supporting details in your response.
   b) Infer two other benefits to the community that this project will provide.

21. List three strategies that have been developed to reduce the amount of VOCs and other pollutant emissions from cars and trucks in Canada.

Communication

22. In a paragraph, explain how particulate matter in the air can be ingested. In your response, include at least one way to avoid ingesting particulate matter.

23. A college student is viewing a furnished room to rent in the basement of a private home for the next semester. She would have to share the bathroom, kitchen, and living room with two other students. The homeowner has a small dog in the house. Before the student leaves the home, she becomes congested, and she sneezes and coughs a lot. Should this student rent this room? Provide at least three supporting details in your response.

24. Create a comic strip, storyboard, or poster that clearly demonstrates to a Grade 6 student the difference between ground ozone and stratospheric ozone.

25. Electrostatic precipitators stop up to 99% of particulates from entering the air. These particles must be collected and then disposed of either as solid waste or hazardous solid waste in appropriate landfills. If the particles were not trapped, they would be released into the atmosphere. Either way, the particles are still on the planet. Explain why it is still beneficial to use electrostatic precipitators to reduce particulate matter emissions.

26. When individuals are asked what they can do to help reduce air pollution, most answer that carpooling or driving their cars less frequently is the best approach. However, carpooling might not be practical for many people who live in rural areas. Create a list of at least five ways that most Canadians can realistically reduce air pollution emissions.

Application

27. Compact fluorescent lights (CFLs) contain mercury vapour. Many stores and schools have collection boxes for the safe disposal of these lights.
   a) If many homeowners dispose of the CFLs in their household garbage, and if the garbage is incinerated, how could these CFLs affect air quality?
   b) If you broke a CFL bulb on the floor at home, would it be safe for you to stay in the room to clean up the mess it created? What advice would you give to someone who needs to clean up a broken CFL so that this person will be safe?

28. The bar graph shows sources and quantities of particulate matter emissions in Canada in 2010. The proportion in percent of total particulate matter emissions is also shown. Based on the information, suggest three ways that Canadians could help to reduce particulate matter emissions.

Particulate matter emissions by source, Canada, 2010

Proportion of national total particulate matter (TPM) emissions

Total particulate matter (TPM)
Respirable particulate matter (PM$_{10}$)
Fine particulate matter (PM$_{2.5}$)

Emissions in kilotonnes

29. Cities across Canada are able to set by-laws regarding the use of CO detectors. Many city by-laws require there to be at least one CO detector in the home.
   a) Why is it important to have a working CO detector in the home? Include a description of at least two physical characteristics of CO in your response.
   b) If a homeowner purchases only one CO detector, where should it be placed?

30. **Did You Know?** Reread the Carl Sagan quotation on page 295. Use the title, "True Priorities", to express your ideas about this quotation."

Pause and Reflect

How could you incorporate what you have learned in this chapter into your daily actions or choices?

CHAPTER 10

Water Quality and Environmental Health

Shrimp boats like this one were used to deploy booms to help contain oil during the BP Deepwater Horizon disaster. After the well was finally capped, more than 365 km of booms were recycled to make air deflectors for the Chevy Volt.

In what other parts of the world is offshore deepwater oil drilling taking place, and what measures are in place if a disaster occurs?

A Clean-up Simulation

This activity will help you appreciate the challenges in cleaning oil that has spilled into marine (saltwater) or freshwater systems.

1. Get a beaker half-filled with salt water or tap water. Add about 5 mL of vegetable oil.

2. Brainstorm ways to separate and remove the oil from the water. Materials you could use include filter paper, detergents, and absorbent materials such as sand, sawdust, and cat litter.

3. Choose one idea to try.
   a) Write a brief procedure to show your teacher.
   b) Include a list of the safety equipment, safety procedures, and protective clothing that you will need and follow.
   c) With your teacher's permission, carry out your method.

4. When you are done, clean up your work area as your teacher instructs.

5. Record the results of your method, and summarizes how effective it was.

6. Explain how the simulation helps you appreciate the challenges involved in removing oil after a real spill.

In this chapter, you will

- identify types of water pollutants, their sources, and their effects on human and environmental health
- describe what can happen when water pollutants enter the bodies of organisms
- explain how legislation, technology, and personal decisions can improve water quality

Quantity, Uses, and Quality of Water

The Distribution of Water on Earth

Figure 10.1 Water is visible from space as both the blue of the oceans and the white vapour of clouds.

People who have seen views of Earth from space, such as the one shown in **Figure 10.1**, marvel at its beauty. Astronaut Edgar Mitchell put it this way: "Suddenly from behind the rim of the moon, in long, slow-motion moments of immense majesty, there emerges a sparkling blue and white jewel, a light, delicate sky-blue sphere laced with slowly swirling veils of white, rising gradually like a small pearl in a thick sea of black mystery. It takes more than a moment to fully realize this is Earth … home."

Water is the key to life on Earth. Most of the planet's millions of species make their home in oceans, lakes, and rivers. Earth's total water supply is estimated at nearly 1.4 billion cubic kilometres. Just one cubic kilometre of water is equivalent to one trillion litres. Most people cannot imagine *that* amount, let alone the staggering amount of all Earth's water.

The great majority of Earth's water—about 97.6%—is the salt water of the oceans and seas. The remaining 2.4% is fresh water. Most of this fresh water is locked up in glaciers and ice caps, so it is not available to be used by living things. Use **Figure 10.2** to help you appreciate the amount of drinkable fresh water there is available to life on Earth.

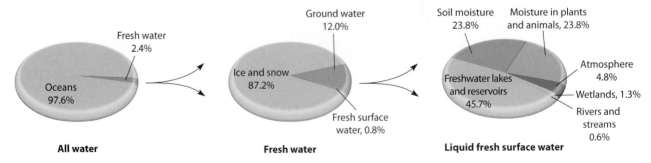

Figure 10.2 Less than 1% of fresh water is the liquid surface water on which all life depends.

Applying How does the increasing population of the world affect the availability of freshwater resources?

The Role of Water in Life

Water occurs as a gas, a solid, and a liquid in Earth's atmosphere and on its surface. It is liquid water, though, that makes life on Earth possible. Most organisms contain over 50% liquid water by weight. This water fills cells and supports their structure. Within organisms, liquid water provides a medium in which life-sustaining chemical reactions can take place. Often, water plays an active role in these reactions, as well as in many other body processes, including regulating body temperature.

As well as keeping your body (and the bodies of other organisms) functioning, water also provides many of the other essentials for your survival. In addition, many of the products and processes described below would disappear without water.

- **Food** Without water, food would never make it to your plate. The plants and animals we eat need water to grow, thrive, and reproduce. Fish also rely on it for support and oxygen. The amount of water used to produce food often surprises people. For instance, it requires 40 L to make one slice of bread, 200 L to produce one egg, and 3500 L to produce 1 kg of chicken.

- **Shelter and Energy** The wood used to build homes and other buildings comes from trees that would not exist without water. The production of steel, cement, and other building materials also requires water. Similarly, most fuels used to heat homes use some water in their production process, with hydroelectricity topping the list.

- **Health and Hygiene** We use water to clean our food, our clothes, and ourselves. Human waste removal and sewage treatment also rely heavily on water. Without water, disease would be a much greater concern. Water is also important in producing the pharmaceutical and plant-based medications that many people rely on for their health.

- **Manufacturing** Most of the products people use each day require water in some stage of their manufacturing processes. Product manufacturing uses even more water than the growth and production of food. For instance, it takes thousands of litres of water to make a single pair of jeans and hundreds of thousands of litres to produce one car. When items are recycled, it takes more water to break them down for re-use.

- **Transportation** It takes water to make cars and to drive them. As a motor vehicle runs, water in the radiator cools the engine. Most fuel production also uses water. Tankers transport much of the oil used in motor vehicles across water. Boats and ferries also transport people and goods.

- **Recreation and Beauty** A dip in a swimming pool, a camping trip, a game of hockey, and a ski vacation all require water. Even simple enjoyment of gardens and tree-lined streets would not be possible without this life-giving substance.

Mini-Activity 10-1 **The Roles of Water**

With a partner or small group, brainstorm the roles that water can play in people's lives. Use your discussion to answer the question, "What roles does water play in my life?"

Use the following questions to guide your discussion.

- Which of the activities that you do or that you take part in each day involve water in any of its three states—solid, liquid, or gas?

- Which of these activities are essential, and which are non-essential (for example, for convenience or pleasure)? How easy or difficult is it to make this distinction?

Present the results of your discussion in the format of your choice. Examples include a slideshow, a video, a dramatic or musical performance, or a blog.

Types of Water Pollution

Water pollution is any physical, biological, or chemical change in water quality that has an adverse effect on organisms or that makes water unsuitable for desired uses. Water may be polluted by natural sources such as volcanoes and landslides. In this chapter, however, you will consider the effect of human activities on water quality.

Point sources of water pollution, such as the one shown in **Figure 10.3**, include factories, power plants, sewage treatment plants, and oil wells. These sources are fairly easy to monitor and regulate. Nonpoint sources of water pollution (see **Figure 10.4**) include run-off from farm fields and feedlots, lawns, construction sites, logging areas, roads, and parking lots. Pollution from nonpoint sources tends to be periodic. The irregular timing of events such as heavy rainfalls and spring thaws washes pollutants from and into many different locations. As a result, it is more difficult to monitor, regulate, and treat nonpoint sources of water pollution.

Perhaps the most challenging nonpoint source to address is the atmosphere itself. Pollutants may be carried great distances by winds and then eventually fall to the ground in rain or snow. As a result, pollutants from one location may turn up in watersheds and surface waters far from their point of origin.

Water pollutants may be classified in a variety of ways because there are so many of them. **Table 10.1** lists many of the most common types of water pollutants.

Figure 10.3 Industrial pipes that discharge waste are a point source of water pollution.

Figure 10.4 Run-off from farm fields is a nonpoint source of water pollution.

Mini-Activity 10-2   **Create Your Own Case Study**

Choose one water pollutant that interests you from Table 10.1. Use the Internet and any other available resources to find out more about this pollutant. Find a specific instance where it has caused a problem in Canada as well as in another part of the world. Using the case studies in this textbook as a model, use the information you gathered to write your own case study.

At a minimum, your case study should include

- a description of the problem, including the pollutant involved, its source, and its effect on the environment and/or human health

- how the problem was or is being dealt with

Table 10.1 Types, Sources, and Effects of Water Pollutants

Pollutant	Sources	Effects on Humans	Effects on Aquatic Ecosystems
Pathogens (includes microscopic disease-causing bacteria, viruses, and parasites)	Dumping of raw and partially treated sewage; run-off of animal wastes from feedlots	Increased costs of water treatment; death and disease; reduced availability and contamination of fish, shellfish, and associated species	Reduced survival and reproduction of aquatic organisms due to disease
Plant Nutrients and Other Organic Matter (includes urine and feces from animals as well as nitrates and phosphates from fertilizers)	Run-off from agricultural fields, pastures, and livestock feedlots; landscaped urban areas; dumping of raw and treated sewage and industrial discharges; phosphate detergents	Increased water treatment costs; reduced availability of fish, shellfish, and associated species; colour and odour associated with algal growth; impairment of recreational uses	Algal blooms occur; death of algae results in low oxygen levels and reduced diversity and growth of large plants; reduced diversity of animals; fish kills
Heavy Metals (inorganic chemicals, including lead, mercury, arsenic, and cadmium)	Atmospheric deposition; road run-off; discharges from sewage treatment plants and industrial sources; creation of reservoirs; acidic mine effluents	Increased costs of water treatment; disease and death; reduced availability and healthfulness of fish and shellfish; biomagnification	Lower fish population due to failed reproduction; death of invertebrates leading to reduced prey for fish; biomagnification
Other Inorganic Chemicals (includes acids, bases, and salts)	Atmospheric deposition; mine drainage; run-off from roads treated for removal of ice or snow; irrigation run-off; brine produced in oil extraction	Reduced availability of fish and shellfish; increased heavy metals in fish; reduced availability of drinking water	Death of sensitive aquatic organisms; increased release of trace metals from soils, rock, and metal surfaces, such as water pipes
Sediment	Run-off from agricultural land and livestock feedlots; logged hillsides; degraded stream banks; road construction; other improper land use	Increased water treatment costs; reduced availability of fish, shellfish, and associated species; filling of lakes, streams, and artificial reservoirs and harbours, requiring dredging	Covering of spawning sites for fish; reduced numbers of insect species; reduced plant growth and diversity; reduced prey for predators; clogging of gills and filters
Petrochemicals and Other Organic Chemicals	Run-off from agricultural fields and pastures, landscaped urban areas, and logged areas; discharges from chemical manufacturing and other industrial processes	Increased costs of water treatment; reduced availability of fish, shellfish, and associated species; odours	Reduced oxygen; fish kills; reduced numbers and diversity of aquatic life
Heat	Urban heat that is transferred to water; solar heating of reservoirs; warm-water discharges from power plants and industrial facilities	Reduced availability of fish	Elimination of cold-water species of fish and shellfish; less oxygen; heat-stressed animals susceptible to disease

Pause and Reflect

1. What percentage of water on Earth is available as fresh water?

2. Describe how water is required for our shelter and energy requirements.

3. **Critical Thinking** Explain how a golf course is a nonpoint source of pollution.

Water Pollution, Organisms, and Human Health

Organisms play several roles related to water pollution.

- Organisms such as bacteria and parasites can act as pollutants themselves. When they do, they affect human health directly.

- Organisms can take up water pollutants. These pollutants become concentrated in the organisms that consume them.

- Water pollution can cause an overgrowth of some organisms. These organisms can produce environmental conditions and toxins that harm other organisms.

Organisms Act as Water Pollutants

Bacteria, viruses, and parasites found in water can endanger human health when they are ingested. On a global scale, these pathogens are among the most dangerous water pollutants. They are responsible for diseases such as cholera, typhoid fever, infectious hepatitis, and polio. Water-related diseases are responsible for over 25 million deaths around the world each year.

Pathogens enter waterways through improperly treated human sewage or run-off contaminated with animal wastes. In Canada, the number of pathogens that enter drinking water is small due to sewage treatment and water chlorination programs. Water quality is also monitored by a series of checks that water quality professionals perform regularly. However, it is time-consuming and expensive to monitor all pathogens. As a result, water is tested for the presence of a few special bacteria that live in human and animal intestines. Because the most common of these is *E. coli*, the bacteria are referred to as **coliform bacteria**. A positive test for these bacteria indicates that fecal contamination and other harmful pathogens may be present. In addition to drinking water, water at public beaches and swimming pools is also tested for coliform bacteria. The process used to test for coliform bacteria is described in **Figure 10.5**.

coliform bacteria
intestinal bacteria whose presence in a water sample indicates fecal contamination

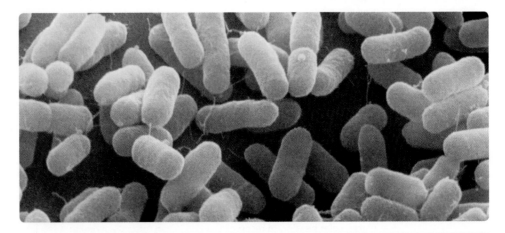

Figure 10.5 *E. coli* bacteria are one of several coliform bacteria species that are tested for in water samples.

Testing for Coliform Bacteria

1. A water sample is taken from a specific source, such as a well, river, beach, or pool. Alternatively, the sample is passed through a filter that is used for testing.
2. The water sample or filter is placed on a growth medium that contains all the nutrients required for the growth of coliform bacteria.
3. After 24 hours, the medium is examined for bacterial growth. By this time, any bacteria in the sample will have multiplied to form small bacterial colonies.
4. Growth of any colonies in the medium indicates that coliform bacteria contaminate the sample. It is unsafe to drink unless properly disinfected.
5. Growth of a limited number of colonies is considered safe for swimming. If the number exceeds a set maximum, the swimming area is closed. The maximum is 200 colonies per 100 mL.

Organisms Magnify Water Pollutants

Some organic chemicals are resistant to degradation, and so they persist in the environment for a long time. These include the pesticide DDT, heavy metals such as mercury, and PCBs. PCBs are organic chemicals that were used in hundreds of different products, from plastics to paints. Production of both DDT and PCBs is now banned in Canada. Why is this the case?

DDT and PCBs, along with heavy metals, enter waterways and are taken up by microscopic organisms such as phytoplankton and bacteria. The pollutants collect in the cells and tissues of these organisms. This process is called *bioaccumulation*. When predators, such as zooplankton and fish, eat these organisms, the pollutants build up further in the predators' fatty tissues. Because predators consume a lot of prey, the pollutants become concentrated in their tissues. This effect is called *biomagnification*. **Figure 10.6** explains how these two processes occur. Notice how the organisms that are highest in the food chain accumulate the greatest levels of the pollutant. Large fish such as sharks, fish-eating birds, and humans all fall into this category. At high levels, these pollutants are extremely harmful.

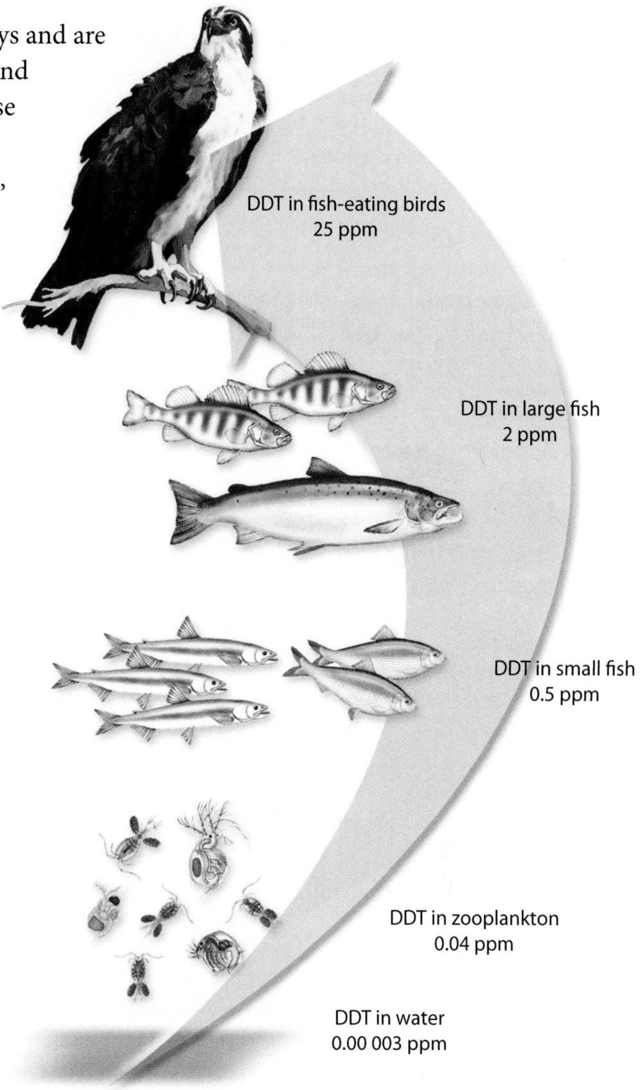

DDT in fish-eating birds
25 ppm

DDT in large fish
2 ppm

DDT in small fish
0.5 ppm

DDT in zooplankton
0.04 ppm

DDT in water
0.00 003 ppm

Figure 10.6 Bioaccumulation and biomagnification work together to magnify certain water pollutants in large predators. The unit *ppm* means "parts per million" of the pollutants. One part per million (1 ppm) is like one second in 11.5 days, or one minute in two years, or one car in bumper-to-bumper traffic from Regina, Saskatchewan, to St. Johns, Newfoundland and Labrador.

Analyzing Would someone who eats small fish such as sardines consume as much DDT as someone who eats large fish such as tuna. Why or why not?

Pause and Reflect

4. What are coliform bacteria?

5. Explain how bioaccumulation and biomagnification are similar and how they are different.

6. Critical Thinking How might biomagnification affect Aboriginal peoples who eat a diet that is high in seafood and marine mammals? Explain your reasoning.

Mini-Activity 10-3 **Pass It On**

Design an activity to explain bioaccumulation and biomagnification to a class of younger students. Create your activity, and then test it on a group of students. Improve the activity based on any feedback you receive.

Organism Overgrowth and Water Pollution

The growth of plants and algae in aquatic ecosystems is limited by the amount of available nutrients, such as phosphorus and nitrogen. As noted in Table 10.1, these nutrients can enter waterways as pollutants. When enough of these nutrients are present, plant and algal growth is stimulated.

Figure 10.7 shows that the overgrowth of algae can occur naturally or as a result of human activity. Whether it happens slowly or quickly, naturally or otherwise, the overgrowth is called an *algal bloom*, and the process that results in its formation is called **eutrophication**. During this process, bacteria break down the organic matter and experience their own overgrowth, which uses up much of the oxygen in the water. This causes the death of fish and other aquatic life. Over time, the water becomes brackish and cloudy, and unpleasant tastes and odours develop. Purification of the water is needed to make the water drinkable.

eutrophication a process in which bodies of water receive excess nutrients that stimulate growth of algae

Figure 10.7 Natural eutrophication **(A)** takes place over hundreds of years. Eutrophication from human activity **(B)** develops much more quickly—within years or decades.

Analyzing *Eutrophication from human activity is also called cultural eutrophication. Explain why you think this is, or is not, an appropriate term for this process.*

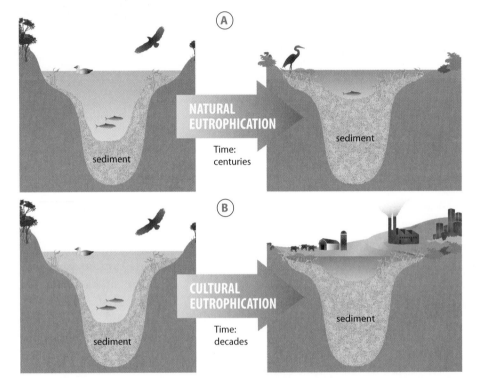

Some algae also produce toxins that can seriously harm animals and human beings. Shellfish can become contaminated with these toxins. Eating them can cause serious illness or, in extreme cases, death. Many consumers are unaware that the bivalve shellfish they buy in grocery stores and at local markets can harbour these harmful contaminants. Bivalve shellfish include clams, mussels, scallops, and oysters. These animals take in nutrients by filtering bacteria and other microscopic organisms from the water. When they feed, they also take up algal toxins. These toxins can build up to dangerous levels in shellfish tissues. As a result, eating even a small amount can cause illness.

Once consumed, the toxins attack an organism's nervous system within a matter of minutes or hours. Loss of muscle coordination, breathing difficulties, headaches, dizziness, and confusion may soon follow. Nausea, cramping, and vomiting are also common. Anyone experiencing these symptoms after eating shellfish should get immediate medical attention.

To help ensure that shellfish are safe to eat, the federal government has set up the Canadian Shellfish Sanitation Program. This program is a joint initiative that involves three government agencies.

InquiryLab 10A, Determining Water Quality, on page 326

Case Study   What's Been Going on at Grassy Narrows?

First Nations March through Downtown Toronto
April 7, 2010

First Nations from Grassy Narrows Reserve travelled more than 1800 km to march in protest through the streets of downtown Toronto en route to Queen's Park. They were joined by members of several other organizations, including Greenpeace and Amnesty International. As they marched, they held banners and carried poles with brightly painted fish. Others carried long pieces of blue fabric to represent a flowing river. Their goal: to bring attention to a problem that's been going on for more than 40 years.

Government Appeals Ontario Superior Court Decision
January 14, 2013

The Government of Ontario today begins an appeal to overturn a decision by the Ontario Superior Court of Justice. That ruling, made in August 2011, said that the government did not have the power to authorize logging and mining development on lands belonging to the Grassy Narrows First Nation. For more than 10 years protesters have set up blockades to keep logging trucks from their traditional lands. Protesters contend that logging activities interfere with their forest-based way of life and would further poison the rivers and fish they depend on for their food and livelihood.

Will They Eat the Fish?
June 7, 2012

An unusual sight greeted visitors on the lawn of the Ontario legislature at Queen's Park this week: a table, decked with a white tablecloth, set for outdoor dining. Name cards designated places for mothers and elders from Grassy Narrows First Nation, alongside places for the Premier and Ministers of Natural Resources, Northern Development, and the Environment. The occasion was a traditional fish fry, and the featured entree was fish from the river systems that run through Grassy Narrows Reserve. The question on everyone's mind: Would the ministers attend, and would they eat the fish?

Research and Analyze

1. Do research to find out what's been going on at Grassy Narrows. What caused this situation? What effects has it had on the Grassy Narrows First Nation? What is the current state of affairs, and who are the stakeholders involved?

2. Public protests are often used to draw attention to a specific issue or concern. Research how protests have been used to shed light on the Grassy Narrows situation. What kinds of actions were taken, and how effective were they?

Communicate

3. The situation in Grassy Narrows is not an isolated case, either in Canada or in the rest of the world. What other environment-related incidents involving First Nations have you heard about? What about Aboriginal peoples in other parts of the world? What kinds of decisions do you think could resolve these types of incidents?

Summary

- Water is essential for life on Earth. Almost 98% of Earth's water is in the form of salt water. Only a small portion of the fresh water is available for use by living things.

- Water is a key component of all cells. It is also needed to make and use products such as food, shelter, medications, and motor vehicles.

- Point sources of water pollution include sewage treatment plants and factories that dump pollutants into waterways. Nonpoint sources of water pollution include run-off from farms and roads.

- Pollutants include pathogens, such as bacteria and viruses, animal feces and urine, nitrates and phosphates, heavy metals such as lead and mercury, petrochemicals, and other organic chemicals.

- Pathogens in waterways can cause diseases such as cholera, typhoid, and hepatitis. Therefore, monitoring the quality of drinking water is performed by water quality professionals. One type of pathogen that is routinely tested for is coliform bacteria.

- Toxic chemicals in water, such as PCBs and DDT, can accumulate in the tissues of organisms. When they are eaten by a predator, concentrations of the toxic substances increase in that predator, which is called biomagnification.

Review Questions

1. Draw a bar graph that represents the relative amounts of fresh water and salt water from natural bodies of water on Earth. Make sure your bar graph has the following information:
   - a title
   - labels for the types of water
   - labels that show the relative amounts of each type of water, expressed as a percentage K/U C T/I

2. Give one example for each of the following forms of water on Earth. T/I
   a) solid water
   b) liquid water
   c) gaseous water

3. In 1798, Samuel Taylor Coleridge published a poem called "The Rime of the Ancient Mariner." The poem is based on a mariner (sailor) who tells the story of his long voyage at sea. In one part of the poem, the Mariner's ship is stranded in a silent, motionless sea. The Mariner recalls:

   Water, water, every where,
   And all the boards did shrink;
   Water, water, every where,
   Nor any drop to drink.

   a) Explain the third and fourth lines.
   b) People have been converting sea water to water that people can drink for centuries. Describe a place in the world where this process is especially important. Explain why you chose this place.
   c) How does water pollution affect the amount of useful water available for organisms? K/U T/I A

4. Water is essential for many processes and products people use every day. K/U A T/I
   a) List three examples of how you depend on water.
   b) How would not having water for these examples have affected what you did today? What, if any, alternative actions or products were possible?

5. Copy the table below into your notebook. Complete it by describing the effects of each pollutant. K/U

Pollutant Effects

Pollutant	Effects on Humans	Effects on Aquatic Ecosystems
Pathogens		
Animal feces and phosphates from fertilizers		
Heavy metals		
Petrochemicals		

6. Describe a source of each pollutant in question 5 and state whether it is a point source or nonpoint source of pollution, or both. Explain your reasoning. K/U C

7. Dehydration is a condition in which the body does not have enough water. Why is severe dehydration life-threatening? A T/I

8. What are coliform bacteria? Why is drinking water tested for it? Draw a flowchart that outlines the testing process for coliform bacteria. K/U C A

9. Would biomagnification be possible without bioaccumulation occurring? Explain. T/I

10. In your own words, describe what occurs during the process of eutrophication. K/U C

Managing Water Resources

Government Management of Water Resources

The clean-up and rehabilitation of Nova Scotia's Sydney Tar Ponds (see **Figure 10.8**) depended on collaboration between the federal and provincial governments. In fact, in Canada, several branches of government work together to manage water resources. Branches at the municipal level also get involved. As well, territorial and Aboriginal governments play an active role in managing this resource, as does the private sector.

The management of water resources focusses on three main goals: to reduce water use, to reduce water pollution, and to clean polluted water. The federal Clean Water Act outlines water resource management in Canada. This Act requires that federal and provincial governments consult each other on water-related issues that affect agriculture and health. Any issues that are important at the national level also must involve both levels of government. The federal and provincial governments co-operate on many initiatives that regulate water resources and support their sustainable use. For example, Environment Canada monitors water quality at the provincial and federal levels. Health Canada works with health departments in Canadian provinces and territories to monitor water-borne pathogens.

Despite this co-operation, some aspects of water resource management and protection are the sole responsibility of the provinces and territories. In some cases, responsibility is passed on to municipal governments. In most urban areas, for example, municipalities monitor and treat drinking water so it is safe to consume. Many municipalities also collect and treat waste water.

Others types of water management are carried out by the federal government alone. For example, fisheries are managed on a national level. Oceans and other bodies of water that are shared with the United States are also managed by the federal government, as are water resources in national parks.

Figure 10.8 Remediation of the Sydney Tar Ponds began in 2007. By 2012, the first and second phases were complete, and work had begun on the third.

Analyzing What water quality issues might arise during clean-up of a toxic site such as the Sydney Tar Ponds?

Mini-Activity 10-4 WHMIS and Water Quality

WHMIS (Workplace Hazardous Materials Information System) regulations provide guidelines for the safe use and handling—which includes proper disposal—of certain regulated substances that may be dangerous to environmental or human health. Each regulated substance has a material safety data sheet (MSDS) that provides this information. Anyone who handles regulated substances must follow WHMIS regulations. For instance, both you and your teacher must follow WHMIS regulations when handling these substances in science class.

Choose one of these WHMIS-regulated substances:

- acetone
- sodium hypochlorite (bleach)
- benzene
- toluene
- ethylene glycol (antifreeze)

Use information from your teacher or online resources to design an information pamphlet (print or digital) that includes the following information about your substance:

1. what it is used for
2. chemical and physical properties that suit its purpose
3. why it is regulated by WHMIS
4. how it should be handled safely
5. proper disposal methods

Explain why you do or do not think that WHMIS regulations are effective for keeping the substance out of waterways.

Using Technology to Protect Water Resources

Technology plays an important role in preventing water pollution and improving water quality. Many types of high-tech computer and machine-based technologies are used to achieve these goals around the world. However, low-tech options can be just as effective in some cases, depending on the water issue being addressed. **Figure 10.9** shows some examples of high- and low-tech solutions to protecting water resources.

Figure 10.9 Computer technology and other high-tech ideas help protect water resources and maintain water quality.

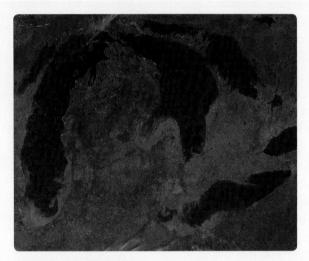

Earth Observation (EO)
These technologies use satellites to monitor algal blooms (green in this image), sedimentation (light blue), and other threats to water quality. EO can also monitor point source pollution discharge into bodies of water, such as sewage and industrial outflow.

UV Water Sterilization
UV water sterilization uses ultraviolet light to kill or deactivate bacteria and other pathogens in water. UV water sterilization is used to clean drinking water from private wells, as well as water used in the food industry and agriculture. It can also treat waste water to remove pathogens.

Living Machines
Living machines combine high-tech computerized monitoring systems and living plants to clean waste water. These systems have been used in a variety of settings, from urban corporate headquarters to science camps for youth. After it is cleaned, the water is used for many purposes, such as flushing toilets, filling fish ponds, and watering plants.

Using Accessible, Low-Tech Approaches to Manage Water

Low-tech options enable people to take a hands-on approach to conserving water and reducing pollution. These methods are more affordable and usually make use of readily available materials. Sometimes a low-tech option is all that is needed to achieve a water quality goal. **Figure 10.10** explores some of the low-tech methods that help protect and improve water quality or provide clean, drinkable water.

Figure 10.10 Low-tech methods can be just as effective as high-tech methods in protecting and preserving water quality.

Fencing in Animals Protects Water Resources

Sometimes a fenced-off stream is the most effective way to keep animal manure out of streams on farmland. The fence keeps animals out of the stream and allows natural vegetation to thrive. The vegetation reduces the amount of waste that rainwater and irrigation wash into the stream from the farm. It also keeps sediment out of the stream.

Solar Ovens Provide Drinking Water

In many countries, people have limited access to sources of clean drinking water. For populations that live near a coast, a solar oven provides a low-tech, cost-effective way to make drinking water from salt water. Water evaporates when solar energy heats salt water in the oven. The evaporated water condenses and provides pure drinking water that is free of salt. A solar oven requires no electricity and can provide several litres of drinking water each day.

Wetlands Clean Waste Water

Natural wetlands are a low-tech version of living machines. These ecosystems both break down and take up nutrients, pathogens, and other water pollutants. They also filter sediment. Wetlands can be constructed artificially, and they are often built on farms to clean water that has been polluted by livestock. Constructed wetlands are also used to treat municipal waste water in Canada and around the world.

Pause and Reflect

7. Describe the three goals of water resource management.

8. Think Critically Explain why low-tech approaches to protecting water resources can sometimes be better than high-tech approaches. When might high-tech approaches be more appropriate?

9. Think Critically Identify at least one limitation to low-tech methods of treating water, and explain your reasoning.

 ThoughtLab 10B, Water Resources and Personal Actions, on page 329

Mini-Activity 10-5 **Exploring Low-tech Solutions**

What ideas do you have for designing any of the low-tech devices listed on the right? Discuss your ideas in a small group. Create a brief presentation to explain how the device works and how it helps improve water quality or conserve water resources. Some choices are more challenging than others and may require online research.

- a low-flush toilet
- a portable water filter
- a rain barrel
- a green roof
- a composting toilet

Case Study Constructed Wetlands

Constructed wetlands use the processes carried out by natural wetland plants and micro-organisms to treat wastewater. Depending on climate and the type of wastewater, these systems may have a combination of ponds, open-water marsh, and subsurface flow areas. They are suitable for treating storm water, farm run-off, sewage from small communities, and some industrial or mining wastewaters.

Study the diagram of a constructed pond-wetland system, and answer the questions.

Overhead View of Constructed Pond-Wetland System

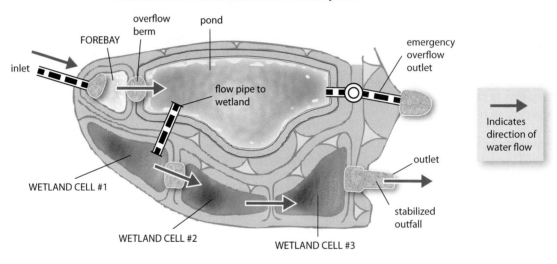

Cross-section View of Wetland Cell

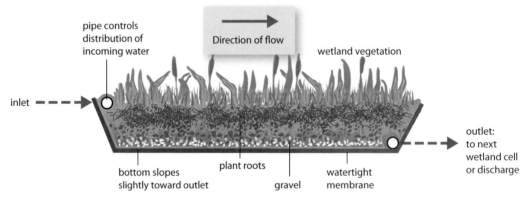

Analyze and Conclude

1. Describe the path of the water and the changes it undergoes from inlet to outlet.

2. What do you think is the purpose of the forebay, pond, and emergency outlet? How would the water in wetland cell #1 compare to wetland cell #3? Explain your answers.

Communicate

3. Wetlands have been described as "nature's kidneys." Explain why this is or is not a suitable analogy.

4. Constructed wetlands are used across Canada. Find a constructed wetland near where you live, and explain why it is located there.

Water Resources and Grassroots Initiatives

Many grassroots initiatives work to prevent water pollution, improve water quality, and conserve water resources. While most are local, some have evolved to work at the provincial, national, or international level. **Figures 10.11** to **10.13** explore several grassroots initiatives that act on these levels.

Acting at a Local Level: The Friends of Medway Creek

This grassroots group in London, Ontario, has been working with students from a local school to restore Medway Creek. Planting native plants on the stream banks reduces sediment pollution and improves water quality in this local creek. Medway Creek flows into the Great Lakes. As a result, restoration of this creek helps keep sediment and other pollutants out of Canada's largest lakes.

Figure 10.11 Medway Creek in London, Ontario

Working on a Provincial/Territorial Level: The Fundy Baykeeper

This New Brunswick grassroots initiative works to protect the water quality and marine environment of the Bay of Fundy. The Baykeeper patrols the bay by boat to monitor environmental threats. When threats occur, they are investigated and action is taken to deal with them. The program also runs a telephone hotline that citizens can use to report pollution problems. Environmental law enforcement and education are top priorities.

Figure 10.12 The Fundy Baykeeper's public paddle in the protected Musquash Estuary

Co-operating on a National/International Level: The Yukon River Inter-tribal Watershed Council

This grassroots organization is made up of members from 70 First Nations and Tribes. Its goal is to protect the Yukon River Watershed and its water quality. To achieve this goal, the Council works with Yukon First Nations and Alaskan Tribes. In addition to protecting the watershed, the Council works to promote the health and wellness of Aboriginal peoples who live within the watershed. Education and awareness of watershed issues play a large role in their work.

Figure 10.13 Yukon First Nations and the Yukon River Watershed

---

Mini-Activity 10-6 Growing Deep Roots

Complete one or both of the activities below, which explore grassroots initiatives related to water resource management.

1. Find a grassroots initiative in your town, city, or region that interests you. The group must be involved with water quality or conservation. Create a list of five or more questions you would like answered about this initiative. Find the answers to your questions through various resources (online, in an interview, and so on). Share your findings with your class.

2. Make a plan to carry out a grassroots initiative that addresses a water quality or conservation issue of your choice. Your plan must explain
   - how you would set up your initiative
   - the goals of your initiative
   - how you will carry out these goals
   - how you will assess your results

Case Study   Improving Water Quality: The Yellow Fish Road Program

Storm drains are openings covered by heavy metal grates along city streets. The drains collect run-off and excess rainwater, but did you know that this storm water goes directly into local lakes and rivers with no treatment? In 1991, the non-profit organization Trout Unlimited established Yellow Fish Road, a national program to conserve freshwater systems and educate the public about storm-water pollution. There are now more than 40 Yellow Fish Road partners across Canada. These include municipalities, conservation authorities, and community groups.

Volunteers with Yellow Fish Road raise awareness by painting a yellow fish symbol on the road near storm drains. They also distribute fish-shaped door hangers to residents to inform them about the issue, and give presentations to classrooms and community groups. Their main message is twofold:

1. Anything that enters a storm drain flows untreated directly to the local water body.
2. There are simple things people can do to prevent storm-water pollution.

What Can You Do to Help?

To help prevent storm-water pollution, always use environmentally responsible products and cleaners. Wash vehicles at a car wash and keep them maintained to prevent leaks. Any hazardous household wastes should be disposed of at proper facilities. In the garden, use natural methods instead of synthetic pesticides and fertilizers. Pick up litter and pet wastes around your property or neighbourhood. Set up a rain barrel to collect rainwater for use on gardens and lawns. Lastly, get involved by joining or creating a local Yellow Fish Road group.

Analyze

1. Create a flowchart to trace the path of at least three common pollutants from their source all the way to local bodies of water.

2. Copy the table into your notebook. Work in a group to list actions people can take to prevent storm-water pollution from each pollutant. Include actions from this case study and some of your own.

Pollution Prevention

Type of Pollution	Human Sources	Actions to Prevent Pollution
Organic debris	• plant matter • soil	
Inorganic debris	• construction wastes • sand and gravel • litter	
Petroleum products	• oil • gasoline	
Phosphates and nitrates	• detergents • fertilizers	
Toxins	• hazardous wastes • pesticides • herbicides	
Harmful bacteria	• animal wastes	

Communicate

3. You are a member of Yellow Fish Road. You have been asked to give a short talk about the group's work at a school assembly. Write the introductory paragraph to your speech and outline four or five points you would make.

Summary

- Managing and protecting water resources in Canada involves federal, provincial, and municipal levels of government. Different agencies in these governments are responsible for different aspects of water management.
- People have designed high-tech ways to monitor and improve water quality, such as satellites and UV sterilization. There are also low-tech approaches, such as solar ovens.
- There are many ways that individuals can help improve water quality, such as not dumping chemicals or dangerous materials down drains and limiting their use of polluting products.
- There are many grassroots initiatives aimed at improving water quality and water conservation. These can act at the local, provincial, national, and international levels.

Review Questions

1. What are the three main goals of managing water resources? K/U

2. The management of water resources is carried out by different levels of government and various agencies. K/U C T/I

   a) In a short paragraph, summarize who is involved in managing Canada's water resources and how they work together to accomplish this.

   b) Describe one advantage and one disadvantage of having multiple agencies involved in managing water resources. Hint: Think about what might happen if different groups at school had to agree on how to handle one issue.

3. Copy the table below in your notebook. Complete it by summarizing, in one or two sentences, three high-tech approaches to protect and improve water quality. K/U

   Water Protection and Improvements

High-tech Approaches	Description

4. According to the World Health Organization, there are still hundreds of millions of people in the world who do not have access to safe drinking water. There are a number of strategies to help address this. K/U T/I A

   a) Describe one technique to improve access to drinking water for people who live near a sea or ocean.

   b) Many people in developing countries must use water sources that are contaminated with different pollutants from various sources. Describe a method of improving water quality that is appropriate for such regions. Provide a reason for why you chose that method.

5. In a table, provide two advantages and two disadvantages of low-tech methods and high-tech methods of protecting and improving water quality. C T/I

6. Some people throw out unused medicines (prescription and non-prescription) by pouring them down the drain or flushing them down the toilet. However, sewage and water treatment plants are not equipped to remove these substances. T/I A

   a) Explain why disposing of drugs in this way is an environmental issue.

   b) How can a local pharmacy play a role in helping to dispose of medications?

7. Describe how fencing off a stream, as shown below, acts to protect water quality. K/U T/I

8. In 2010, Canadian federal regulations came into effect that banned the sale of household dishwashing detergents, laundry detergents, and other cleaning products that contained more than 0.5% phosphates by weight.

   a) How does this action help improve the quality of our waterways, such as lakes and rivers?

   b) What do high levels of phosphates cause in waterways such as lakes and ponds? T/I A

Case Study   Mercury: A Global Concern

Mercury is a heavy-metal element that has been, and continues to be, used in a variety of industrial processes and consumer products. Mercury is toxic. Regulations and bans on its use have decreased mercury pollution in many regions. However, challenges remain in some industries that still use mercury, especially in developing countries.

Mercury in the Environment

Once released into the environment, mercury persists for a long time. Mercury can be transported long distances and can accumulate in water, soil, and in the bodies of organisms far from its original source.

There are three main uses of mercury in industry: small-scale gold mining, compact fluorescent lighting, and the production of certain plastics. Unintended mercury emissions also occur as a result of burning coal, as well as in the production of metals and cement. (Mercury occurs naturally in many kinds of rock, including those containing metals and coal.)

How mercury can enter our environment

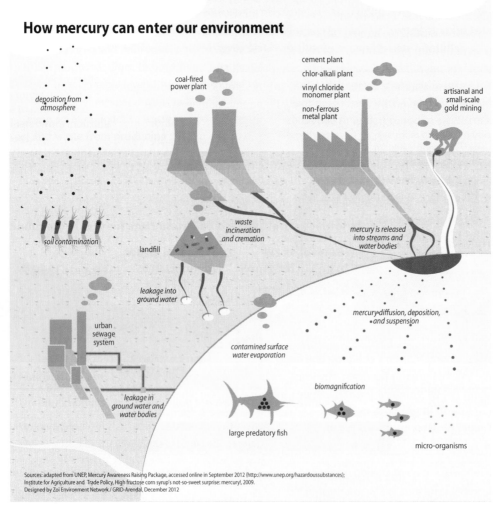

Sources: adapted from UNEP, Mercury Awareness Raising Package, accessed online in September 2012 (http://www.unep.org/hazardoussubstances);
Institute for Agriculture and Trade Policy, High fructose corn syrup's not-so-sweet surprise: mercury!, 2009.
Designed by Zoï Environment Network / GRID-Arendal, December 2012

This graphic shows the many ways that mercury can enter the environment and eventually accumulate in organisms.

Mercury in Daily Life and Effects on Human Health

People can be exposed to mercury in different ways. Those who work directly with the substance, such as small-scale gold miners, are most in danger of exposure. Also at risk are people whose main food source has been contaminated by mercury through pollution of the aquatic environment. However, many common products in the home contain mercury and have the potential to affect human health. This can occur through direct exposure by using the products or from improper disposal of mercury-containing items. Symptoms range in severity depending on the amount of exposure and the form of the mercury. The most toxic form is methylmercury, which can cause symptoms such as impaired balance, numbness in the hands and feet, general muscle weakness, and sensory impairment. In extreme cases, insanity, paralysis, coma, and death can result.

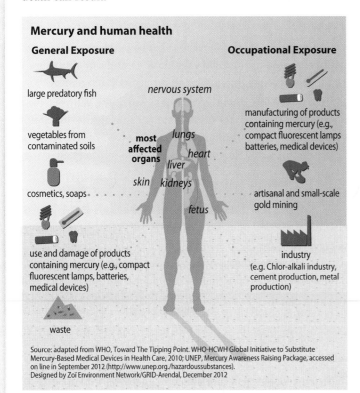

Mercury and human health

General Exposure

large predatory fish

vegetables from contaminated soils

cosmetics, soaps

use and damage of products containing mercury (e.g., compact fluorescent lamps, batteries, medical devices)

waste

nervous system

most affected organs lungs

heart

liver

skin kidneys

fetus

Occupational Exposure

manufacturing of products containing mercury (e.g., compact fluorescent lamps, batteries, medical devices)

artisanal and small-scale gold mining

industry (e.g. Chlor-alkali industry, cement production, metal production)

Source: adapted from WHO, Toward The Tipping Point. WHO-HCWH Global Initiative to Substitute Mercury-Based Medical Devices in Health Care, 2010; UNEP, Mercury Awareness Raising Package, accessed on line in September 2012 (http://www.unep.org./hazardoussubstances). Designed by Zoï Environment Network/GRID-Arendal, December 2012

This graphic shows ways that people can be exposed to mercury.

Who Needs to Act?

In January 2013, officials from around the world met in Geneva, Switzerland, to discuss how to put mercury regulations and bans into place on a wide scale. This set of regulations, agreed to by more than 140 countries, is known as the Minamata Convention. It aims to define measures to reduce the supply and demand for mercury, while managing its use and ensuring proper disposal by industries that continue to use it. A lasting solution to the dangers of mercury contamination will only come about through the combined efforts of governments, organizations, industries, and individuals.

Research and Analyze

1. What is the significance of the name of the Convention agreed to in Switzerland?

2. What common products in the home contain mercury? Find and name as many as you can. (There are many more than you might think, even if you do not include pharmaceutical products—which you should.) Organize your findings in a graphic organizer.

Communicate

3. Examine the graphics used in this case study. Convert the information presented to another form in order to demonstrate your understanding of what they communicate. For example, you could "translate" the information into several descriptive paragraphs, create a photo slideshow with captions, or develop a short story or skit that includes all the information.

4. Design an information campaign to alert your friends, family, and community to the threats posed by mercury and provide guidance on how to minimize exposure. (Assume that you do not have permission to reproduce the graphics from this case study, in whole or in part. Therefore, you will have to find a different way to communicate any information from them.)

Safety Precautions

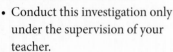

- Conduct this investigation only under the supervision of your teacher.

- Always remain in shallow, calm water when collecting samples.

- Be careful when handling living organisms.

- Use appropriate protective equipment such as apron, safety goggles, and gloves as well as taking any other safety precautions that are stated in associated Material Safety Data Sheets (MSDS).

- Be sure to wash your hands when you are finished the Lab.

Materials

- rubber boots or hip waders
- safety waist tether
- 1.2 m³ nylon screen or mesh net with a metal rim
- pan
- turkey baster or plastic forceps
- illustrated classification keys to the macroinvertebrates
- hand lens
- thermometer
- pH paper
- water quality test kits

Determining Water Quality

Chemical indicators of water quality include measurements of dissolved oxygen, phosphates, and nitrate levels, as well as the temperature and pH of the water. Biological indicators are organisms that reflect how polluted a water system is. Macroinvertebrates are organisms that are visible to the unaided eye and lack a backbone. A survey of the number and type of macroinvertebrates found in a waterway can provide a general indication of the presence of water pollutants and quality of water. It is important to keep in mind that organisms found in conditions of poor water quality may be found in any type of water, whereas organisms that are representatives of good-quality water are only found in water of good quality.

Pre-Lab Questions

1. How can water quality affect biodiversity in a water resource?
2. Name seven types of water pollutants and describe the effects each can have on an aquatic ecosystem.

Question

How can you assess water quality using biological and chemical indicators?

Procedure

Working in small groups and under the supervision of your teacher, you will visit a local stream to identify biological and chemical indicators of the quality of water.

Part A: Determining Biological Indicators

1. Collect a sample of organisms from the river or stream bottom by placing the nylon screen or mesh net against the bottom and kicking against the bottom upstream of the net for at least 1 minute. You should overturn and scrape any rocks that are present. Be sure that your net is placed to intercept all of the floating debris stirred up by the kicking.
   Note: If you are testing a pond or lake where there is no current, use the net with a metal rim to scoop material from the bottom mud, especially around the base of any weedy areas.

2. Examine the larger bits of wood that are disturbed by your kicking, since some of the organisms you are attempting to collect may be stuck to the underside of the wood.

3. Wash away the mud and dirt by shaking the screen or net while holding it partly under the surface of the water.

4. Using the turkey baster or plastic forceps, transfer any collected organisms to the pan or paper plate, and group them by shape.

5. Once out of the water, use a classification key and hand lens to identify as many of the organisms as you can. Macroinvertebrates that you might see in waterways near you include crustaceans (such as crayfish), molluscs (such as clams and mussels), gastropods (such as snails), oligochaetes (such as worms), and insects. The illustrations on page 328 will help.

6. As you identify an organism in your sample, record it in a table like the one below. Use the chart on the next page to assign each organism a number of points. Collect at least 10 organisms.

Organisms Collected

Organisms	Points	Organisms	Points

7. Return the organisms to the water as close as possible to where you collected them.

Part B: Determining Chemical Indicators

1. Create a table like the one below.

Data

	Measurements and Observations
Air temperature	
Water temperature	
Appearance of the water	
Water pH	
Dissolved oxygen*	
Dissolved phosphates*	

*Your teacher may perform these tests as a demonstration.

2. Measure the air and water temperatures, and record them in your table.

3. Determine the pH of the water by dipping a piece of pH paper into a small sample of water on your plate or pan.

4. If your teacher approves, follow the instructions on the water quality test kits to determine the dissolved oxygen and phosphates in the water. Alternatively, your teacher may perform these tests as a demonstration and provide you with the data.

5. Wash your hands once you are finished collecting the samples and performing any measurements.

6. Draw or photograph the collection site. Note any evidence of human activity near the waterway. Also describe the weather conditions when collecting samples.

Analyze and Interpret

1. Determine a water quality rating based on the points you assigned to the organisms collected. Use the following guide to help determine the quality:

   Excellent (>23)   Good (17–22)   Fair (11–16)   Poor (<11)

2. What water quality did the test kits indicate?

Conclude and Communicate

3. Did all groups identify the same water quality? If not, what differences were there?

4. Why were you asked to return the organisms to their original location?

Continued on next page >

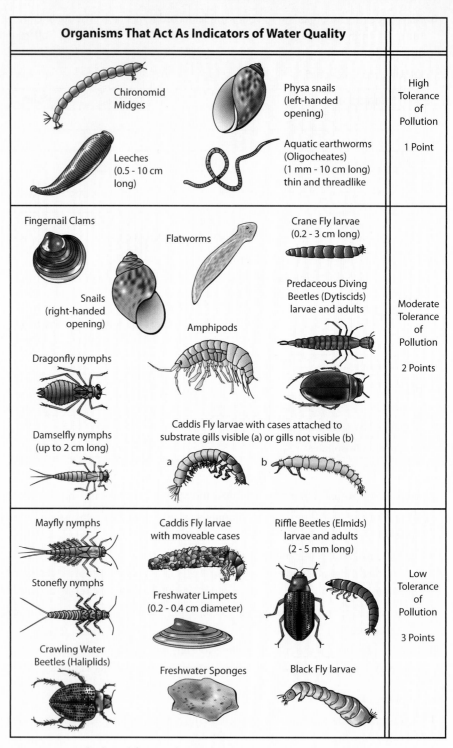

Organisms That Act As Indicators of Water Quality	
Chironomid Midges — Physa snails (left-handed opening) — Leeches (0.5 - 10 cm long) — Aquatic earthworms (Oligocheates) (1 mm - 10 cm long) thin and threadlike	High Tolerance of Pollution 1 Point
Fingernail Clams — Flatworms — Crane Fly larvae (0.2 - 3 cm long) — Snails (right-handed opening) — Amphipods — Predaceous Diving Beetles (Dytiscids) larvae and adults — Dragonfly nymphs — Damselfly nymphs (up to 2 cm long) — Caddis Fly larvae with cases attached to substrate gills visible (a) or gills not visible (b) a b	Moderate Tolerance of Pollution 2 Points
Mayfly nymphs — Caddis Fly larvae with moveable cases — Riffle Beetles (Elmids) larvae and adults (2 - 5 mm long) — Stonefly nymphs — Freshwater Limpets (0.2 - 0.4 cm diameter) — Crawling Water Beetles (Haliplids) — Freshwater Sponges — Black Fly larvae	Low Tolerance of Pollution 3 Points

Note: Because the larval forms of most insects are the most numerous, they are usually the focus of most surveys.

Water Resources and Personal Actions

Compared with the people in most other countries, Canadians are fortunate to have so much water that is readily available. Could such easy access be causing us to take water for granted? Canadian households use more water, per person, than most other households in the world. Responsibility for managing water resources is not limited to government, industry, scientists, and technologists. There are many ways that you, and your family, can conserve water resources.

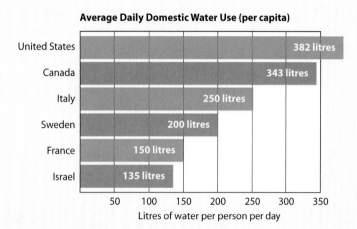

Average Daily Domestic Water Use (per capita)

Litres of water per person per day

United States — 382 litres
Canada — 343 litres
Italy — 250 litres
Sweden — 200 litres
France — 150 litres
Israel — 135 litres

Pre-Lab Questions

1. According to the graph above, what is the average amount of water that each Canadian uses in their home per day?

2. Besides your home, where else do you use water?

Question

What personal actions can you take to conserve water?

Procedure

1. With a partner, brainstorm ideas about actions each of you and your families can take to reduce the water that your household uses. Think about the activities that go on in your home that involve using water.

2. Use your ideas to create a water reduction plan that you and your family can use.

3. Carry out your plan and assess its success.

Analyze and Interpret

1. Was your plan successful? Describe the factors you used to assess its success.

2. Were you able to carry out some actions more easily than others? Explain why.

Conclude and Communicate

3. How did your household water use compare with the average given in the graph above?

4. How did your water reduction plan compare with others in the class? Account for any differences.

Chapter 10 SUMMARY

Section 10.1 Quantity, Uses, and Quality of Water

Water is one of our most important resources. It is necessary for organisms to live and supports our lives in many ways. Since only a small amount of fresh water is available, it is important to protect water quality against pollutants such as pathogens and chemicals.

Key Terms
coliform bacteria
eutrophication

Key Concepts
- Water is essential for life on Earth. Almost 98% of Earth's water is in the form of salt water. Only a small portion of the fresh water is available for use by living things.
- Water is a key component of all cells. It is also needed to make and use products such as food, shelter, medications, and motor vehicles.

- Point sources of water pollution include sewage treatment plants and factories that dump pollutants into waterways. Nonpoint sources of water pollution include run-off from farms and roads.
- Pollutants include pathogens, such as bacteria and viruses, animal feces and urine, nitrates and phosphates, heavy metals such as lead and mercury, petrochemicals, and other organic chemicals.
- Pathogens in waterways can cause diseases such as cholera, typhoid, and hepatitis. Therefore, monitoring the quality of drinking water is performed by water quality professionals. One type of pathogen that is routinely tested for is coliform bacteria.
- Toxic chemicals in water, such as PCBs and DDT, can accumulate in the tissues of organisms. When they are eaten by a predator, concentrations of the toxic substances increase in that predator, which is called biomagnification.

Section 10.2 Managing Water Resources

Both low-tech and hi-tech methods are used to manage water quality. Protecting and improving water quality is the responsibility of governments at all levels, as well as of businesses and industries. In addition, individual members of society as well as groups of citizens working together can protect and improve water quality.

Key Concepts
- Managing and protecting water resources in Canada involves federal, provincial, and municipal levels of government. Different agencies in these governments are responsible for different aspects of water management.

- People have designed high-tech ways to monitor and improve water quality, such as satellites and UV sterilization. There are also low-tech approaches, such as solar ovens.
- There are many ways that individuals can help improve water quality, such as not dumping chemicals or dangerous materials down drains and limiting their use of polluting products.
- There are many grassroots initiatives aimed at improving water quality and water conservation. These can act at the local, provincial, national, and international levels.

Chapter 10 REVIEW

Knowledge and Understanding

Choose the letter of the best answer below.

1. Which one of the following statements about water is *incorrect*?
   a) Most species on Earth live in aquatic environments.
   b) All of the fresh water on Earth is available for use by living organisms.
   c) All cells of living organisms contain water.
   d) Water is needed in chemical reactions in the cells of organisms.
   e) Water is needed to provide us with food.

2. An example of a nonpoint source of water pollution is
   a) sewage treatment plants
   b) power plants
   c) paper mill factory
   d) run-off from farm fields
   e) oil wells

3. When an industrial facility releases warm water into a waterway,
   a) more oxygen is available for fish and other aquatic organisms near this discharge area
   b) cold-water fish species will usually move near the warm water discharge point to raise their body temperature
   c) some aquatic organisms will be stressed by the heat, making them vulnerable to disease
   d) reproductive rates of cold-water species will likely increase in the warm water area
   e) it is beneficial to the nearby aquatic ecosystem

4. After a heavy rainfall, a public beach might be closed by the public health authority due to
   a) an excessive amount of acid rain in the water
   b) the presence of sediment from run-off
   c) the possibility of cholera spreading in the human population
   d) an increase in the number of cold-water fish species
   e) the presence of coliform bacteria in quantities that are beyond the acceptable range

5. Which statement about organic chemicals such as DDT and PCBs in waterways is *incorrect*?
   a) Since organic chemicals resist breaking down, they can last a long time in the environment.
   b) Sharks and birds that eat fish that live in the waterways will accumulate high levels of DDT and PCBs in their tissues.
   c) Biomagnification occurs when DDT and PCBs build up in the cells of phytoplankton.

d) Bioaccumulation can occur when heavy metals enter the cells of organisms that are lowest in the food chain.
   e) The production of DDT and PCBs is banned in Canada.

6. Two signs of eutrophication in a waterway are
   a) algal blooms and high levels of oxygen in the water
   b) a decrease in the number of algae and low levels of oxygen in the water
   c) high levels of phosphorus entering the water and the presence of algal blooms
   d) a decrease in bacteria and an increase in fish populations in the waterway
   e) clear water with a slight odour to it

7. Which of the following technologies is used to protect water resources?
   a) Earth Observation technologies that are used to treat algal blooms.
   b) Earth Observation technologies that monitor nonpoint sources of pollution discharge.
   c) Ultraviolet light being used to kill bivalve shellfish.
   d) Living machines using live plants and computerized monitoring systems to clean wastewater.
   e) UV water sterilization in private homes to remove sedimentation.

8. Solar ovens are a cost-effective way to make drinking water from salt water because
   a) the salt will evaporate from the oven, leaving fresh water behind to drink
   b) the solar oven uses only a small amount of electricity to work
   c) the evaporated water from the heated saltwater can be condensed to provide salt-free water
   d) the solar heat melts the salt and only fresh water remains
   e) each solar oven is able to produce several thousand litres of drinking water per day

Answer the questions below.

9. Water is used in many ways in daily life. Briefly describe three ways that the use of cars depends on water.

10. Pathogens are one type of water pollutant.
    a) What does the term *pathogen* refer to?
    b) Name three human diseases that are caused by drinking water that is contaminated with pathogens.

11. Why are bivalve shellfish that are contaminated with toxins from algae a potential threat to human health? Provide two examples of bivalve shellfish that could be a problem, and describe the health problems that contaminated ones can cause in people.

12. Federal, provincial, and municipal levels of government in Canada are involved in the management of water resources. Describe three examples of how water resources are managed by government.

13. *E. coli* bacteria live in the intestines of humans and other animals. Most strains of *E. coli* do not cause any harm. Some, however, do make people sick.

a) Describe two ways that water may be contaminated by *E. coli*.

b) What are some effects of *E. coli* contamination of water on humans and aquatic ecosystems?

c) Describe one method for preventing contamination of water sources by *E. coli*.

d) What are coliform bacteria? Describe how water is tested for coliform bacteria.

14. Explain how a fenced-off stream can help to protect water resources in agricultural areas. Is this an example of high-tech or low-tech water resource management? Explain your choice.

15. Describe two ways that constructed wetlands can be used to treat wastewater.

16. Describe three ways that you can demonstrate responsible water resource use at home.

17. Explain why you should not pour common household chemicals such as bleach or motor oil down the drain or into a storm sewer.

18. Describe two ways that the presence of heavy metals in waterways can affect organisms.

Thinking and Investigation

19. As a water quality professional, you receive the following information in a water quality report: 50 colony-forming units per 100 millilitres (50 CFU/100 mL).

a) What actions would you recommend if this report came from a well that supplies drinking water to a community, and why?

b) What actions would you recommend if this report came from a local swimming pool, and why?

20. Measuring contaminants in water requires determining very small amounts of materials, especially if they are very toxic substances. Often the units used to report concentration of contaminants are parts per million (ppm) and parts per billion (ppb).

a) Which represents the higher concentration of water pollutant: 1 ppm or 1 ppb? Why?

b) If the water contaminant measured was lead, would UV radiation be an effective way to treat the water? Explain your answer.

21. The diagram below shows the increasing concentration of a pollutant within an aquatic food chain. Explain the information presented in this diagram.

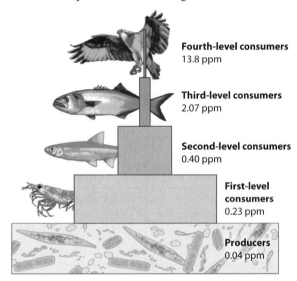

Fourth-level consumers
13.8 ppm

Third-level consumers
2.07 ppm

Second-level consumers
0.40 ppm

First-level consumers
0.23 ppm

Producers
0.04 ppm

22. An estuary is an area of coastal wetland that occurs where the fresh water of a river meets the salt water of an ocean. Minerals and other nutrients that come from the land, rivers, and the ocean accumulate in estuaries. This makes them ideal habitat for both plants and animals. The nutrients are distributed throughout the estuaries by winds, currents, and tides. The action of tides also flushes pollutants and debris out of estuaries.

a) Based on this description, infer why estuaries are rich environments for biodiversity.

b) How do you think activities such as building homes, commercial shipping, and sports fishing could affect this biodiversity?

23. How would planting native plants on stream banks help to reduce sediment pollution? Would it be as beneficial to plant non-native plants on the stream banks? Explain your answer.

24. How can pesticides that are applied to plants as a spray become a water pollutant?

Communication

25. You see an ad stating that a company is selling "pure" water that they collect directly from mountain streams. Write a letter to the editor of the newspaper explaining why the water is not pure, even though it is good-quality drinking water.

26. You are attending a community meeting about a local company that has been dumping their waste into a nearby river. At the meeting, a company representative proposes an alternative method of disposal, which is to pump the waste into the ground. Compose an argument for or against this proposed change that you could stand up and present at the meeting. Keep in mind that you have less than 5 minutes to speak, so your argument must be brief and to the point.

27. A community relies a great deal on tourism due to its lakes and other waterways. Prepare a summary about what algal blooms are and how they could affect tourism in the community, and present the information in a format of your choice. Keep in mind that the information must be understood by the general public.

28. Explain the difference between bioaccumulation and biomagnification so that students in Grade 7 can understand. Include a diagram to help support your explanation.

Application

29. Boil water advisories are announcements telling people they should boil their tap water for a minimum of 1 minute before using it for drinking, preparing food, and brushing their teeth.
    a) What type of water pollutant is this type of treatment for?
    b) Describe a situation or condition that would require a boil water advisory to be issued.
    c) Why does Health Canada recommend that infant formulas should always be prepared using tap water that has been boiled for a minimum of 1 minute?

30. Before the 1950s, lead pipes were used to deliver municipal water to homes. Also, until 1990, lead solder was used by plumbers when sealing, connecting, or repairing pipes.
    a) Why are lead pipes and lead solder for plumbing no longer allowed to be used?
    b) Research what your local public health office recommends that people do if they have lead pipes in their homes.

31. Chlorination acts as a disinfectant—it kills or inhibits growth of live pathogens. Drinking water is treated with chlorine. Also, swimming pools are chlorinated in order to maintain a healthy swimming environment.
    a) You hear someone say "Better safe than sorry, I'll throw in extra chlorine in my pool to make sure I am protected from diseases." Why should someone not do this?
    b) According to information on Health Canada's website, what chemical by-products can be formed in chlorinated water? What is an alternative to using chlorine for disinfecting municipal drinking water?

32. Trucks like the one shown here apply salt to roads in winter to reduce ice on the roads.

    a) What happens to the salt that is applied to the roads?
    b) Briefly describe how run-off in the spring from roads that have been treated with salt over the winter might affect plants near the roads.
    c) Is this a point source or nonpoint source of pollution? Explain your answer.
    d) Perform research to identify alternatives that have been used or proposed for de-icing roadways. Are these as effective?

33. Make a list of three issues regarding water pollution that might be reported to an environmental hotline by the general public.

34. **Did You Know?** Reread the Benjamin Franklin quotation on page 309. Use the title "The Worth of Water" to express your thoughts and feelings about this vital resource.

Pause and Reflect

How could you incorporate what you have learned in this chapter into your daily actions or choices?

Kim McKay-McNabb: Protecting Environmental Health

The lives of Canada's First Nations people are closely linked to the land. Diets include traditional foods such as fish and game that are vulnerable to pollutants in air, water, and soil. Kim McKay-McNabb, of the Sakimay First Nation in Saskatchewan, has made it her life's work to ensure that First Nations people have the tools they need to protect themselves against environmental contaminants.

"My background is in holistic health and in the idea of all things being connected," Kim says, "which I learned from our Elders and knowledge keepers throughout my First Nations teachings." She adds that the field of environmental health was "a good fit for me because, as with the four directions and the medicine wheel teachings, the physical, mental, emotional, and spiritual aspects all relate to Mother Earth."

Since 2005 Kim has been coordinator of the National First Nations Environmental Contaminants Program (NFNECP), which informs communities south of the 60th parallel about possible threats, helps them make better decisions on ways to protect their health, and advises them on the action they can take to protect their communities. The NFNECP and its Drinking Water Quality Program work to support communities doing their own research, combining scientific and traditional ways of investigating their surroundings. Kim is also a member of the advisory group of the First Nations Environmental Health Innovations Network. This Web-based network promotes this traditional knowledge in environmental health research and provides a place for communities to exchange information with one another and with other groups interested in the environment.

Kim teaches in the science department at First Nations University in Regina, and she completed her Doctorate in Clinical Psychology at the University of Regina in 2012. She says her interest in psychology is closely linked to her work in environmental health: "There are significant emotional reactions to our Earth becoming sick. When First Nations communities are living on 'boil water' advisories as their lands and waters are being contaminated, it has a direct result on the mental health of the communities."

Environmental Science at Work

Focus on Air and Water Quality and Environmental Health

Occupational Therapist

Environmental Lawyer

Air Quality Specialist

Air Quality Engineer

Air and Water Quality and Environmental Health

Environmental Auditor

Limnologist

Limnologists monitor and protect freshwater resources and may be employed by universities, government agencies, or industry. They investigate abiotic factors such as water chemistry and biotic factors such as vegetation, microbes, and invertebrates.

Compliance Promotion Specialist

Compliance promotion specialists provide technical, scientific, and management advice to industries to help them comply with federal acts and regulations. They may also be involved in increasing awareness, education, and promoting best practices.

Air Quality Technician

Air quality technicians monitor, measure, and report on air pollutants in both urban and rural areas so that their health effects can be properly assessed. They may also work in environmental emergency situations to determine the effects of events such as fires or chemical spills.

For Your Consideration

1. What other jobs and careers do you know or can you think of that involve air and water quality and environmental health?

2. Research a job or career in this field that interests you. What essential knowledge, skills, and aptitudes are needed? What are the working conditions like? What attracts you to this job or career?

Building Improvements for Better Health

Air and water pollution is not just a concern when outdoors. Today, health standards for buildings are stricter than in the past because there is greater awareness of the negative effects that some materials have on indoor air and water quality. In this project, you will investigate health standards for buildings and improvements that can be made to reduce the negative effect that a material can have on human health.

In the past, building insulation contained asbestos, which is a health hazard. Cellulose insulation that is made from recycled paper products is now available.

Question

What improvement can be made to a building to help reduce the negative impact it has on human health?

Initiate and Plan

1. In a group, select what type of building you want to consider. Examples include a home, office building, or warehouse.

2. Brainstorm a list of changes to the building you could make that would improve the interior air or water quality. Consider the insulation material, indoor paints, material used for plumbing pipes, air filtration systems, and water filtration systems.

3. Make a research plan based on the topics outlined in step 6. Your research plan should include
   • using a variety of sources of information, such as government, academic, and industry
   • assigning each group member a topic to research

4. Have your teacher approve your research plan before proceeding.

Perform and Record

5. Describe the structural change your group has chosen and explain why the alteration was developed.

6. Research the technological, societal, and economical aspects of the alteration. Consider the following:
   • What current building health standards is the alteration designed to meet? What improvement to human health is the alteration supposed to make?
   • In new buildings, is the alteration required or only recommended? Also, does the alteration only apply to new buildings, or must it be made in all structures?
   • What are the estimated costs of making this alteration? Are there programs offering financial assistance for the costs of making the alteration?
   • What is the lifespan of the alteration? Does it only need one-time action for the life of the structure? If it requires continual replacement of materials, what costs are associated with this?
   • Are different types of materials to make the alteration available? What are the differences between these products or approaches?

7. As a group, combine your research findings.

Analyze and Interpret

1. Using an appropriate graphic organizer, summarize the advantages and disadvantages of making the alteration.

2. Complete a risk-benefit analysis for the alteration.

Communicate Your Findings

3. Present your findings to another group in the class. Record the suggestions for improvement they provide.

Assessment Checklist

Review your project when you complete it. Did you ...

☑ **K/U** choose a building and an alteration that is designed to improve air or water quality?

☑ **T/I** assess your sources for credibility and bias?

☑ **A** implement your research plan, with teacher approval?

☑ **A** research the technological, societal, and economical aspects of the alteration?

☑ **C** cite your research sources using proper academic format?

☑ **C** choose an appropriate format for presenting your findings, keeping your audience in mind?

Analyzing an Environmental Issue from a Certain Perspective

People help to preserve and protect our environment in many ways. As a result, there are many professions related to environmental science, each with its own role and contributions. In this project, you will choose an environmental issue that you are interested in and assume the role of someone whose profession is associated with that issue. As part of your analysis, you will consider the contributions someone in that role makes to the environmental issue.

Question

What is the current status of the environmental issue you chose, and what contributions do certain individuals make in addressing that issue?

Initiate and Plan

1. Based on the environmental issues you have learned about in this course, choose one that interests you.

2. Consider the different people who make contributions to address the environmental issue you chose. You may need to research some of this information. Some suggestions are
   - lawyer
   - head of a non-profit organization
   - lobbyist
   - research scientist
   - educator
   - volunteer coordinator
   - fundraiser
   - professional writer

   Choose one of these roles or another one you have thought of.

3. Have your teacher approve the environmental issue and the role you have chosen.

Perform and Record

4. Research how someone who works in the profession you have chosen would contribute to the environmental issue that interests you. Answer the following questions as part of your research:
   - What educational background and skills are required for the job?
   - What is the current status of the environmental issue? What progress has been made in addressing the issue?
   - In what particular way does an individual in your chosen profession help address the environmental issue?
   - What biases are associated with addressing the environmental issue in the role you chose?

Analyze and Interpret

1. Summarize your research findings about the current status of the environmental issue and how the person in the role you chose contributes to addressing the issue.

2. What factors or sides of the issue are not considered or addressed by an individual in the profession you chose?

Communicate Your Findings

3. In your role, share your findings with the class as though you are making a presentation at a career day at your school. Use a format of your choice, keeping in mind your audience and the purpose of the presentation.

Assessment Checklist

Review your project when you complete it. Did you ...

☑ **K/U** have your teacher approve your choice of environmental issue and profession?

☑ **T/I** research the requirements for the profession?

☑ **T/I** research the current status of the environmental issue or program that you chose?

☑ **A** prepare a presentation that is appropriate for someone at a high school career day?

☑ **C** communicate your findings with your audience in mind?

Unit 5 REVIEW

Knowledge and Understanding

Choose the letter of the best answer below.

1. Which of the following statements is correct?
   a) Air pollutants only enter the body through the nose.
   b) Pollutants are substances that are harmful or could be harmful to organisms.
   c) Contaminants are also considered pollutants.
   d) Pollutants are substances that occur in concentrations higher than expected.
   e) Water pollutants can only enter the body if the water is ingested.

2. An example of nonpoint source pollution is
   a) a can of VOC-containing paint.
   b) a contaminated water well.
   c) a drain pipe carrying waste from an industry to a river.
   d) emissions from the smokestack of a factory.
   e) emissions from transportation vehicles.

3. Which of the following is a list of primary air pollutants?
   a) carbon monoxide, nitrogen oxides, sulfur oxides
   b) nitrogen oxides, sulfur oxides, photochemical smog
   c) photochemical smog, particulate matter, nitrogen oxides
   d) bacteria, particulate matter, radon
   e) None of the above is correct.

4. Which of the following statements about coliform bacteria is *incorrect*?
   a) They are considered a pathogen.
   b) Their presence indicates the water is contaminated with fecal matter.
   c) Water sources tested include drinking water and water at public beaches.
   d) They consist only of *E. coli* bacteria.
   e) They are tested for by placing a water sample in a medium that supports bacterial growth.

5. Which of the following is used to protect and improve water quality?
   a) wetlands
   b) solar ovens
   c) Earth Observation technologies that involve satellite monitoring
   d) UV sterilization
   e) All of the above are.

Answer the questions below.

6. Describe three ways that pollutants can enter the body of an organism.

7. Several types of water pollutants have certain effects on humans and aquatic ecosystems.
   a) Name two sources of sediment pollution in a waterway. Describe three ways that sediment pollution can affect organisms.
   b) Why is heat considered a water pollutant? What effects can it have on an aquatic ecosystem?
   c) How do bioaccumulation and biomagnification contribute to the effects of some water pollutants?

8. Photochemical smog is a harmful air pollutant in large cities.
   a) Briefly describe what it is and how it forms.
   b) How have catalytic converters helped to reduce the amount produced?

9. Describe three indoor air pollutants and their effects on human health.

10. Water is one of our most important resources.
    a) Describe three products or processes that we rely on water for.
    b) What effects can pathogen water pollutants have on humans? Describe a test used to monitor pathogen contamination of water.

Thinking and Investigation

11. Particulate matter is a major source of air pollution.
    a) Why might particulate matter aggravate respiratory system conditions, such as bronchitis or asthma?
    b) What guidelines have been developed to help reduce particulate matter pollution?
    c) Describe one technology that industries have developed to reduce particulate matter emissions. Explain how it works.

12. The graph below shows the distribution of fresh water on Earth.

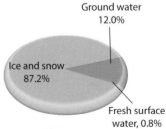

Fresh water

a) Describe the information represented in this graph.
b) If there is 2.4 L of fresh water for every 1.0×10^2 L of total water on Earth, what volume of that fresh water is available as liquid surface water?

13. Air quality can be described using the Air Quality Health Index (AQHI).

   a) What is the AQHI? What groups are responsible for developing it?

   b) If the AQHI is 8 and you have a cold, would the air quality likely affect you when exercising outdoors? Explain.

   c) Why is the health of young children and seniors adversely affected by air pollution?

14. An estuary is an area of coastal wetland. Nutrients from the area accumulate in estuaries, and winds and current distribute them. This makes estuaries ideal habitats for both plants and animals. The action of tides also flushes pollutants out of estuaries.

   a) Infer why estuaries are rich environments for biodiversity.

   b) How do you think activities such as building homes, commercial shipping, and sports fishing could affect this biodiversity?

Communication

15. Study the diagram below. In a table, record all sources of pollution shown, whether each source is point or nonpoint, and a suggestion for how each pollution problem may be solved.

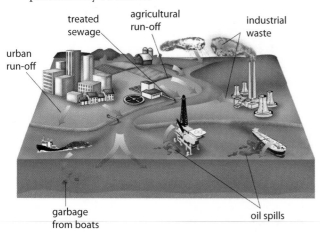

16. Draw a diagram with descriptive captions that could be used to inform the general public about algal blooms. Include the following information: what they are, why they form, their effects on aquatic life, and the quality of the water where they occur.

17. Develop a one-page bulletin that could be posted in your school describing three steps each student can take to help conserve water.

Application

18. The term *biodegradable* refers to anything that bacteria breaks down into biomass, water, and carbon dioxide gas. The breakdown of biodegradable soap occurs in soil and can take months or even years.

   a) Many hikers and campers believe it is safe and "green" to bathe in rivers and lakes if the soap they use is biodegradable. Explain why this belief could be incorrect.

   b) Suggest an alternative to soap that hikers and campers could use to clean themselves and objects such as dishes and clothing.

19. The Gulf of St. Lawrence is home to beluga whales, shown below. There were once thousands of belugas in this population, but now there are fewer than 1000 whales. High levels of harmful organic chemicals have been found in the bodies of dead whales. Describe how bioaccumulation and biomagnification have contributed to harming the beluga whales.

20. Many school buses have pollution control devices to help reduce the harmful effects of exhaust fumes. Describe three strategies that can be used in the schoolyard to help reduce the possibility of emissions being inhaled by people near the buses.

21. The Drive Clean program in Ontario ensures that Ontarians maintain their vehicles so that fewer pollutants will be emitted into the environment.

   a) Some people might have a negative view of this program if it results in having to pay for car repairs. What could you say to convince them that the program is worthwhile.

   b) Why are people who live in Northern Ontario excluded from the program? Should the program eventually cover Northern Ontario as well? Explain.

22. We are exposed to primary air pollutants on a daily basis.

   a) Describe three primary air pollutants that you are commonly exposed to and their sources.

   b) What can you do to reduce these pollutants or your exposure to them?

Guide to the Appendices

Guide to the Appendices

Green Chemistry

Chemistry is going green! Green chemistry is a sustainable way of looking at the production, use, and disposal of chemical products or chemical processes. In the past, chemists tried to reduce the environmental impact of chemical products and processes by focussing on one area, such as cleaning up toxic waste. The goals of green chemistry include making all steps as sustainable as possible, including planning, synthesis, re-use, disposal, and clean-up. To increase sustainability, the principles of green chemistry look for ways to reduce energy use, resource use, and waste generation, while also improving safety and efficiency.

The 12 Principles of Green Chemistry*

1. Prevention
   It is better to prevent waste than to treat or clean up waste after it has been created.

2. Atom Economy
   Synthetic methods should be designed to maximize the incorporation of all materials used in the process into the final product.

3. Less Hazardous Chemical Syntheses
   Wherever practical, synthetic methods should be designed to use and generate substances that possess little or no toxicity to human health and the environment.

4. Designing Safer Chemicals
   Chemical products should be designed to effect their desired function while minimizing their toxicity.

5. Safer Solvents and Auxiliaries
   The use of auxiliary substances (for example, solvents, separation agents, etc.) should be made unnecessary whenever possible and innocuous when used.

6. Design for Energy Efficiency
   Energy requirements of chemical processes should be recognized for their environmental and economical impacts and should be minimized. If possible, synthetic methods should be conducted at ambient temperature and pressure.

7. Use of Renewable Feedstocks
   A raw material or feedstock should be renewable rather than depleting whenever technically and economically practicable.

8. Reduce Derivatives
   Unnecessary derivatization (modifications that temporarily change a compound for various purposes) should be minimized or avoided if possible, because such steps require additional reactants and can generate waste.

9. Catalysis
   Catalytic reactants (as selective as possible) are superior to stoichiometric reactants.

10. Design for Degradation
    Chemical products should be designed so that at the end of their function they break down into innocuous degradation products and do not persist in the environment.

11. Real-Time Analysis for Pollution Prevention
    Analytical methodologies need to be further developed to allow for the real-time, in-process monitoring and control prior to the formation of hazardous substances.

12. Inherently Safer Chemistry for Accident Prevention
    Substances and the form of a substance used in a chemical process should be chosen to minimize the potential for chemical accidents, including releases, explosions, and fires.

* Anastas, P. T. and Warner, J. C. Green Chemistry: Theory and Practice. Oxford University Press: New York, 1998, p. 30.

Instant Practice

Using print and/or Internet resources, describe how the principles of green chemistry have been applied to the life cycle of a chemical process or chemical product in Canada. What improvements, if any, occurred when these principles were applied? How can you use the principles of green chemistry in your labwork?

Analyzing STSE Issues

What STSE issue have you heard about lately? Did it concern banning pesticide use, for example? Or maybe it was about the possible health risks of eating undercooked hamburger or growing genetically modified crops.

STSE stands for **S**cience, **T**echnology, **S**ociety, and the **E**nvironment. STSE issues involve people talking and making decisions about events that affect them and others—either directly or indirectly, either now or in the future. An issue is anything that causes different people to engage in a debate. They can debate the usefulness of something or how to solve a problem. Regardless of the topic, analyzing STSE issues involves several important steps:

- teach yourself about the issue
- think about different ways to solve it
- think about the consequences that each solution might have
- decide on what you think is the best solution at a particular time
- communicate your decision respectfully

The flowchart below breaks down the process of analyzing an STSE issue into a few steps. You can use a flowchart like this to analyze STSE issues.

What Comes Next?

Present your decision in a respectful manner. Make an action plan to set out the specific steps that need to be done. Include a timeline in your plan that states when each step will be completed.

Look at your solution again once some time has passed. All the parts of STSE issues—**S**cience, **T**echnology, **S**ociety, and the **E**nvironment—are constantly changing. A decision made today might not apply in months or years to come, so you might need to re-evaluate your plan at various times in the future.

1. Define the Issue
People often have strong feelings or opinions about an issue. In order to analyze any issue, it is important to make sure that everyone agrees on what it is they have feelings or opinions about.

2. Gather Information about the Issue
Sources of information include the library, the Internet, and newspapers or magazines. If the issue you are analyzing is local, you can talk directly to people involved or attend a community debate. All of these parties have a stake (an interest) in the outcome to this problem. They are known as "stakeholders"—and you may be one of them.

3. Identify Possible Solutions
Record each solution that has been proposed. If you have an idea about a solution that has not been suggested yet, include that, too.

4. Identify Consequences That Could Result from Each Solution
Identify the pros and cons of each solution. (Some people call this doing a risk-benefit analysis.) Set up a chart to record your points. Things to consider include effects on people, other living things, and the environment. If possible, also consider what financial costs might be involved.

5. Make a Decision
Look at each pro and con and decide if it is a major or a minor point. For example, one solution might have two positive points and three negative points. If the positives are major and the negatives are minor, the solution could still be a good choice. Make your decision.

Think about how the different stakeholders will react to the decision. Also think about how permanent your solution is. Will it fix the problem forever or only for now? Finally, make sure your decision is supported by facts and logic. If you think you were influenced too much by emotion, complete steps 4 and 5 again.

A Sample STSE Issue: Smoke-Free Restaurant Patios

Your town council is considering a new bylaw that would ban smoking on all restaurant patios.

1. Define the Issue

Patios are open spaces but people are still seated close to each other. Restaurant servers have to work on the patio to do their jobs. This new smoking bylaw would make the patio area smoke-free for everyone.

2. Gather Information about the Issue

Health Canada and the Canadian Lung Association report that second-hand smoke increases the risk of lung disease by 25% and heart disease by 10%. When a person is smoking, two thirds of the smoke enters the air around them and may be inhaled by people nearby. The negative effects of breathing in second-hand smoke accumulate over time.

For this issue, there are five main groups of people involved (stakeholders): restaurant patrons who smoke, restaurant patrons who do not, servers, restaurant owner, pedestrians who may be near the patio at any time.

The new bylaw would take away the rights of smokers who want to smoke while they are at the restaurant. However, allowing some patrons to smoke takes away the rights of non-smoking patrons who do not want to be exposed to cigarette smoke. It also exposes the restaurant employees who have to work on the patio to second-hand smoke. The restaurant owner could lose business if the restaurant no longer offers smokers a place to smoke.

4. Identify Consequences That Could Result from Each Solution

Solution	Consequences
Smoking is banned from patios.	• The restaurant owner must obey the law. • No more smoking will be allowed on the patio. • The law will apply to all restaurants with patios, so the owner might not lose too much business.
Construct a barrier on the patio to separate smoking and non-smoking sections.	• Non-smoking patrons will be able to eat on the patio in a smoke-free area. • Smokers will still have a place to smoke on the patio. • It might cost the restaurant owner a lot of money to install the barrier.
Servers can choose not to work on the patio.	• The rights of the servers to work in a smoke-free environment are protected. • Servers could lose tips, which make up a good portion of their wages.

5. Make a Decision

Here is one possible decision about the issue.

Sample Decision:
The bylaw should be passed so that no smoking is allowed on restaurant patios. This will take away smokers' rights to smoke, but it is more important to protect the health of the non-smoking patrons and servers who also share the patio. The owner should not lose much business, since the law applies to all patios in the city.

3. Identify Possible Solutions

Outcome of Bylaw	Solution	Pros	Cons
Bylaw passes.	Smoking is banned from patios.	• All non-smokers are protected from second-hand smoke.	• Rights of smokers are taken away. • Restaurant could lose business.
Bylaw does not pass Smoking is still allowed on patios.	Construct a barrier on the patio to separate smoking and non-smoking sections.	• Non-smoking patrons are protected from second-hand smoke. • Smoking patrons have a place to smoke. • Restaurant owner does not lose business.	• Non-smoking servers have to work in the smoking section. • Cost of the barrier could be expensive for the restaurant owner.
Bylaw does not pass. Smoking is still allowed on patios.	Servers can choose not to work on the patio.	• Servers have a choice to avoid the patio. • Smoking patrons have a place to smoke. • Restaurant owner does not lose business.	• Non-smoking patrons are not protected from second-hand smoke. • Servers who choose not to work on the patio may lose tips and make less money.

Scientific Inquiry

Scientific inquiry is a process that involves several steps. These include:

- making observations
- asking questions
- conducting a lab
- drawing conclusions

These steps may not always happen in the same order. Here is one model of the scientific inquiry process:

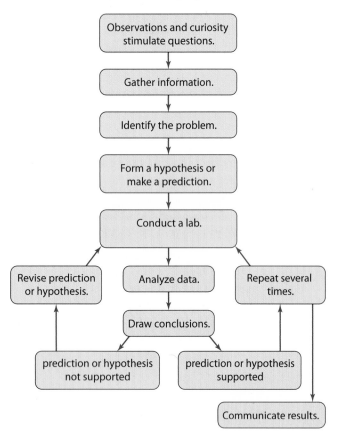

The Scientific Inquiry Process

Making Observations and Asking Questions

The rain has stopped, and the Sun is out. You notice that a puddle of water has disappeared from the sidewalk. You know that the water has evaporated, but how can you verify that? You can carry out a scientific inquiry.

Gathering Information and Identifying the Problem

First, you might observe what happens to some other puddles. You would watch them closely until they disappeared and record what you observed.

You might observe, "The puddle is almost gone." This is a **qualitative observation**, which is an observation in which numbers are not used.

Later, you might say, "It took five hours for the puddle to disappear completely." This is a **quantitative observation**, which is an observation that uses numbers.

Although the two puddles were the same size, one evaporated faster than the other. Your quantitative observations tell you that one evaporated in 4 h, and the other in 5 h. Your qualitative observations tell you that the puddle that evaporated in 4 h was in the sunlight. The other puddle was in the shade. You have now identified a problem to solve: Does water always evaporate more quickly in the sun than in the shade?

Stating a Hypothesis

Now you are ready to make a **hypothesis**. A hypothesis is a statement about an idea that you can test. Testing a hypothesis involves comparing two things to find the relationship between them. You know that the Sun is a source of heat, so your hypothesis might be: "If a puddle of water is in sunlight, then the water will evaporate faster than if the puddle is in the shade."

Making a Prediction

A **prediction** is a forecast about what you expect to observe. In this case, you might predict that three puddles in the sunlight (A, B, and C) will dry up more quickly than three puddles in the shade (X, Y, and Z).

Conducting a Lab

Conducting a Lab involves these steps:

- identifying variables
- controlling variables for a fair test
- recording and organizing data
- analyzing and presenting data

Identifying Variables

Suppose there was a strong breeze when you made your observations. The breeze is one factor that might affect evaporation. Energy from the Sun is another factor. Any factor that might affect a test is called a **variable**. Scientists try to identify every variable that could affect tests they conduct.

You need to control variables by changing only one at a time. The variable that is changed is the **independent variable**. In this case, the independent variable is the condition under which you observe the puddle. One such variable would be heat from the Sun. Another would be air moving across the puddle.

According to your hypothesis, adding heat will change the time it takes for the puddle to evaporate. The time in this case is the **dependent variable**.

The diagram below shows some examples of variables. Keep in mind that many Labs also have a **control**. A control is a test that you carry out with *no* variables so that you can observe whether your independent variable actually does cause a change.

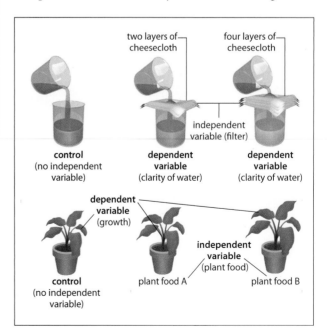

Appendix A

Identifying Variables

For each question, state your control, your independent variable, and your dependent variable.

1. Does adding compost to soil promote vegetable growth?
2. How effective are various kinds of mosquito repellent?

Controlling Variables for a Fair Test

A **fair test** is one that is valid and unbiased. To have a fair test, it is important to test only one variable at a time. If you do not, you will not know whether the breeze or the Sun's heat caused the water to evaporate.

How can you conduct a fair test on puddles? How can you be sure the puddles are the same size? In situations like this, scientists often use **models**. A model can be a picture, a diagram, a working model, or even a mathematical expression. To make sure your test is fair, you can prepare model puddles that you know are all exactly the same.

Before you start a laboratory investigation, review safety procedures and identify what safety equipment you may need. Refer to page xii in this textbook for more information on safety.

Recording and Organizing Data

Another step in conducting a laboratory investigation is recording and organizing data. Often you can record your data in a table like the one shown below.

Puddle Evaporation Times

Puddle	Evaporation Time (min)
A	37
B	34
C	42
X	100
Y	122
Z	118

Analyzing and Presenting Data

After recording your data, you will need to present the data in a format that you can analyze. Often, scientists make a graph, such as the bar graph below.

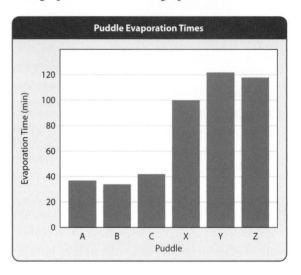

Forming a Conclusion

In complex Labs, there are many possibilities for error. That is why it is so important to record careful qualitative and quantitative observations.

After you have completed all your observations and analyzed your data, you are ready to draw a **conclusion**. A conclusion is a statement that says whether your results support or do not support your hypothesis. If your results do not support your hypothesis, you can use what you have learned in the Lab to come up with a new hypothesis to test. (Keep in mind that scientists often set up Labs without knowing what will happen. Sometimes they set out to find out whether something will *not* happen.)

A note about **causality** and conclusions: One thing to avoid is implying a causal relationship where none exists. In other words, even though two variables may show a relationship, it does not necessarily follow that changes in one variable *caused* the changes to the other variable. Unless you conduct a controlled experiment and replicate the results, or can show a definitive connection between the two variables, or eliminate all other variables as having an influence, you cannot imply causation in any conclusion you make.

Organizing Data in a Table

An important part of any successful scientific inquiry is recording and organizing your data. Often, scientists create tables in which to record data.

Planning to Record Your Data

Suppose you are investigating the water quality of a stream that runs near your school. You take samples of the numbers and types of organisms at two different locations along the stream. You need to decide how to record and organize your data. First, list what you need to record:

- the sample site
- the pH of the water at each sample site
- the types of organisms found at each sample site
- how many of each type of organism you collected

Creating Your Data Table

To record your data neatly and clearly, you will need to create a data table with these features:

- headings to show what you are recording
- columns and rows that you will fill with data
- enough cells to record all the data
- a title for the table (to appear above the table)

In your data table, every row representing a sample site will have at least three rows associated with it for the different organisms. You can leave space at the bottom of your table, in case you find more than three organisms at a sample site. (If you use the extra rows, be sure to identify which sample site the extra data are from.) Your data table might look like the one on the right above.

Observations Made at Two Sample Stream Sites

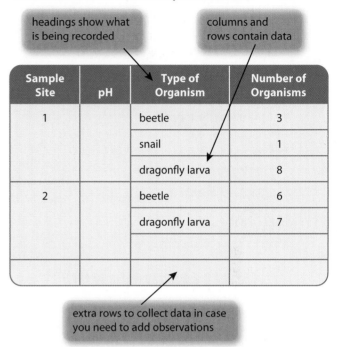

headings show what is being recorded

columns and rows contain data

Sample Site	pH	Type of Organism	Number of Organisms
1		beetle	3
		snail	1
		dragonfly larva	8
2		beetle	6
		dragonfly larva	7

extra rows to collect data in case you need to add observations

Instant Practice

Organizing Data in a Table

You are interested in how weeds grow in a garden. You decide to collect data from your garden every week for a month. You will identify the weeds and count how many there are of each type of weed. Design and draw a data table that you could use to record your data.

Copy the following data table into your notebook and fill in the missing title and headings. The Lab tests the effect of increased fertilizer on plant height. There are four plants, and measurements are being taken every four days.

Day 1	Plant 1	5 mL	
	Plant 2	10 mL	10 cm
		15 mL	
		20 mL	

Communicating Your Labwork in a Lab Report

To communicate their discoveries, scientists submit papers to scientific journals for publication. Other scientists then verify or refute the results and use these results for further investigations of their own. Your Lab reports are similar to scientific papers. The sample Lab report below shows how you can use the headings in the Labs in this textbook to help you organize your report.

Give your Lab report a title. Write the title at the top of the page.

Write the question that you tried to answer when you conducted the Lab.

Make a hypothesis and write it out in a full sentence.

Your hypothesis must be something that can be tested. It should also be directly related to your question.

Make a list of all the materials that you used in the Lab. If specific quantities of substances were used, include the amounts.

Refer to the Procedure for the Lab on the relevant page in your textbook. Write out only modifications to the steps in the Procedure (if there are any). If you collected data, include the data in a table. If you made a graph from your data, include the graph.

Acids and Bases in the Home

Question
How can you determine if a household liquid will be an acid or a base based on how it is used?

Hypothesis
It is not possible to predict whether a household liquid is an acid or a base.

Materials
8 test tubes	dilute acid
test tube rack	dilute base
glass stirring rod	vinegar
medicine dropper	mayonnaise
labelling tape	liquid drain cleaner
paper towels	window cleaner
red litmus paper	cola soft drink
blue litmus paper	liquid laundry soap
water	

Procedure
CAUTION: Do not let any of the liquids get on your skin. If any liquid does get on your skin, rinse your skin with plenty of water.
1. Prepare a data table with the following headings: Sample, Blue Litmus Paper, Red Litmus Paper. Make eight rows and write the name of the sample in the first column. Record the data in the other columns.
2. Put the eight test tubes in the test tube rack. Label each test tube with tape by writing the name of each of the eight liquids in the materials list on the tape.
3. With the medicine dropper, put about 1 mL of each liquid into each of the test tubes with the correct label. Rinse the medicine dropper with water between each type of liquid.
4. Place eight small pieces of red litmus paper and eight small pieces of blue litmus paper on a paper towel.

5. Use the medicine dropper to transfer one drop of the dilute acid to a piece of red litmus paper. Transfer another drop of dilute acid to a piece of blue litmus paper.
6. Record the final colour of each piece of litmus paper.
7. Rinse the medicine dropper with water and dry it with some paper towel.
8. Repeat steps 4 through 6 for each of the other seven liquids.

Litmus Paper Tests Used with Acids and Bases

Sample	Blue Litmus Paper	Red Litmus Paper
dilute acid	turned red	stayed red
dilute base	stayed blue	turned blue
vinegar	turned red	stayed red
mayonnaise	turned red	stayed red
liquid drain cleaner	stayed blue	turned blue
window cleaner	stayed blue	turned blue
cola soft drink	turned red	stayed red
liquid laundry soap	stayed blue	turned blue

Analyze and Interpret

1. When acid was placed on red litmus paper, it stayed red. When acid was placed on blue litmus paper, it turned red.
2. When base was placed on red litmus paper, it turned blue. When base was placed on blue litmus paper, it stayed blue.
3. The following table shows the results for the household liquids.

Results for Household Liquids

Red litmus paper stayed red. Blue litmus paper turned red.	Red litmus paper turned blue. Blue litmus paper stayed blue.
vinegar	drain cleaner
mayonnaise	window cleaner
cola soft drink	liquid laundry soap

> Examine your data and look for relationships between variables or other factors. The questions in the "Analyze and Interpret" section of the Lab you conducted will guide you through this section. If you made a data table or a graph, explain its meaning.

Conclude and Communicate

4. Acid turns blue litmus paper red but red litmus paper stays red.
5. Base turns red litmus paper blue but blue litmus paper stays blue.
6. Vinegar, mayonnaise, and cola soft drinks are acids.
7. Drain cleaner, window cleaner, and liquid laundry soap are bases.
8. The hypothesis is not correct. It is possible to predict whether a household liquid is an acid or a base. Cleaning liquids are usually bases and food liquids are usually acids.

> Compare your results with your hypothesis. State whether the results supported the hypothesis or not. State any conclusions you made. The questions in the "Conclude and Communicate" section of the Lab you conducted will guide you in writing this section.

Appendix A

Constructing and Interpreting Graphs

Data can be presented visually using graphs. Three of the most common types of graphs are **line graphs**, **bar graphs**, and **circle graphs**. Each type of graph represents information about a relationship between variables in different ways, as you can see below.

Type of Graph	Description	Example
Line Graph	• Used to show changes in data over a period of time • The peaks and dips in this graph allow you to see how the data change.	**Whitehorse Average Maximum Temperature** (line graph showing Temperature (°C) vs Month: J F M A M J J A S O N D) This line graph shows changes in temperature over a period of one year in Whitehorse in Canada's Yukon Territory.
Bar Graph	• Used to compare data among different groups or categories • The height of a bar represents how large or small a value is.	**Gadgets Owned by a Grade 8 Class** (bar graph showing Number of Gadgets vs Gadgets: Calculator, Cell Phone, Computer, MP3 Player) This bar graph compares numbers of different types of gadgets owned by a class of Grade 11 students.
Circle Graph	• Used for comparing parts to the whole using percentages. • The size of each part of the circle shows you how much of the whole that part makes up.	**Andre's Weekly Activities (25 h)** Reading 15%, Television 20%, Playing Drums 20%, Playing Sports 25%, Internet/Video Games 20% This circle graph compares the amount of time a person spends doing various activities each week.

Line Graphs

A line graph has a horizontal *x*-axis that shows the independent variable. (The independent variable is the variable that the experimenter controls.) It also has a vertical *y*-axis that shows the dependent variable. (The dependent variable is the variable that changes, because it "depends" on the independent variable.) Data are plotted as points on the graph. A line of best fit connects the points as smoothly as possible. In the graph above that shows changes in temperature in Whitehorse, the line of best fit is curved, but in some line graphs it is straight.

Sample Line Graph

Brenda has a summer job working for a tree planting company. She gets $10 for every 100 trees she plants. That means if she plants

• 200 trees she gets $20

• 300 trees she gets $30

• 400 trees she gets $40

In this example, the independent variable (on the *x*-axis) is the number of trees Brenda plants. The dependent variable (on the *y*-axis) is how much money Brenda earns. The line graph of Brenda's earnings looks like this.

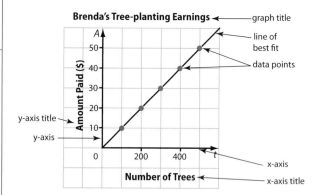

Instant Practice

How much money would Brenda earn if she planted 600 trees?

Bar Graphs

A bar graph has an *x*-axis (horizontal) that shows the different categories of data collected. It also has a *y*-axis (vertical) that shows the dependent variable (the variable that changes because it "depends" on which category it belongs to). Data are plotted as vertical bars on the graph.

Sample Bar Graph

A provincial park has 338 species of birds that live in various habitats in the park. The categories of data on the *x*-axis are the four habitats: marsh, meadow, lakeshore, and woods. The *y*-axis shows the number of bird species living in each habitat. The bar graph of number of bird species by habitat is shown on the following page.

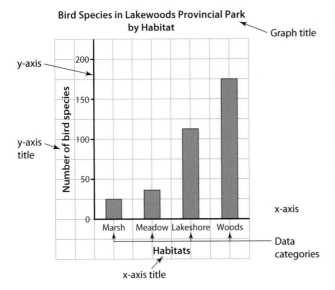

Bird Species in Lakewoods Provincial Park by Habitat

Graph title · y-axis · y-axis title · x-axis · Data categories · x-axis title

Instant Practice

How many bird species live in the lakeshore habitat?

Circle Graphs

A circle graph compares parts to a whole using percentages. This kind of graph is like a pie with different-sized pieces, or sections. The size of each piece shows what percentage of the whole that category represents. All the categories must add up to 100%.

To construct a circle graph, first calculate what percentage each category represents. Second, calculate what angle in the circle each percentage takes up. See how this sample circle graph showing students' breakfast choices has been constructed.

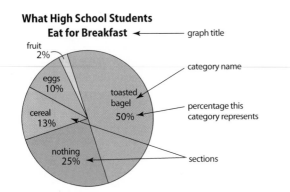

What High School Students Eat for Breakfast ← graph title

fruit 2% · eggs 10% · cereal 13% · nothing 25% · toasted bagel 50% · category name · percentage this category represents · sections

Constructing the Sample Circle Graph

A total of 1500 high school students across Canada were asked what they eat for breakfast. The results are shown in the table of this page.

Calculate the percentages as follows:

(Number of students in category) ÷ (Total number of students surveyed) × 100

For example, 30 students eat fruit for breakfast.

$$30 \div 1500 \times 100 = 2\%$$

2% written as a decimal is 0.02. This is known as the "decimal equivalent" of 2%.

Calculate the angle a category takes up in the circle as follows:

(Decimal equivalent of %) × 360° = angle for that category

For the fruit category, multiply 0.02 by 360° (the number of degrees in a circle). The fruit category has an angle of approximately 7°.

$$0.02 \times 360° = 7.2° \approx 7° \text{ (the value is rounded to 7°)}$$

Students' Breakfast Choice Survey Results

Breakfast Food	Number of Students	Percentage of Total (%)	Decimal Equivalent	Angle
Fruit	30	2	0.02	$0.02 \times 360° = 7.2°$ $\approx 7°$
Eggs	150	10	0.10	$0.10 \times 360° = 36°$
Cereal	195	13	0.13	$0.13 \times 360° = 46.8°$ $\approx 47°$
Nothing	375	25	0.25	$0.25 \times 360° = 90°$
Toast/ Bagel	750	50	0.50	$0.50 \times 360° = 180°$

The angles calculated in the table are used to draw the graph, as shown below. You can use a protractor to help you draw the angles accurately in a circle graph.

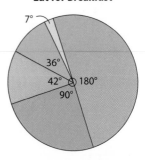

What High School Students Eat for Breakfast

7° · 36° · 42° · 90° · 180°

Instant Practice

What percentage of students eats either eggs or cereal for breakfast?

Appendix A

Measurement

The Metric System

The metric system has a **base unit** for each type of quantity being measured. For example, the base unit of length is the metre and the base unit of mass is the gram. Table 1 shows the most common metric units for various quantities, along with the symbol used to represent each unit. For example, the symbol for litre is L and the symbol for second is s.

The metric system of measurement is based on multiples of 10. This makes it easier to convert from one unit to another. To get larger units of a quantity, you multiply by a factor of 10 ($\times$ 10, $\times$ 100, $\times$ 1000, etc.). To get smaller units of a quantity, you divide by a factor of 10 ($\div$ 10, $\div$ 100, $\div$ 1000, etc.). Each unit that is larger or smaller than the base unit has a prefix to indicate what multiple or division of 10 it represents. For example, **kilo-** means "multiplied by 1000." Thus, one kilometre is 1000 metres.

$$1 \text{ km} = 1000 \text{ m}$$

The prefix **milli-** means "divided by 1000." So, one milligram is one thousandth of a gram.

$$1 \text{ mg} = \frac{1}{1000} \text{ g}$$

These prefixes can be used with most types of measurements. Table 2 shows the most common metric prefixes.

Table 1: Commonly Used Metric Quantities, Units, and Symbols

Quantity	Unit	Symbol
Length	nanometre micrometre millimetre centimetre metre kilometre	nm μm mm cm m km
Mass	gram kilogram tonne	g kg t
Area	square metre square centimetre hectare	m² cm² ha (10 000 m²)
Volume	cubic centimetre cubic metre millilitre litre	cm³ m³ mL L
Time	second	s
Temperature	degree Celsius	°C

Table 2: Commonly Used Metric Prefixes

Prefix	Symbol	Relationship to the Base Unit
tera-	T	$10^{12} = 1\ 000\ 000\ 000\ 000$
giga-	G	$10^{9} = 1\ 000\ 000\ 000$
mega-	M	$10^{6} = 1\ 000\ 000$
kilo-	k	$10^{3} = 1\ 000$
hecto-	h	$10^{2} = 100$
deca-	da	$10^{1} = 10$
—	—	$10^{0} = 1$
deci-	d	$10^{-1} = 0.1$
centi-	c	$10^{-2} = 0.01$
milli-	m	$10^{-3} = 0.001$
micro-	μ	$10^{-6} = 0.000\ 001$
nano-	n	$10^{-9} = 0.000\ 000\ 001$
pico-	p	$10^{-12} = 0.000\ 000\ 000\ 001$

Measuring Length

Length is the measurement of how long an object is or the measurement of anything from end to end. You can use a metre stick or a ruler to measure short distances. These tools are usually marked in centimetres (cm) and/or millimetres (mm). When you compare length measurements, make sure the measurements are in the same units.

Instant Practice

Measure and record the length of three objects on your desk in both millimetres (mm) and centimetres (cm). Compare your results with those of a partner.

Calculating Area

Area is the measurement of the size of a surface. For a regular, flat surface the area is calculated by multiplying the length of the surface by its width. Always use the same units for the length and width. Here is how to calculate area for a square and a rectangle.

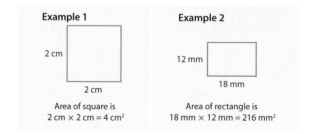

Example 1

2 cm, 2 cm

Area of square is
2 cm × 2 cm = 4 cm²

Example 2

12 mm, 18 mm

Area of rectangle is
18 mm × 12 mm = 216 mm²

Measuring Mass

Mass is the amount of matter in a substance or an object. Mass is measured in milligrams (mg), grams (g), kilograms (kg), and tonnes (t). The gram is the basic unit of mass. Use a balance for measuring mass.

Measuring the Mass of a Solid Object

To measure the mass of a solid object, place the object directly on the balance and read the mass. You could place a sheet of paper under the object to protect the balance. A solid object's mass can be measured on a manual balance or an electronic balance.

Measuring the Mass of a Substance Using a Manual Balance

To measure the mass of a substance, such as table salt, the substance cannot be placed directly on the balance. It will have to be measured in a container, such as a beaker. Follow these steps to find the mass of a substance using a manual balance.

1. Find the mass of the beaker, as shown in photograph A below.

2. Pour the substance into the beaker and find the mass of the beaker and the substance together, as shown in photograph B below.

3. To find the mass of the substance, subtract the beaker's mass from the combined mass of the beaker and the substance.

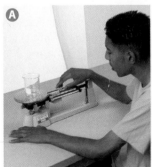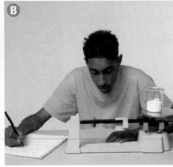

Mass of substance = (mass of beaker with substance) − (mass of empty beaker)

Example:

Mass of beaker with substance	= 325 g
Mass of empty beaker	= 140 g
Mass of substance	= 325 g − 140 g = 185 g

Instant Practice

Use the following information to determine the mass of a quantity of sand. The mass of the beaker is 160 g. The mass of the sand and the beaker together is 270 g.

Measuring the Mass of a Substance Using an Electronic Balance

When you measure the mass of a substance with an electronic balance, the balance will do the calculation for you. Follow these steps to find the mass of a substance using an electronic balance.

1. Place the empty beaker on the balance and hit the "Tare," "Zero," or "Re-Zero" button. This resets the balance to zero, as shown in photograph A below.

2. Add the substance to be measured into the beaker. The balance subtracts the mass of the beaker before the contents are even added, so it reports only the mass of the contents.

3. Read the mass of the substance as shown on the electronic balance, as shown in photograph B below.

Appendix A

Measuring Volume

Calculating the Volume of a Regular Solid

The **volume** of an object is the amount of space that the object occupies. The volume of a regularly shaped solid object can be calculated using well known mathematical formulas. For example, to find the volume of a cube or a rectangular solid, multiply length × width × height (as shown in Diagram A below). The only difference between a cube and a rectangular solid is that for a cube, the length, width, and height are equal. Remember to measure all sides using the same unit.

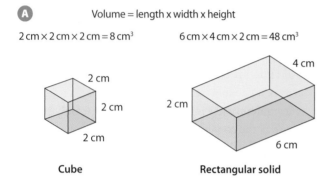

A   Volume = length x width x height

$2\,cm \times 2\,cm \times 2\,cm = 8\,cm^3$ $6\,cm \times 4\,cm \times 2\,cm = 48\,cm^3$

Cube   Rectangular solid

If all the sides of a solid object are measured in millimetres (mm), the volume will be in cubic millimetres (mm³). If all the sides are measured in centimetres (cm), the volume will be in cubic centimetres (cm³). The units for measuring the volume of a solid are called **cubic units**.

Measuring the Volume of a Liquid

The units used to measure the volume of liquids are called **capacity units**. The basic unit of volume for liquids is the litre (L). Recall that 1 L = 1000 mL.

Cubic units and capacity units are interchangeable. For example,

$1\,cm^3 = 1\,mL$

$1\,dm^3 = 1\,L$

$1\,m^3 = 1\,kL$

A graduated cylinder is used to measure the volume of a liquid, as shown in Diagram B on the right above. When a liquid is in a container, it forms a slight curve called a **meniscus**. To measure accurately, always place your eye at the same level as the bottom of the meniscus and read the volume measurement there.

B

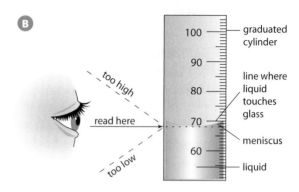

Measuring the volume of a liquid

Instant Practice

Determine the volume of liquids present in the three graduated cylinders shown here.

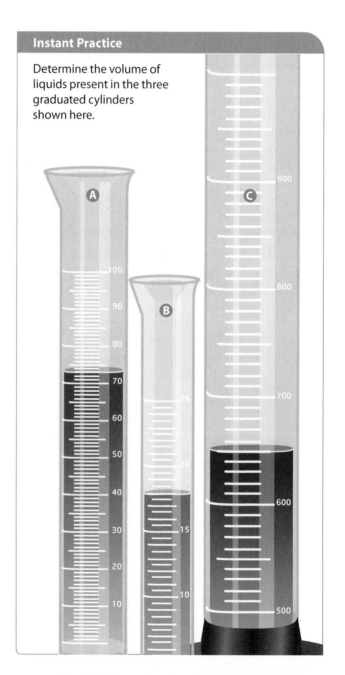

Finding the Volume of an Irregular Solid

The volume of an irregularly shaped solid object must be measured indirectly. This is done by measuring the volume of liquid that the object displaces. Follow these steps to find the volume of an irregular solid.

1. Fill a graduated cylinder partway and record the volume (see step 1).

2. Place the irregular object in the liquid (see step 2). The object will displace some of the liquid and cause the liquid level to rise (see step 3).

3. Read the new volume shown on the cylinder. Calculate the volume of the object by subtracting the starting volume from the new volume.

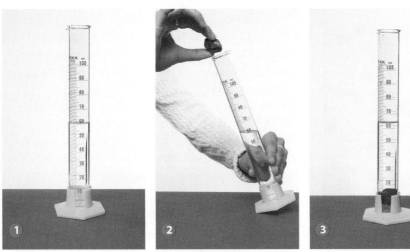

Measuring the volume of an irregularly shaped solid

Volume of object = (volume of liquid with object) − (initial volume of liquid)

Example:

Volume of liquid with object = 98 mL

Initial volume of liquid   = 50 mL

Volume of object   = 98 mL − 50 mL = 48 mL

(This is the same as 48 cm^3)

Instant Practice

Use the following information to determine the volume of an irregular solid. The initial volume of the liquid is 35 mL. The volume of the liquid with the object is 75 mL.

Developing Research Skills

In this course, you will need to conduct research to answer specific questions and to explore broad research topics. The following skills will take you through the research process from start to finish:

- focussing your research
- searching for resources that contain information related to your topic
- evaluating the reliability of your information sources
- gathering, recording, and organizing information in an appropriate format
- presenting your work

Focussing Your Research

- Start by carefully reading your assignment. Pick out key words and phrases, such as *apply*, *analyze*, *argue*, *compare and contrast*, *describe*, *discuss*, *evaluate*, *explain*, *identify*, *infer*, *interpret*, and *predict*. These key words and phrases will guide you on what kind of information you need to collect, and what you need to do with the information.
- Jot down ideas on your own, and then get additional input from others, including your teacher.
- Once you have done some general research, narrow down your topic until you can express it in one specific question. This will help you focus your research.
- Ensure that the question you are researching fulfills the guidelines of the assignment provided by your teacher.

Searching for Resources

- It is important to find reliable resources to help you answer your question. Potential sources of information include print and on-line resources such as encyclopedias, textbooks, non-fiction books, journals, websites, and newsgroups.

- The library and the Internet can both provide information for your search. Whether you are looking at print or digital resources, you need to evaluate the accuracy and objectivity of the information.

Evaluating the Reliability of Your Information Sources

Assess the reliability of your information sources to help you decide whether the information you find is likely to be accurate. To determine the validity of a source, check that the author is identified, a recent publication date is given, and the source of facts or quotations is identified. An author's credentials are important. Look for an indication of educational background, work experience, or professional affiliation. If the information is published by a group, try to find out what interests the group represents. The following guidelines may be helpful in assessing your information sources:

- On-line and print scientific journals provide data that have been reviewed by experts in a field of study (peer-reviewed), so they are usually a reliable source. Be aware, however, that the conclusions in journal articles may contain opinions as well as facts.

- Data on the websites of government statistical departments tend to be reliable. Be sure to read carefully, however, to interpret the data correctly.

- University resources, such as websites ending in ".edu" are generally reliable.

- Reliable experts in a field of study often have a PhD or MSc degree, and their work is regularly cited in other publications.

- Consumer and corporate sources may present a biased view. That is, they may present only data that support their side of an issue. Look for sources that treat all sides of an issue equally and fairly, or that clearly specify which perspective(s) they are presenting.

- Some sources, such as blogs and editorials, provide information that represents an individual's point of view or opinion. Therefore, the information is not objective. However, opinion pieces can alert you to controversy about an issue and help you consider various perspectives. The opinion of an expert in a field of study should carry more weight than that of an unidentified source.

- On-line videos and podcasts can be dynamic and valuable sources of information. However, their accuracy and objectivity must be evaluated just as thoroughly as all other sources.

- A piece of information is generally reliable if you can find it in two other sources. However, be aware that several on-line resources might use the same incorrect source of information. If you see identical wording on multiple sites, try to find a different source to verify the information.

Gathering, Recording, and Organizing Information

- As you locate information, you may find it useful to jot it down on large sticky notes or colour-coded entries in a digital file so you can group similar ideas together. Remember to document the source of your information for each note or data entry.

Avoid Plagiarism Copying information word-for-word and then presenting it as your own work is called *plagiarism*. Instead, you must cite every source you use for a research assignment. This includes all ideas, information, data, and opinions that appear in your work. If you include a quotation, be sure to indicate it as such, and supply all source information. Avoid direct quotations whenever possible—put information in your own words. Remember, though, that even when you paraphrase, you need to cite your sources.

Record Source Information A research paper should always include a bibliography—a list of relevant information sources you have consulted while writing it. Bibliographic entries include information such as the author, title, publication year, name of the publisher, and city in which the publisher is located. For magazine or journal articles, the name of the magazine or journal, the name of the article, the issue number, and the page numbers should be recorded. For on-line resources, you should record the site URL, the name of the site, the author or publishing organization, and the date on which you retrieved the information. Remember to record source information while you are taking notes to avoid having to search it out again later! Ask your teacher about the preferred style for your references.

- You might find it helpful to create a chart to keep track of detailed source information. For on-line searches, a tracking chart is useful to record the key words you searched, the information you found, and the URL of the website where you found the information.

- Write down any additional questions that you think of as you are researching. You may need to refine your topic if it is too broad, or take a different approach if there is not enough information available to answer your research question.

Presenting Your Work

- Once you have organized all of your information, you should be able to summarize your research so that it provides a concise answer to your original research question. If you cannot answer this question, you may need to refine the question or do a bit more research.

- Check the assignment guidelines for instructions on how to format your work.

- Be sure that you fulfill all of the criteria of the assignment when you communicate your findings.

Instant Practice

1. Your assignment asks you to research the effectiveness of some Canadian technologies in providing food for growing populations.
   a. What search terms might you use for your initial research on the Internet or at the library?
   b. How might you narrow down this assignment into a research question?

2. How could you verify the information in an article about genetically modified organisms that you found on a wiki site?

3. Suggest two or more clues that could indicate that the information in an on-line video might not be reliable.

Using Critical Thinking to Analyze and Assess Research

Critical thinking is a valuable decision-making and problem-solving skill that you can apply to almost every field of life. It is an especially important skill in science, because it helps you determine the validity of a claim and identify the implications and consequences of a conclusion. You can also use critical thinking to help you clearly state a goal, examine data, identify assumptions, and evaluate evidence.

The word "critical" in critical thinking does not mean finding fault or being negative. "Critical" refers to using reason and reflection to analyze and evaluate thinking in order to improve it. A critical thinker displays well-developed intellectual values, such as clarity, accuracy, precision, relevance, logic, significance, depth, breadth, and fairness. Developing skills in critical thinking is a lifelong practice that is self-guided and self-disciplined.

The following points and questions can help you develop your critical thinking skills. You can use the skills to critically evaluate both your own and others' work to determine whether a researcher has conducted fair and objective research. You can also use the questions to evaluate investigations, arguments, advertisements, and science resources. The questions are written as if the research is in a written format, but you can also apply them to spoken formats, such as discussions, or to viewed formats, such as science television shows. If you are evaluating your own work, then you are "the researcher" in each of the categories below. The sentences beginning *Look for* will alert you to which values may be displayed. The sentences beginning *Watch out for* point out some possible shortcomings in the research.

1. What is the researcher trying to accomplish?

Look for clarity and precision. The old expression, "Well begun is half done" applies to stating the purpose. The more accurately and precisely the purpose is stated, the easier it is to develop a plan to achieve it and evaluate whether the investigation has succeeded. The researcher should clearly state the purpose or goal of the inquiry or investigation in a direct statement such as, "The purpose of this study is to describe the characteristics of…." If there is not a subheading called "Purpose" or "Goal" then ask yourself, "What is the problem the researcher is trying to solve? What is the decision the researcher wants to make?"

Watch out for a fuzzy or imprecise purpose, such as "The purpose of this study is to find out something more about…"

You should be able to complete this sentence stem for the inquiry or investigation you are evaluating: **The main purpose of this research is …**

2. What question does the researcher want to answer?

Look for clarity, precision, and significance. The question in scientific inquiry is directly related to the purpose. If a question is not explicitly stated, you can often identify it from the hypothesis or from a statement about the problem the researcher investigated. The researcher may have described the concerns or issues that led him or her to undertake the investigation. Once you know what the question is, ask yourself, "Is there more than one right answer to this question? What other ways can I think about the question? If I divided the question into sub-questions, what would they be?" If you are evaluating your own research, you may realize that you could improve your question by going more deeply into the complexity of what you are investigating.

Watch out for questions that require judgement rather than just facts to be answered. Is there a bias toward a certain answer implied in the question? For example, the question "Wouldn't it be a good idea to add acidic fertilizer to basic garden soil?" implies that the "correct" answer would be "yes" and that the answer should be based on judgement of what "good" means as well as on facts.

You should be able to complete this sentence stem for the inquiry or investigation you are evaluating: **The question directing this research is ….**

3. What data does the researcher present?

Look for accuracy and relevance. The researcher should have gathered enough relevant data to respond to the research question and reach a reasonable conclusion. The data section may include descriptive statistics, which summarize the raw data, such as means and standard deviations. You may be able to infer conclusions from data called inferential statistics. The researcher may also mention how the data were screened for errors in data entry, outliers and distribution, etc.

As you review and analyze the data, it is important to think about internal validity, which means whether the changes in the dependent variable were the result of something other than, or in addition to, changes in the manipulated variable. Sometimes, the testing procedure itself can cause changes in the dependent variable.

In some investigations that occur over a long period of time, the dependent variable may change due to aging or deterioration. Studies that include data from a comparable control group help to reduce the problems of internal validity.

Watch out for data that is not relevant to the question or investigation.

You should be able to complete this sentence stem for the inquiry or investigation you are evaluating: **The most important data presented in this research are…**

4. What assumptions are the researcher's point of view based on?

Look for logic, breadth, depth, and fairness. Every researcher has a point of view or frame of reference while conducting research. The point of view is based on major assumptions in the research. Try to recognize unstated assumptions and values as well as those that the researcher has clearly acknowledged. As you read the research, ask yourself, "What assumptions is the researcher making? What is the researcher taking for granted? Which of those assumptions would I question? Does the researcher show open mindedness by recognizing and assessing his or her assumptions? What assumption is the conclusion based on?" It is important that a researcher address the strengths and weaknesses of his or her point of view when evaluating data. The researcher also needs to fairly consider objections that arise from alternative relevant points of view or lines of reasoning and to be aware of the strengths and weaknesses of those points of view.

Watch out for a point of view that does not include other possibilities, i.e., "this is the correct point of view and other points of view are not valid." Notice whether the researcher is distorting ideas to try to strength his or her position or to support a point of view.

You should be able to complete this sentence stem for the inquiry or investigation you are evaluating: **The main assumptions underlying the researcher's point of view are…**

5. How well do the conclusions address the research question?

Look for relevance, logic, significance, breadth, and depth. To reach a well-reasoned conclusion, the researcher interprets, appraises, and evaluates data. The researcher considers what the data imply and how those implications address the research question. The conclusions or discussion should always include a clear reference to the original hypothesis or question. The researcher should also attempt to explain data that is contrary to the main conclusion. The conclusions of a research study often include questions that can be the basis of future research.

When you read conclusions in the research you are evaluating, ask yourself, "How did the researcher arrive at these conclusions? What inferences and interpretations did he or she make? What data did he or she use? How well do the data support the conclusions? Do the data suggest other conclusions than what the researcher has provided? In what other way might the data be interpreted?" In order to accept or reject a researcher's conclusions, you need to have a clear idea of how the research was conducted.

All research has implications, which may be either positive or negative. Has the researcher considered and stated the broader implications of his or her findings?

Watch out for unsupported conclusions that go beyond what the data imply. Sometimes, researchers describe insignificant results as though they were significant, or draw conclusions from future studies. For example, a researcher may suggest that even though there was not a significant difference in this particular study, there probably would be in future studies. This is not a valid or supported conclusion.

You should be able to complete these sentence stems for the inquiry or investigation you are evaluating: **The main conclusions in this research are… The implications of these conclusions are…**

Using Statistics to Test Hypotheses

You may be interested in answering many types of science questions. Is Earth's surface getting warmer? Which air bag most effectively reduces the severity of injuries in a car accident? Does the public prefer a certain colour in a new line of fashion? To answer those questions you need to conduct a study or experiment.

Statistics is the science of conducting studies to collect, organize, summarize, analyze, and draw conclusions from data. Data are the values (measurements or observations) that a variable can assume. A collection of data values is called a data set.

Suppose you wanted to know: Does a new medication affect people's pulse rates? You cannot give the drug to everyone in the population and measure their pulse rate. You need to take a sample. A sample is a group of subjects selected from a population.

If you could sample the entire population, you would know exactly what the effect of the medication is on the pulse rate of the population. Since this is not possible, you use the observations from the sample to approximate the effect on pulse rate of the population.

Outliers Once you have conducted your study and collected the data, you need to check the data for extremely high or low values, called *outliers*. For example, suppose that in your pulse rate study the data you collected included
$$20, 80, 81.5, 82, 82, 82.5, 83$$
The value 20 is an outlier and you might be suspicious of whether it is valid.
- An outlier may have resulted from a measurement or observational error.
- An outlier may have resulted from a recording error.
- An outlier may have been obtained from a subject that is not in the defined population.
- An outlier might be a legitimate value that occurred by chance (although the probability is extremely small).

There are no hard and fast rules on what to do with outliers, nor is there complete agreement among statisticians on ways to identify them. If outliers occurred as a result of an error, you should attempt to correct the error or omit the data value. When outliers occur by chance, you need to make a decision about whether or not to include them in the data set.

Describing the Data

The data you collect in your study is called raw data. To draw conclusions from the data, you need to look at how it is distributed. The two most important statistics for describing the data are the mean and the standard deviation.

- The *mean* describes the average. You find the mean by adding all values of the data and dividing by the total number of values.
- *Standard deviation* describes the spread of the data.

The distribution that is defined by these two parameters from the sample describes the likely effect on the pulse rate in the population. The most likely average pulse rate of the population with the drug is the mean of the sample. However, since you did not measure the pulse rate of the entire population, there is some uncertainty in your measurements. As you get further away from the mean, the likelihood that the true population average has that value diminishes. The shape of the likelihood curve is described by the normal distribution around the mean (the bell curve, shown below).

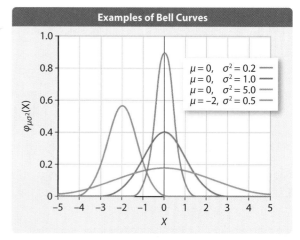

Examples of normal distribution of data

Testing the Research Question

The basic question you are testing in describing your data is: How likely is the chance that the mean pulse rate of the sample with the drug could actually be the same as the population average pulse rate without the drug? The population average pulse rate would have been obtained from a previous study, which also studied a sample and not the entire population.

The statistical method for examining whether there is a difference in pulse rates between the two samples is called hypothesis testing. A hypothesis is a proposed effect of a phenomenon or treatment. Hypothesis testing always include two hypotheses. Note that you only investigate the null hypothesis. The alternative hypothesis is just the complement.

- The *null hypothesis*, symbolized by H_0, states that there is no effect (the drug does not affect the pulse rate).

- The *alternative hypothesis*, symbolized by H_1, states that there is an effect (the drug does affect the pulse rate).

One of these hypotheses needs to be true. The result of a hypothesis test is whether we accept or reject the null hypothesis.

- By accepting the null hypothesis, we say that there is no statistical difference between the mean pulse rate of the sample with the drug and the average pulse rate of the population.
- By rejecting the null hypothesis, we say that there is a statistical difference between the mean pulse rate of the sample with the drug and the average pulse rate of the population.

The decision rule for accepting or rejecting the null hypothesis is defined by the significance level. Since you only measured the pulse rate of the sample and not the entire population, you can never be certain that your measurements actually truly represent the population. The significance level is the probability of a type I error you are willing to accept for your decision. A type I error is rejecting the null hypothesis when it was actually true (i.e., you say there is a difference when there is actually none). Typically, researchers only reject the null hypothesis when the probability of a type I error is smaller than 5%. In other words, they reject when there is only a 1 in 20 chance that the drug does not affect pulse rate.

Representing the Hypotheses Mathematically

You can state the hypotheses using mathematical symbols as shown in the table and examples below.

Common Phrases for Hypothesis-Testing	
>	**<**
Is greater than	Is less than
Is above	Is below
Is higher than	Is lower than
Is longer than	Is shorter than
Is bigger than	Is smaller than
Is increase	Is decreased or reduced from
≥	**≤**
Is greater than or equal to	Is less than or equal to
Is at least	Is at most
Is not less than	Is not more than
=	**≠**
Is equal to	Is not equal to
Is exactly the same as	Is different from
Has not changed from	Has changed from
Is the same as	Is not the same as

Example: Will it change?

Suppose the mean pulse rate for the population under study is 82 beats per minute. The null hypothesis for your pulse rate study states that the medication has no effect on the pulse rate.

$$H_0 \; \mu = 82$$

(The Greek letter μ (mu) is used to represent the mean of a population.)

The alternative hypothesis states that the medication has an effect on the pulse rate.

$$H_1 \; \mu \neq 82$$

Example: Will it increase?

A chemist invents an additive to increase the life of a car battery. The mean lifetime of the battery is 36 months. In this situation, the chemist in only interested in increasing the lifetime of the battery.

The null hypothesis states that the additive has no effect on the battery life or even shortens it.

$$H_0 \; \mu \leq 36$$

The alternative hypothesis is that the additive increases the battery life.

$$H_1 \; \mu > 36$$

Example: Will it decrease?

A contractor wishes to lower heating bills by using a special type of insulation in homes. The average monthly heating bill is $78.00.

The null hypothesis states that the insulation has no effect on the monthly heating bill or even increases it.

$$H_0 \; \mu \geq \$78$$

The alternative hypothesis is that the insulation decreases the monthly heating bill.

$$H_1 \; \mu < \$78$$

Instant Practice

1. State the null hypothesis and alternative hypothesis for each of the following situations. Give your answer in words first and then in mathematical symbols.

   a. A psychologist is curious whether the playing of soft music during a test will change the results of the test. The psychologist is not sure whether the grades will be higher or lower. In the past, the mean of the scores without soft music was 73.

   b. An engineer hypothesizes that the mean number of defects can be decreased during the manufacture of compact discs by using robots instead of humans for certain tasks. The mean number of defective discs in human production per 1000 is 18.

   c. A researcher thinks that if expectant mothers use vitamin pills, the birth weight of the babies will increase. The average birth weight of babies from mothers who are not using vitamin pills is 3.9 kg.

Appendix A

Significant Digits and Rounding

You might think that a measurement is an exact quantity. In fact, all measurements involve uncertainty. The measuring device is one source of uncertainty, and you, as the reader of the device, are another. Every time you take a measurement, you are making an estimate by interpreting the reading. For example, the illustration below shows a ruler measuring the length of a rod. The ruler can give quite an accurate reading, since it is divided into millimetre marks. But the end of the rod falls between two marks. There is still uncertainty in the measurement. You can be certain that if the ruler is accurate, the length of the rod is between 5.2 mm and 5.3 mm. However, you must estimate the distance between the 2 mm and 3 mm marks.

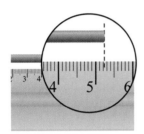

Significant Digits

Significant digits are the digits you record when you take a measurement. The significant digits in a measured quantity include all the certain digits plus the first uncertain digit. In the example above, suppose you estimate the length of the rod to be 5.23 cm. The first two digits (5 and 2) are certain (those marks are visible), but the last digit (0.03) is estimated. The measurement 5.23 cm has three significant digits.

Determining the Number of Significant Digits

The following rules will help you determine the number of significant digits in a given measurement.

1. All non-zero digits (1–9) are significant.
   Examples:
   • 123 m – three significant digits
   • 23.56 km – four significant digits

2. Zeros between non-zero digits are also significant.
   Examples:
   • 1207 m – four significant digits
   • 120.5 km/h – four significant digits

3. Any zero that follows a non-zero digit *and* is to the right of the decimal point is significant.
   Examples:
   • 12.50 m/s^2 – four significant digits
   • 6.0 km – two significant digits

4. Zeros that are to the left of a measurement are not significant.
   Examples:
   • 0.056 – two significant digits
   • 0.007 60 – three significant digits

5. Zeros used to indicate the position of the decimal are not significant. These zeros are sometimes called spacers.
   Examples:
   • 500 km – one significant digit (the decimal point is assumed to be after the final zero)
   • 0.325 m – three significant digits
   • 0.000 34 km – two significant digits

6. In some cases, a zero that appears to be a spacer is actually a significant digit. All counting numbers have an infinite number of significant digits.
   Examples:
   • 6 apples – infinite number of significant digits
   • 125 people – infinite number of significant digits
   • 450 deer – infinite number of significant digits

Instant Practice

Determine the number of significant digits in each measurement.
a. 46 units
b. 2973 L
c. 82.9 cm
d. 9.0034 W
e. 20.380 g
f. 0.0073 mm
g. 0.340 kg
h. 2 s
i. 400 J
j. 439.0001 km

Using Significant Digits in Mathematical Operations

When you use measured values in mathematical operations, the calculated answer cannot be more certain than the measurements on which it is based. Often the answer on your calculator will have to be rounded to the correct number of significant digits.

Rules for Rounding

1. When the first digit to be dropped is less than 5, the preceding digit is not changed.

Example:

6.723 m rounded to two significant digits is 6.7 m. The digit after the 7 is less than 5, so the 7 does not change.

2. When the first digit to be dropped is 5 or greater, the preceding digit is increased by one.

Example:

7.237 m rounded to three significant digits is 7.24 m. The digit after the 3 is greater than 5, so the 3 is increased by one.

3. When the first digit to be dropped is 5, and there are no following digits, increase the preceding number by 1 if it is odd, but leave the preceding number unchanged if it is even.

Examples:

8.345 L rounded to two significant digits is 8.34 L, because the digit before the 5 is even.

8.375 L rounded to two significant digits is 8.38 L, because the digit before the 5 is odd.

Adding or Subtracting Measurements

Perform the mathematical operation, and then round off the answer so it has the same number of significant digits as the value that has the fewest decimal places.

Example:

Add the following measured lengths and express the answer to the correct number of significant digits.

$x = 2.3$ cm $+ 6.47$ cm $+ 13.689$ cm
$= 22.459$ cm
$= 22.5$ cm

Since 2.3 cm has only one decimal place, the answer can have only one decimal place.

Multiplying or Dividing Measurements

Perform the mathematical operation, and then round off the answer so it has the same number of significant digits as the value that has the least number of significant digits.

Example:

Multiply the following measured lengths and express the answer to the correct number of significant digits.

$x = (2.342$ m$)(0.063$ m$)(306$ m$)$
$= 45.149\ 076$ m^3
$= 45$ m^3

Since 0.063 m has only two significant digits, the final answer must also have two significant digits.

Instant Practice

Perform the following calculations, rounding off your answer to the correct number of significant digits.

a. 2.0 cm $+$ 0.52 cm $+$ 3.2 cm

b. 642.0 mg/33.402 mg

c. 23.8 L $\times$ 0.00321 L

d. 8.045 kJ $-$ 3.10 kJ

e. 72.1 mm/0.3 mm $+$ 2.09 mm

Scientific Notation and Logarithms

An exponent is the symbol or number denoting the power to which another number or symbol is to be raised. The exponent shows the number of repeated multiplications of the base. In 10^2, the exponent is 2 and the base is 10. The expression 10^2 means 10×10.

Powers of 10

Digits	Standard Form	Exponential Form
Ten thousands	10 000	10^4
Thousands	1 000	10^3
Hundreds	100	10^2
Tens	10	10^1
Ones	1	10^0
Tenths	0.1	10^{-1}
Hundredths	0.01	10^{-2}
Thousandths	0.001	10^{-3}
Ten thousandths	0.0001	10^{-4}

Why use exponents? Consider this: One molecule of water has a mass of 0.000 000 000 000 000 000 000 029 9 g. Using such a number for calculations would be quite awkward. The mistaken addition or omission of a single zero would make the number either 10 times larger or 10 times smaller than it actually is. Scientific notation allows scientists to express very large and very small numbers more easily, to avoid mistakes, and to clarify the number of significant digits.

Scientific Notation

In scientific notation, a number has the form $x \times 10^n$, where x is greater than or equal to 1 but less than 10, and 10^n is a power of 10. To express a number in scientific notation, use the following steps:

1. To determine the value of x, move the decimal point in the number so that only one non-zero digit is to the left of the decimal point.

2. To determine the value of the exponent n, count the number of places the decimal point moves to the left or right. If the decimal point moves to the right, express n as a positive exponent. If the decimal point moves to the left, express n as a negative exponent.

3. Use the values you have determined for x and n to express the number in the form $x \times 10^n$.

Examples

Express 0.000 000 000 000 000 000 000 029 9 g in scientific notation.

1. To determine x, move the decimal point so that only one non-zero number is to the left of the decimal point:

   2.99

2. To determine n, count the number of places the decimal moved:

   0.000 000 000 000 000 000 000 02.99 g
   3   6   9   12   15   18   21   23

   Since the decimal point moved to the right, the exponent will be negative.

3. Express the number in the form $x \times 10^n$:

   2.99×10^{-23} g

Express 602 000 000 000 000 000 000 000 in scientific notation.

1. To determine x, move the decimal point so that only one non-zero number is to the left of the decimal point:

   6.02

2. To determine n, count the number of places the decimal moved:

   6.02 000 000 000 000 000 000 000.
   23   21   18   15   12   9   6   3

   Since the decimal point moved to the left, the exponent will be positive.

3. Express the number in the form $x \times 10^n$:

   6.02×10^{23}

Rules for Scientific Notation

1. To multiply two numbers in scientific notation, add the exponents.

   Example:

   $(7.32 \times 10^{-3}) \times (8.91 \times 10^{-2})$
   $= (7.32 \times 8.91) \times 10^{(-3) + (-2)}$
   $= 65.2212 \times 10^{-5}$
   $= 6.52 \times 10^{-4}$

   (Remember to report the correct number of significant digits in your answer. Also, you will need to move the decimal point to achieve correct scientific notation.)

2. To divide two numbers in scientific notation, subtract the exponents.

Example:

$(1.842 \times 10^6) \div (1.0787 \times 10^2)$

$= (1.842 \div 1.0787) \times 10^{(6-2)}$

$= 1.707\ 611 \times 10^4$

$= 1.708 \times 10^4$

3. To add or subtract numbers in scientific notation, first convert the numbers so they have the same exponent. Each number should have the same exponent as the number with the greatest power of 10. Once the numbers are all expressed to the same power of 10, the power of 10 can be ignored (neither added nor subtracted) in the calculation.

Example:

$(3.42 \times 10^6) + (8.53 \times 10^3)$

Express 8.53×10^3 as a number multiplied by 10^6. Then perform the addition.

$= (3.42 \times 10^6) + (0.008\ 53 \times 10^6)$

$= 3.428\ 53 \times 10^6$

$= 3.43 \times 10^6$

Logarithms

Logarithms are a convenient method for communicating large and small numbers. The logarithm, or log, of a number is the value of the exponent to which 10 would need to be raised in order to equal this number. Every positive number has a logarithm. Numbers that are greater than 1 have a positive logarithm. Numbers that are between 0 and 1 have a negative logarithm. The number 1 has a logarithm of 0.

Some Numbers and Their Logarithms

Number	Scientific Notation	As a Power of 10	Logarithm
1 000 000	1×10^6	10^6	6
7 895 900	7.8959×10^6	$10^{6.897\ 40}$	6.897 40
1	1×10^0	10^0	1
0.000 001	1×10^{-5}	10^{-5}	−5
0.004 276	4.276×10^{-3}	$10^{-2.3690}$	−2.3690

For logarithmic values, only the digits to the right of the decimal point count as significant digits. The digit to the left of the decimal point fixes the location of the decimal point of the original value. (Notice in the second row of the table at the bottom of the page that 7 895 900, which has 5 significant digits, has a log of 6.897 40. The log has 5 digits to the right of the decimal point.)

Logarithms are especially useful for expressing values that span a range of powers of 10. The Richter scale for earthquakes, the decibel scale for sound, and the pH scale for acids and bases all use logarithmic scales.

Logarithms and pH

The pH of an acid solution is defined as $-\log[H^+]$. (The square brackets mean *concentration*.) Thus, you can find the pH of a solution by taking the negative log of the concentration of hydrogen ions.

Example:

Find the pH of a solution with a hydrogen ion concentration of 0.004 76 mol/L.

$$-\log[0.004\ 76 \text{ mol/L}] = 2.322$$

The pH scale is a negative log scale. Thus, a decrease from pH 7 to pH 4 is actually an increase of 10^3, or 1000 times, in the acidity of a solution. An increase from pH 3 to pH 6 is a decrease in acidity of 10^3 times.

Instant Practice

1. Convert the following values into scientific notation.
   a. 5.319
   b. 0.03
   c. 99 482
   d. 0.000 718
   e. 8 382 441 002

2. Convert the following numbers from scientific notation into ordinary numbers.
   a. 2.3×10^3
   b. 8.003×10^{-2}
   c. 1.3492×10^6

3. Perform the following calculations.
   a. $(5.2 \times 10^5) + (1.32 \times 10^2)$
   b. $(9.231 \times 10^{-3}) - (2.4 \times 10^1)$
   c. $(7.92 \times 10^{-1}) \times (3.05 \times 10^{-3})$
   d. $(8.228 \times 10^4) \times (1.1 \times 10^5)$

4. Determine the pH of a solution, given each hydrogen ion concentration.
   a. 0.00001 mol/L
   b. 1 000 mol/L
   c. 3.28 mol/L

Using Graphic Organizers

Six different graphic organizers are shown here. These are but a few examples of the many types of graphic organizers that can help you to:

- brainstorm
- show relationships among ideas
- summarize a section of text
- record research notes
- review what you have learned before writing a test

PMI Chart

A *PMI chart* has three columns. PMI stands for "Plus," "Minus," and "Interesting." A PMI chart can be used to state the pros and cons related to an issue. The third column in the PMI chart is used to list interesting information related to the issue. PMI charts help you to organize your thinking after reading about a topic that is up for debate or that can have positive or negative effects.

Grocery stores now charge 5¢ for a plastic bag.		
P	**M**	**I**
• This may make people want to use fewer plastic bags. • There will be fewer plastic bags heading to landfills. • People may be more likely to bring reusable cloth bags to the store. • Cloth bags are strong and can carry many groceries. • Cloth bags can easily be washed before being re-used.	• The charge may not be high enough to stop people from using plastic bags. • It is easy to forget your reusable bags at home. • Over time, cloth bags can collect bacteria if they are used to carry meat, fish, dairy, and produce. • It is inconvenient to have to pay for bags or bring bags from home.	• Some stores have plastic bag recycling centres. • If you have a certain spot where you always keep your bags, you'll be more likely to remember them. • Plastic bags were introduced to replace paper bags. At the time, people thought plastic bags were better for the environment than paper ones.

Concept Map

A *concept map* uses shapes and lines to show how ideas are related. Each idea, or concept, is written inside a circle, a square, a rectangle, or another shape. Words that explain how the concepts are related are written on the lines that connect the shapes.

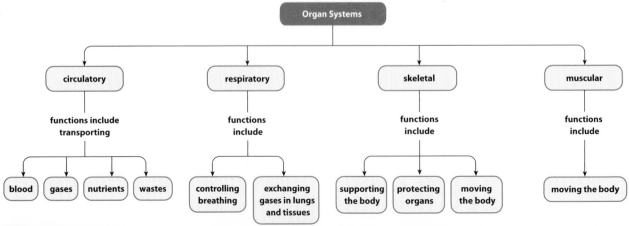

Flowchart

A *flowchart* shows a sequence of events or the steps in a process. A flowchart starts with the first event or step. An arrow leads to the next event or step, and so on, until the final outcome. The events or steps are shown in the order in which they occur.

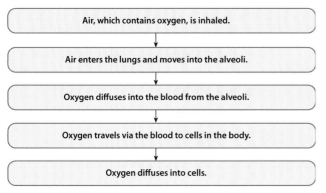

Cycle Chart

A *cycle chart* is a flowchart that has no clear beginning or end. The events are shown in the order in which they occur, as shown by arrows, but there is no first or last event. Instead, the events occur again and again in a continuous cycle.

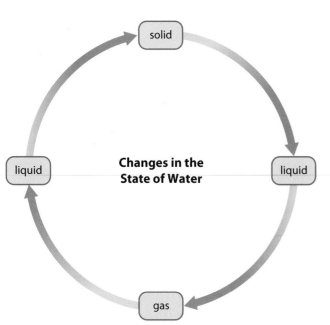

Venn Diagram

A *Venn diagram* uses overlapping shapes to compare concepts (show similarities and differences).

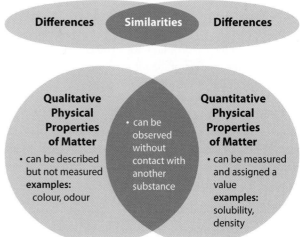

Cause-and-Effect Map

The first cause-and-effect map below shows one cause that results in several effects. The second map shows one effect that has several causes.

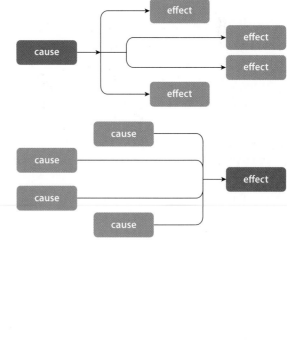

Tips for Answering Written Response Type Exam Questions

Answers to written response type questions on an exam are assessed on the basis of how well you communicate both your understanding of the information presented and your understanding of the applicable science.

Key Exam Skills

Evaluators will be looking for examples of your understanding of scientific principles and techniques. In order to successfully answer the questions, you must be able to do the following.

1. Read critically and identify
   - key words, phrases, and data that deliver useful information
   - distractor information and data that can be ignored because they do not have any bearing on the answer to the question
   - if the question is an open-response style that requires a unified response, or if it is a closed-response style that requires a more analytical approach
   - precisely what the question is asking
     - pay close attention to the process words (see below)
     - pay close attention to the directing words (see list following) to determine how you should answer the question. The directing words are sometimes highlighted in boldface type.
   - the scientific concept(s) that you should include in your answer
   - any formulas that you should include in your answer

2. Interpret and Analyze
   - process words and directing words
   - information including the key words, phrases, and data presented in the information box
   - information that is presented in charts, tables, and graphs

3. Communicate
   - conclusions by making a formal statement
   - results in the form of charts, graphs, or diagrams
   - ideas or answers to questions in the form of complete sentences, paragraphs, or short essays

4. If you are asked to perform an experiment, write the experimental design as follows.
   - State the problem or questions to be answered.
   - Formulate the hypothesis or make a prediction.
   - Identify the independent and dependent variables if required.
   - Provide a method for controlling variables.
   - Identify the required materials clearly.
   - Describe any applicable safety procedures.
   - Provide a sketch of the apparatus to help make the set-up clear.
   - Provide a method for collecting and recording pertinent data—include the units for the data being collected.

Process Words

You may find this list of words helpful in both understanding what the questions are asking and what is expected in your answer.

Hypothesis: A single proposition intended as a possible explanation for an observed phenomenon, e.g., a possible cause for a specific effect

Conclusion: A proposition that summarizes the extent to which a hypothesis and/or theory has been supported or contradicted by evidence

Experiment: A set of manipulations and/or specific observations of nature that allow the testing of hypotheses and/or generalizations

Variables: Conditions that can change in an experiment. Variables in experiments are categorized as

  Independent variable (manipulated variable): Condition that was deliberately changed by the experimenter

  Controlled variables: Conditions that could have changed but did not, because of the intervention of the experimenter

  Dependent variable (responding variable): Condition that changed in response to the change in the independent variable

Directing Words

Algebraically Using mathematical procedures that involve letters or symbols to represent numbers

Analyze To make a mathematical, chemical, or methodical examination of parts to determine the nature, proportion, function, interrelationship, etc. of the whole

Compare Examine the character or qualities of two things by providing characteristics of both that point out their similarities and differences

Conclude State a logical end based on reasoning and/or evidence

Contrast/ Distinguish Point out the differences between two things that have similar or comparable natures

Criticize Point out the demerits of an item or issue

Define Provide the essential qualities or meaning of a word or concept; make distinct and clear by marking out the limits

Describe Give a written account or represent the characteristics of something by a figure, model, or picture

Design/Plan Construct a plan, i.e., a detailed sequence of actions for a specific purpose

Determine Find a solution to a specified degree of accuracy, to a problem by showing appropriate formulas, procedures, and calculations

Enumerate Specify one-by-one or list in concise form and according to some order

Evaluate Give the significance or worth of something by identifying the good and bad points or the advantages and disadvantages

Explain Make clear what is not immediately obvious or entirely known; give the cause of or reason for; make known in detail

Graphically Use a drawing that is produced electronically or by hand and that shows a relation between certain sets of numbers

How Show in what manner or way, with what meaning

Hypothesize Form a tentative proposition intended as a possible explanation for an observed phenomenon i.e., a possible cause for a specific effect. The proposition should be testable logically and/or empirically.

Identity Recognize and select as having the characteristics of something

Illustrate Make clear by giving an example. The form of the example must be specified in the question i.e., word description, sketch, or diagram

Infer Form a generalization from sample data; arrive at a conclusion by reasoning from evidence

Interpret Tell the meaning of something; present information in a new form that adds meaning to the original data

Justify/ Show How Show reason for or give facts that support a position

Model Find a model that does a good job of representing a situation

Outline Give, in an organized fashion, the essential parts of something. The form of the outline may be specified in the question, e.g., list, flow chart, or concept map.

Predict Tell in advance on the basis of empirical evidence and/or logic

Prove Establish the truth or validity of a statement for the general case by giving factual evidence or logical argument

Relate Show logical or causal connection between things

Sketch Provide a drawing that represents the key features of an object or graph

Solve Give a solution of a problem, i.e., explanation in words and/or numbers

Summarize Give a brief account of the main points

Trace Give a step-by-step description of the development

Verify Establish, by substitution for a particular case or by geometric comparison, the truth of a statement

Why Show the cause, reason, or purpose

The Classification of Organisms

Classification systems are used to organize items into categories so it is easier to see how the items relate to each other—how they are the same and how they are different. For example, imagine there are just four sports in the world: hockey, soccer, tennis, and golf. Any sports competition could be classified in one of these four categories. The simple classification system in (A) is called an *un-nested system*. There are no intermediate categories between the main category (sports) and the individual types of sports (hockey, soccer, tennis, and golf).

Un-nested Classification

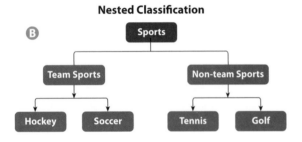

Nested Classification

The classification system shown in (B) is called a *nested system*, because more than one category level is used. A nested system always starts with the most general category at the top (Sports). The system continues with more specific categories (Team Sports; Non-team Sports) as you move down to lower levels of the classification chart. This kind of arrangement (from general to more specific) is called a hierarchy and nested systems are known as *hierarchical classification*.

How Categories Are Used To Classify Organisms

Biologists arrange organisms in a hierarchical classification to help understand the great diversity of life on Earth. A species is classified by placing it in eight nested categories: Domain, Kingdom, Phylum, Class, Order, Family, Genus, and Species. These general category names are called *ranks*. The specific category into which an organism is classified in each rank is called its *taxon*. The table shows how the grey wolf (*Canis lupus*) is classified using these categories.

Classification of the Grey Wolf

A domain is the broadest of the ranks and represents the greatest number of species. For example, the grey wolf, along with all organisms composed of eukaryotic cells, is placed in the domain Eukarya. There are millions of organisms, from animals to plants to fungi, in this domain.

Down the classification table, the ranks become more specific and the number of species in each taxon decreases. For example, the grey wolf belongs to the mammal class, *Mammalia*, a group of warm-blooded animals with specific characteristics. This class has approximately 5000 species. Once the wolf's genus rank (*Canis*) is reached, there are only seven species included. Finally, we reach the species level for the grey wolf: *Canis lupus*. At this rank, only the grey wolf species is left. (Species names always have two parts: the genus name followed by the species name. These two parts are always in italics. The first letter of the genus name is uppercase ; the first letter of the species name is lowercase.)

Rank (Taxonomic Category)	Grey Wolf Taxon	Number of Species in Taxon	Examples of Species in Taxon
Domain	Eukarya	4–10 million	
Kingdom	Animalia	2 million	
Phylum	Chordata	50 000	
Class	Mammalia	5 000	
Order	Carnivora	270	
Family	Canidae	34	
Genus	*Canis*	7	
Species	*Canis lupus*	1	

The Circulatory System

The circularly system interacts directly with all other systems in the human body. For instance, all living cells in the body need water, nutrients, and oxygen to survive. The circulatory system delivers these to the cells in each system as it transports blood. It also carries away wastes produced by the cells. Other components in the blood defend the body systems against disease and injury. The water in blood also holds heat. As a result, the circulatory system helps keep all systems at the right temperature as it pumps blood through the body.

Integumentary System

Blood vessels deliver nutrients and oxygen to skin; carry away wastes; blood clots if skin is broken.

Skin prevents water loss; helps regulate body temperature; protects blood vessels.

Skeletal System

Blood vessels deliver nutrients and oxygen to bones; carry away wastes.

Rib cage protects heart; red bone marrow produces blood cells.

Muscular System

Blood vessels deliver nutrients and oxygen to muscles; carry away wastes.

Muscle contraction keeps blood moving in heart and blood vessels.

Nervous System

Blood vessels deliver nutrients and oxygen to neurons; carry away wastes.

Brain controls nerves that regulate the heart and width of blood vessels.

Endocrine System

Blood vessels transport hormones from glands; blood services glands.

Hormones and other chemicals help regulate blood volume and blood cell formation.

How the Circulatory System Works with Other Body Systems

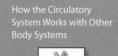
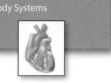

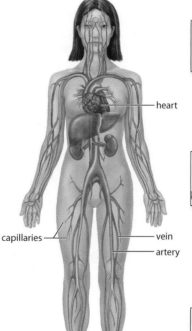

heart

capillaries

vein

artery

Lymphatic System/Immune System

Blood vessels transport white blood cells.

Lymphatic vessels collect excess tissue fluid and return it to blood vessels; the spleen filters blood.

Respiratory System

Blood vessels transport gases to and from lungs; blood services respiratory organs.

Gas exchange in lungs rids body of carbon dioxide, helping to regulate the pH of blood.

Digestive System

Blood vessels transport nutrients from digestive tract to body; blood services digestive organs.

Digestive tract provides nutrients for blood cell formation; liver detoxifies blood, makes plasma proteins, destroys old red blood cells.

Urinary System

Blood vessels deliver wastes to be excreted; blood pressure aids kidney function; blood services urinary organs.

Kidneys filter blood and excrete wastes; maintain blood volume, pressure, and pH.

The Respiratory System

The respiratory system contributes to homeostasis in many ways. For example, body cells depend on the respiratory system for gas exchange. The respiratory system also keeps the pH of the blood within healthy limits by changing levels of carbon dioxide in the blood. As well, the cilia and mucus of the respiratory tract help to defend against bacteria and other disease-causing microbes by trapping and removing them from respiratory surfaces.

Integumentary System

Gas exchange in lungs provides oxygen to skin and rids body of carbon dioxide from skin.

Skin helps protect respiratory organs and helps control body temperature.

Skeletal System

Gas exchange in lungs provides oxygen and rids body of carbon dioxide.

Rib cage protects lungs and helps breathing; bones provide sites of attachment for muscles involved in breathing.

Digestive System

Gas exchange in lungs provides oxygen to the digestive organs and rids body of carbon dioxide produced by those organs.

Breathing is possible through the mouth, because the digestive and respiratory tracts share the pharynx.

Nervous System

Lungs provide oxygen for nerve cells and rid the body of carbon dioxide produced by nerve cells.

Respiratory centres in the brain control breathing rate.

How the Respiratory System Works with Other Body Systems

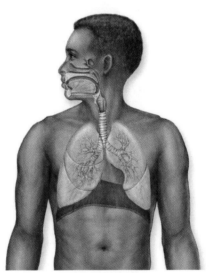

Endocrine System

Gas exchange in lungs provides oxygen and rids body of carbon dioxide.

Hormones help promote expanding of bronchioles during breathing; proteins control production of red blood cells that carry oxygen.

Circulatory System

Gas exchange in lungs provides oxygen and rids body of carbon dioxide, which also helps control pH of blood.

Blood vessels transport gases to and from lungs; blood carries oxygen to tissues of respiratory system.

Lymphatic System/Immune System

Gas exchange in lungs provides oxygen and rids body of carbon dioxide.

Immune cells provide protection against microbes and toxins and aid in tissue repair.

Muscular System

Lungs provide oxygen for contracting muscles and rid the body of carbon dioxide from contracting muscles.

Muscle contraction helps breathing; physical activity can improve efficiency of inhaling and exhaling.

Urinary System

Gas exchange in lungs provides oxygen and rids body of carbon dioxide; helps maintain pH balance of blood.

Lungs remove carbon dioxide.

The Digestive System

The digestive system works with the circulatory system when the nutrients it digests are absorbed into the blood. The nutrient-rich blood is transported by blood vessels to the cells in the body. However, as blood absorbs the nutrients and cells take them up, the water-to-nutrient balance in the body changes. The urinary system helps to restore this balance by removing excess water and nutrients from the blood through the kidneys.

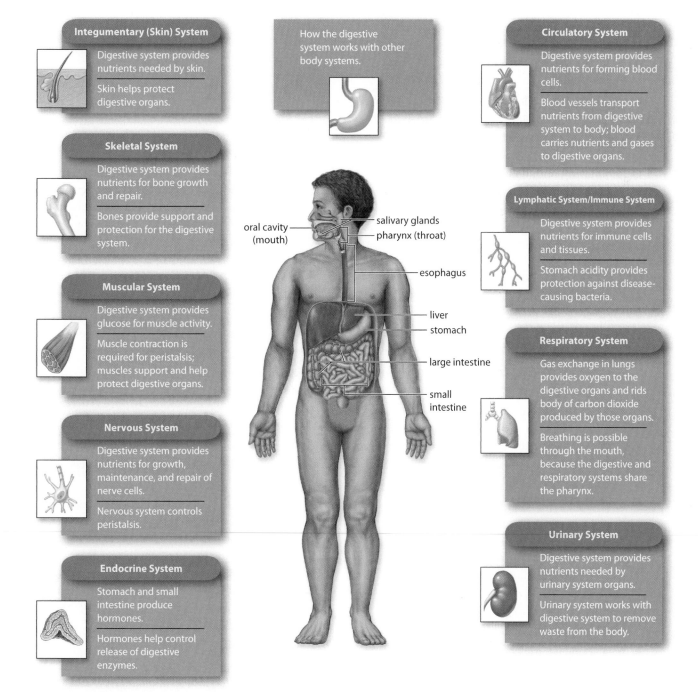

Integumentary (Skin) System

Digestive system provides nutrients needed by skin.

Skin helps protect digestive organs.

Skeletal System

Digestive system provides nutrients for bone growth and repair.

Bones provide support and protection for the digestive system.

Muscular System

Digestive system provides glucose for muscle activity.

Muscle contraction is required for peristalsis; muscles support and help protect digestive organs.

Nervous System

Digestive system provides nutrients for growth, maintenance, and repair of nerve cells.

Nervous system controls peristalsis.

Endocrine System

Stomach and small intestine produce hormones.

Hormones help control release of digestive enzymes.

How the digestive system works with other body systems.

oral cavity (mouth)
salivary glands
pharynx (throat)
esophagus
liver
stomach
large intestine
small intestine

Circulatory System

Digestive system provides nutrients for forming blood cells.

Blood vessels transport nutrients from digestive system to body; blood carries nutrients and gases to digestive organs.

Lymphatic System/Immune System

Digestive system provides nutrients for immune cells and tissues.

Stomach acidity provides protection against disease-causing bacteria.

Respiratory System

Gas exchange in lungs provides oxygen to the digestive organs and rids body of carbon dioxide produced by those organs.

Breathing is possible through the mouth, because the digestive and respiratory systems share the pharynx.

Urinary System

Digestive system provides nutrients needed by urinary system organs.

Urinary system works with digestive system to remove waste from the body.

Assessing Scientific Validity of Advertisements

Manufacturers often make claims of "scientific" or "clinical" evidence for their product's effectiveness. How valid are such claims? Use the following questions to assess the validity of scientific evidence claims.

1. **Control Groups** Control groups provide a reference with which to compare experimental groups to determine if there is an effect caused by the treatment. They provide baseline values for comparison to the measures of the experimental group.
   a. Was a control group used in the investigation?
   b. What is the importance of a control group to a scientific investigation?

2. **Size of Sample** Samples of a population vary with respect to a particular variable—that is, they have a degree of error associated with them. Including many subjects in each study group minimizes the possibility that the differences between groups are due to chance differences between individuals assigned to groups. The greater the sample size, the more variation that is accounted for, and the less error that is associated with them. Studies with larger sample sizes increase the ability to determine a difference between two groups, if one exists.
   a. How many subjects were included in the study?
   b. Why is it important to have as many subjects as possible in an investigation?

3. **Grouping** Subjects must be placed into study groups (control/experimental) in such a way that there is no systematic difference (bias) between them that could affect the results of a study. This makes comparisons between the groups valid.
   a. How were the subjects assigned to the study groups?
   b. Why is the manner in which subjects are assigned to study groups important to the validity of scientific investigations?

4. **Controlling Variables** Controlling for variables reduces the likelihood of alternative explanations and provides for clear interpretation of the results.
   a. What variables were controlled for in the investigation?
   b. Why is it important to the validity of the results that particular variables be controlled for?

5. **Data** Science is a method of inquiry that is based on a critical evaluation of evidence (data). Data must be made available to public scrutiny to be considered valid.
   a. What data are presented?
   b. Why is providing the results of the investigation important?

6. **Replicating Results** Only results that can be reproduced are scientifically valid. If others can reproduce the experiment, they can attempt to verify the data. The methods, as well as the results, of the study are subject to public scrutiny.
   a. Is enough information provided so that the investigation can be replicated?
   b. Why is providing this information vital to a scientific claim?

7. **Analyzing Data** Because most scientific studies use samples, which have an associated degree of error, procedures are needed to determine if the results are significantly different. Statistical procedures can account for the error of samples and determine if significant differences exist between study groups at a given level of confidence.
   a. Is the data statistically analyzed to draw conclusions?
   b. Why is it important that statistical tests be used to analyze experimental results?

8. **Anecdotal Evidence** Anecdotal evidence represents the observations of a limited number of subjects and is of limited scientific value because these subjects are often not representative of the study population.
   a. Is anecdotal evidence (testimonials) presented?
   b. What are the limitations/weaknesses of anecdotal evidence?

9. **Critical Evaluation** A critical evaluation of scientific claims by the peer review process is a hallmark of science as a method of inquiry. A critical evaluation of both the methods and the results of scientific investigations helps to establish the validity of scientific findings. The integrity of science rests on the exposure of new ideas and findings.
   a. What questions would you like to ask the researchers?
   b. What is the role of peer review in the science?

Adapted from: Rutledge, M.L. (2005). Making the nature of science relevant: Effectiveness of an activity that stresses critical thinking skills. The American Biology Teacher, *67(6), 325-329*

A Quick Chemistry Reference

Matter is anything that takes up space and has mass. All matter can be classified as mixtures or pure substances. Pure substances are either elements or compounds. An element is a form of matter that cannot be broken down further by chemical or physical methods. A compound is made up of two or more elements that have been chemically combined. Mixtures contain two or more pure substances.

Like all matter, the matter in living organisms—whether it is a pure substance or a mixture—is made up of atoms. An ordinary chemical reaction cannot destroy, create, or split an atom. Atoms are made up of subatomic particles—particles that are smaller than an atom.

The Structure of Atoms

To understand and explain the properties of matter and the nature of chemical reactions, you need to know about the subatomic particles called *protons, neutrons,* and *electrons.* Their properties are summarized in the following table.

Protons, Neutrons, and Electrons

Subatomic Particle	Symbol	Type of Charge	Amount of Charge	Mass (u)
Proton	p^+	Positive	+1	1.0
Neutron	n^0	Neutral	0	1.0
Electron	e^-	Negative	−1	0.000 55

In atoms, subatomic particles are arranged in a characteristic structure, as shown in the diagram below. The protons and neutrons are clustered together in the nucleus, which contains over 99 percent of an atom's mass but makes up less than 1 percent of its volume. The electrons surround the nucleus in regions called shells. Electrons make up less than 1 percent of an atom's mass, although the shells they occupy make up over 99 percent of its volume.

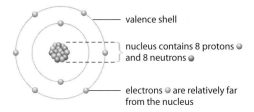

This model of an oxygen atom shows the arrangement of its subatomic particles. Fixed numbers of electrons occupy regions called shells. The outermost shell is called the valence shell.

Different elements, such as hydrogen and oxygen, are distinguished from one another by the number of protons their atoms contain. All atoms of the same element contain the same number of protons. In a neutral atom, the number of its electrons always equals the number of its protons. The periodic table lists and provides information about all the known elements. For a given element, the number of neutrons may vary from one atom to another.

Covalent Bonds

A *covalent bond* forms when the electron shells of two atoms overlap so that the valence electrons of each atom are shared between both atoms. A covalent bond between the two hydrogen atoms in a hydrogen molecule is shown below. Atoms of different elements can also form molecules held together by covalent bonds. Compounds that are made of molecules are called *molecular compounds.*

Two hydrogen atoms that share a pair of electrons form a molecule (H_2) with a single covalent bond. The structural formula to show the single covalent bond between the hydrogen atoms is H–H.

The Tendency Toward Stability

The noble gases (Group 18 on the periodic table) are so chemically stable that they are unlikely to take part in chemical reactions. When atoms bond, they share, give up, or gain electrons to achieve the same arrangement of valence electrons as that of the noble gas to which they are closest in the periodic table.

The maximum number of electrons that can occupy the first valence shell outside a nucleus is two (the valence-shell arrangement of the noble gas helium). In a hydrogen molecule, which is made up of two hydrogen atoms, each atom achieves a stable valence-shell arrangement by sharing a valence electron with the other atom. For elements with atomic numbers 3 to 20, the maximum number of electrons that can occupy the valence shell of each atom is eight (the valence-shell arrangement of the other noble gases).

Appendix A

Double and Triple Bonds

The bond between atoms in a hydrogen molecule is a single covalent bond, since it involves a single pair of shared electrons. However, atoms can also share two pairs of electrons or three pairs of electrons in a covalent bond. In a double covalent bond, two atoms share two pairs of electrons. This is shown using two oxygen atoms in the illustration below.

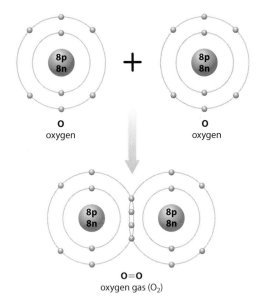

By sharing two pairs of electrons (in a double bond), each of the oxygen atoms has access to eight electrons in its valence shell. This gives each oxygen atom the same stable valence-shell arrangement as the noble gas closest to it in the periodic table, which is neon.

Carbon dioxide (CO_2) is an example of a three-atom molecule held together by double covalent bonds. Examine the CO_2 molecule in the following Lewis structure. Look for evidence that each atom in the molecule has access to a stable valence-shell arrangement.

$$\ddot{\text{O}}\colon \colon\text{C}\colon \colon\ddot{\text{O}}\colon$$

This Lewis structure illustrates a carbon dioxide molecule. The structural formula of the molecule is O=C=O.

In a triple covalent bond, two atoms share three pairs of electrons, as shown in the following Lewis structure for a molecule of nitrogen (N_2).

$$\colon\text{N}\colon\colon\colon\text{N}\colon$$

This Lewis structure illustrates molecular nitrogen, which consists of two nitrogen atoms joined by a triple bond. The structural formula of the molecule is N–N.

Polar Covalent Bonds and Polar Molecules

Most of the biochemical reactions in a living cell take place in a water solution.

$$\underset{\displaystyle :\ddot{\text{O}}\colon\text{H}}{\overset{\displaystyle \text{H}}{}}$$

This Lewis structure is a model of a water molecule.

The Lewis structure above shows that a water molecule is held together by covalent bonds. What the diagram does not show is that the oxygen atom attracts electrons with greater force than the hydrogen atoms do, because the oxygen atom has more protons in its nucleus. This type of covalent bond, in which the electrons are unequally shared between atoms, is called a *polar covalent bond*.

There are two polar covalent bonds in water. Because of the bent shape of the water molecule, it has what can be thought of as a "hydrogen end" and an "oxygen end." Due to the polar bonds, the oxygen end of a water molecule has a slight negative charge, while the hydrogen end of the molecule has a slight positive charge. Water is considered a polar molecule. The following diagram provides a representation of the polarity in a water molecule.

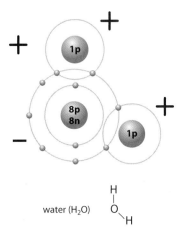

water (H_2O)

Even though a water molecule is held together with polar covalent bonds and is itself polar, it is electrically neutral overall.

Water and Hydrogen Bonds

When water molecules or other polar molecules are near each other, the slight negative charge on one end of a molecule attracts the slight positive charge on the other end of another molecule.

The attraction that forms between water molecules is called a *hydrogen bond*. Hydrogen bonds are much weaker than covalent bonds, but they are strong compared to other bonds that form between molecules. Hydrogen bonds are responsible for some of the unique properties of water. For example, hydrogen bonding allows water molecules to "stick together" as they are pulled up the trunk of a tall tree. The illustration below shows the pattern of attractions that forms between liquid water molecules as a result of hydrogen bonding. Hydrogen bonds are found not just in water—they are also in other important biological molecules, such as DNA and proteins.

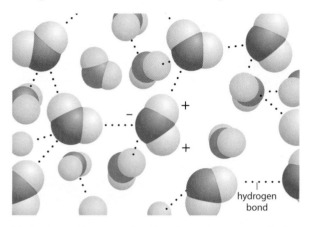

The polarity of water molecules allows attractions called hydrogen bonds to form between the water molecules. The dotted lines represent hydrogen bonds. Dotted lines are used to indicate the weakness of hydrogen bonds relative to covalent and ionic bonds.

Ionic Bonds

Atoms can also form ionic bonds. Ionic bonds form when atoms or groups of atoms transfer electrons rather than share electrons (as in a covalent bond). When an atom or group of atoms gains or loses electrons, it acquires an electric charge and becomes an *ion*. When the number of electrons is less than the number of protons, the ion is positive (and called a cation). When the number of electrons exceeds the number of protons, the ion is negative (and called an anion). Ions can be composed of only one element, such as the hydrogen ion (H^+), or of several elements, such as the bicarbonate ion (HCO_3-). The attraction between oppositely charged ions is called an *ionic bond*.

Forming Ionic Compounds

Ionic bonds hold positively and negatively charged ions together within an *ionic compound*. When solid sodium, $Na(s)$, is exposed to chlorine gas, $Cl_2(g)$, there is an explosive reaction that releases both heat and light. In this reaction, electrons are transferred from the sodium atoms to the chlorine atoms. Sodium achieves the stable electron configuration of the noble gas neon, while chlorine achieves the configuration of the noble gas argon.

As shown in the diagram below, two ions are formed simultaneously: Na^+ and Cl^-.

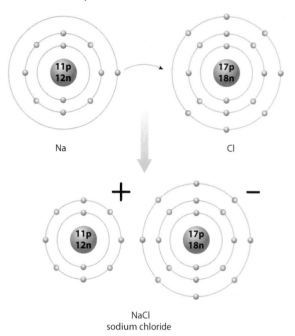

NaCl
sodium chloride

Sodium chloride is created when an ionic bond forms between Na^+ and Cl^-.

The ions align themselves into a regular, repeating pattern based on the size of the individual ions, the amount of charge they carry, and the type of charge they carry (either positive or negative). The ionic compound formed by sodium and chlorine is sodium chloride, $NaCl(s)$, or table salt.

Ionic Compounds in Solution

Table salt and many other ionic compounds dissolve in water. What happens to the ions in sodium chloride when the compound dissolves in water? Attraction by the charged poles of the surrounding water molecules pulls the ions away from the compound and into solution. The following diagram shows how the polar water molecules interact with and surround the sodium and chloride ions.

Notice the orientation of the water molecules around the sodium ion and the chloride ion.

Once dissolved in water, the sodium and chloride ions are free to move about and collide with other particles. This makes the ions mobile enough to carry an electric current from one location to another. So, like many ionic compounds, sodium chloride is an electrolyte. An *electrolyte* is a substance that, when dissolved in water, enables the solution to carry an electric current.

The Biological Significance of Ions

Dissolved ions play a vital role in the chemistry of living cells and body systems. The following table identifies the significance of the important ions in your body.

Significant Ions in the Body

Name	Symbol	Special Significance
Bicarbonate	HCO_3^-	important in acid-base balance
Calcium	Ca^{2+}	found in bones and teeth; important in muscle contraction
Chloride	Cl^-	found in body fluids; important in maintaining fluid balance
Hydrogen	H^+	important in acid-base balance
Hydroxide	OH^-	important in acid-base balance
Phosphate	PO_4^{3-}	found in bones, teeth, and the high-energy molecule that cells use for energy
Potassium	K^+	found primarily inside cells; important in muscle contraction and nerve conduction
Sodium	Na^+	found in body fluids; important in muscle contraction and nerve conduction

Understanding pH

Biological processes take place within specific limits of acidity and basicity. If an environment becomes too acidic or too basic for a process to continue at optimum levels, the organism that depends on that process may suffer and die. Freshwater fish, for example, cannot survive in water that is too acidic. Pitcher plants, sundews, and many other plants that grow in acidic soils cannot tolerate basic conditions.

Whether an environment is acidic or basic depends on its pH. The pH scale is used to measure the acidity or basicity of a solution. The scale, shown at the top of the page, goes from 0 to 14, with 0 representing an extremely acidic solution and 14 representing an extremely basic one. The midpoint of the scale, 7, represents a neutral

solution—one that is neither acidic nor basic. Each change in number up or down the scale represents a tenfold increase or decrease in the acidity of the solution.

Measuring pH

For a relative indication of the pH level, a sample can be tested with litmus paper. This simple test will determine whether a solution is acidic or basic. The pH can also be tested by adding an acid-base indicator such as bromothymol blue. The resulting colour is then compared to a colour chart that indicates relative pH. More precise readings can be determined using a pH meter or probe.

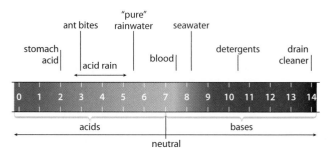

The average pH values for various substances are indicated on this pH scale.

Periodic Table of the Elements

Key (example):
- Atomic number: 26
- Electronegativity: 1.8
- Element Symbol: Fe
- Name of element: iron
- Relative atomic mass (u): 55.85
- Common ion charges: 3+, 2+

Legend

Element categories	State
Alkali metals	Solid
Alkaline earth metals	Liquid
Transition metals	Gas
Other metals	No stable isotopes
Other non-metals	
Halogens	
Noble gases	
Lanthanoids	
Actinoids	
Metalloids	

main-group elements

transition elements

inner transition elements

*Temporary names

Group 1
- 1 / 2.2 / H / 1.01 / hydrogen / 1+,1−
- 3 / 1.0 / Li / 6.94 / lithium / 1+
- 11 / 0.9 / Na / 22.99 / sodium / 1+
- 19 / 0.8 / K / 39.10 / potassium / 1+
- 37 / 0.8 / Rb / 85.47 / rubidium / 1+
- 55 / 0.8 / Cs / 132.91 / cesium / 1+
- 87 / 0.7 / Fr / (223) / francium / 1+

Group 2
- 4 / 1.6 / Be / 9.01 / beryllium / 2+
- 12 / 1.3 / Mg / 24.31 / magnesium / 2+
- 20 / 1.0 / Ca / 40.08 / calcium / 2+
- 38 / 1.0 / Sr / 87.62 / strontium / 2+
- 56 / 0.9 / Ba / 137.33 / barium / 2+
- 88 / 0.9 / Ra / (226) / radium / 2+

Group 3
- 21 / 1.4 / Sc / 44.96 / scandium / 3+
- 39 / 1.2 / Y / 88.91 / yttrium / 3+
- 57–71 La–Lu lanthanoids
- 89–103 Ac–Lr actinoids

Group 4
- 22 / 1.5 / Ti / 47.87 / titanium / 4+,3+
- 40 / 1.3 / Zr / 91.22 / zirconium / 4+
- 72 / 1.3 / Hf / 178.49 / hafnium / 4+
- 104 / Rf / (261) / rutherfordium / —

Group 5
- 23 / 1.6 / V / 50.94 / vanadium / 5+,4+,3+,2+
- 41 / 1.6 / Nb / 92.91 / niobium / 5+,3+
- 73 / 1.5 / Ta / 180.95 / tantalum / 5+
- 105 / Db / (262) / dubnium / —

Group 6
- 24 / 1.6 / Cr / 52.00 / chromium / 3+,2+
- 42 / 1.8 / Mo / 95.94 / molybdenum / 6+
- 74 / 1.7 / W / 183.84 / tungsten / 6+
- 106 / Sg / (266) / seaborgium / —

Group 7
- 25 / 1.5 / Mn / 54.94 / manganese / 2+,4+,7+
- 43 / 1.9 / Tc / (98) / technetium / 7+
- 75 / 1.9 / Re / 186.21 / rhenium / 7+
- 107 / Bh / (264) / bohrium / —

Group 8
- 26 / 1.8 / Fe / 55.85 / iron / 3+,2+
- 44 / 2.2 / Ru / 101.07 / ruthenium / 3+
- 76 / 2.2 / Os / 190.23 / osmium / 4+
- 108 / Hs / (277) / hassium / —

Group 9
- 27 / 1.8 / Co / 58.93 / cobalt / 2+,3+
- 45 / 2.2 / Rh / 102.91 / rhodium / 3+
- 77 / 2.2 / Ir / 192.22 / iridium / 4+
- 109 / Mt / (268) / meitnerium / —

Group 10
- 28 / 1.8 / Ni / 58.69 / nickel / 2+,3+
- 46 / 2.2 / Pd / 106.42 / palladium / 2+,4+
- 78 / 2.2 / Pt / 195.08 / platinum / 4+,2+
- 110 / Ds / (271) / darmstadtium / —

Group 11
- 29 / 1.9 / Cu / 63.55 / copper / 2+,1+
- 47 / 1.9 / Ag / 107.87 / silver / 1+
- 79 / 2.4 / Au / 196.97 / gold / 3+,1+
- 111 / Rg / (272) / roentgenium / —

Group 12
- 30 / 1.7 / Zn / 65.41 / zinc / 2+
- 48 / 1.7 / Cd / 112.41 / cadmium / 2+
- 80 / 1.9 / Hg / 200.59 / mercury / 2+,1+
- 112 / Cn / (285) / copernicium / —

Group 13
- 5 / 2.0 / B / 10.81 / boron / 3+
- 13 / 1.6 / Al / 26.98 / aluminum / 3+
- 31 / 1.8 / Ga / 69.72 / gallium / 3+
- 49 / 1.8 / In / 114.32 / indium / 3+
- 81 / 1.8 / Tl / 204.38 / thallium / 1+,3+
- 113 / Uut* / (284) / ununtrium / —

Group 14
- 6 / 2.6 / C / 12.01 / carbon / 4+
- 14 / 1.9 / Si / 28.09 / silicon / 4+
- 32 / 2.0 / Ge / 72.64 / germanium / 4+
- 50 / 2.0 / Sn / 118.71 / tin / 4+,2+
- 82 / 1.8 / Pb / 207.2 / lead / 2+,4+
- 114 / Uuq* / (289) / ununquadium / —

Group 15
- 7 / 3.0 / N / 14.01 / nitrogen / 3−
- 15 / 2.2 / P / 30.97 / phosphorus / 3−
- 33 / 2.2 / As / 74.92 / arsenic / 3−
- 51 / 2.1 / Sb / 121.76 / antimony / 3+,5+
- 83 / 1.9 / Bi / 208.98 / bismuth / 3+,5+
- 115 / Uup* / (288) / ununpentium / —

Group 16
- 8 / 3.4 / O / 16.00 / oxygen / 2−
- 16 / 2.6 / S / 32.07 / sulfur / 2−
- 34 / 2.6 / Se / 78.96 / selenium / 2−
- 52 / 2.1 / Te / 127.60 / tellurium / 2−
- 84 / 2.0 / Po / (209) / polonium / 2+,4+
- 116 / Uuh* / (292) / ununhexium / —

Group 17
- 9 / 4.0 / F / 19.00 / fluorine / 1−
- 17 / 3.2 / Cl / 35.45 / chlorine / 1−
- 35 / 3.0 / Br / 79.90 / bromine / 1−
- 53 / 2.7 / I / 126.90 / iodine / 1−
- 85 / 2.2 / At / (210) / astatine / 1−

Group 18
- 2 / He / 4.00 / helium / —
- 10 / Ne / 20.18 / neon / —
- 18 / Ar / 39.95 / argon / —
- 36 / Kr / 83.80 / krypton / —
- 54 / Xe / 131.29 / xenon / —
- 86 / Rn / (222) / radon / —
- 118 / Uuo* / (294) / ununoctium / —

Lanthanoids (period 6)
- 57 / 1.1 / La / 138.91 / lanthanum / 3+
- 58 / 1.1 / Ce / 140.12 / cerium / 3+
- 59 / 1.1 / Pr / 140.91 / praseodymium / 3+
- 60 / 1.1 / Nd / 144.24 / neodymium / 3+
- 61 / — / Pm / (145) / promethium / 3+
- 62 / 1.2 / Sm / 150.36 / samarium / 3+,2+
- 63 / — / Eu / 151.96 / europium / 3+,2+
- 64 / 1.2 / Gd / 157.25 / gadolinium / 3+
- 65 / — / Tb / 158.93 / terbium / 3+
- 66 / 1.2 / Dy / 162.50 / dysprosium / 3+
- 67 / 1.2 / Ho / 164.93 / holmium / 3+
- 68 / 1.2 / Er / 167.26 / erbium / 3+
- 69 / 1.3 / Tm / 168.93 / thulium / 3+
- 70 / — / Yb / 173.04 / ytterbium / 3+,2+
- 71 / 1.0 / Lu / 174.97 / lutetium / 3+

Actinoids (period 7)
- 89 / 1.1 / Ac / (227) / actinium / 3+
- 90 / 1.3 / Th / 232.04 / thorium / 4+
- 91 / 1.5 / Pa / 231.04 / protactinium / 5+,4+
- 92 / 1.7 / U / 238.03 / uranium / 6+,4+
- 93 / 1.3 / Np / (237) / neptunium / 5+
- 94 / 1.3 / Pu / (244) / plutonium / 4+,6+
- 95 / — / Am / (243) / americium / 3+,4+
- 96 / — / Cm / (247) / curium / 3+
- 97 / — / Bk / (247) / berkelium / 3+,4+
- 98 / — / Cf / (251) / californium / 3+
- 99 / — / Es / (252) / einsteinium / 3+
- 100 / — / Fm / (257) / fermium / 3+
- 101 / — / Md / (258) / mendelevium / 2+,3+
- 102 / — / No / (259) / nobelium / 2+,3+
- 103 / — / Lr / (262) / lawrencium / 3+

Although Group 12 elements are often included in the transition elements, these elements are chemically more similar to the main-group elements.

Although Group 12 elements are often included in the transition elements, these elements are chemically more similar to the main-group elements.

Any value in parentheses is the mass of the least unstable or best known isotope for elements that do not occur naturally.

Risk and Economics

What Is Risk?

Risk is the probability that a condition or action will lead to an injury, damage, or loss. When considering any activity or situation that poses a risk, we generally think about three factors:

- the probability of a bad outcome
- the consequences of a bad outcome
- the cost of dealing with a bad outcome

Probability is a mathematical statement about how likely it is that something will happen. For example, "The probability of developing a particular illness is 1 in 10 000," or "The likelihood of winning the lottery is 1 in 5 million." Probability is not the same as possibility. When we say something is possible, we are just saying that it could occur. It is a very inexact term. Probability specifically defines how likely it is that a possible event will occur.

The consequences of a negative outcome may be minor or major. For example, ammonia is a common household product. One hundred percent of people will get watery eyes and other symptoms when exposed to ammonia. The probability that a person will be exposed to ammonia and have an adverse effect are high; however, there are no lasting effects after the person recovers. Therefore, we are willing to use ammonia in our homes and accept the high probability of an exposure. On the other hand, if a large dam were to fail, it could cause extensive property damage and the deaths of thousands of people downstream. The consequences of a dam failure are great. Therefore we insist on very high engineering standards so that the probability of a failure is extremely low.

In accepting risk, we must understand that there will likely be an economic cost if a negative outcome occurs. For example, there will be health care costs if people become ill or are injured. If a dam fails and a flood occurs downstream, there will be an economic cost associated with the loss of life and property. To properly assess and manage risk we must understand probability and the consequences of our decisions. **Table 1** gives examples of the risk of dying of certain causes.

Table 1 Risk of Dying of Select Causes

Cause of Death	Risk of Dying
Motor vehicle accident	1 in 5814
Cancer	1 in 34 483
Homicide	1 in 40 000
Cardiovascular Disease	1 in 58 824
Accidental poisoning	1 in 76 923
Drowning	1 in 142 857
Flu and pneumonia	1 in 333 333
Tuberculosis	1 in 1 000 000

Data for 15–19 year olds from Statistics Canada report, "Mortality, Summary List of Causes," 2006.

Risk and Economics

Most decisions in life involve an analysis of two factors: risk and cost. We commonly ask such questions as "How likely is it that someone will be hurt?" and "What is the cost of this course of action?" Furthermore, these two factors are often interrelated. When we make economic decisions, we may be risking our hard earned money. There are also costs associated when someone gets physically hurt, such as medical care costs or legal fees. Environmental decision-making is no different. In environmental issues there is the cost of the environmental risk and the economic cost of eliminating the conditions that pose the risk.

Making an informed decision about environmental matters involves finding a balance between these two kinds of cost. For example, the government might introduce a new air pollution regulation to reduce the public's exposure to a chemical that causes disease in a small percentage of people. Companies will have to pay a considerable amount of money to meet the standards of the new regulation, which will reduce their profitability. Citizens may have to pay increased taxes so that the government can monitor companies and make sure that the regulations are followed. On the other side, the new regulation may reduce the risk of illness (and therefore reduce health care costs) for people who live in areas affected by the pollutant.

Assessing Environmental Risk

Environmental risk assessment uses facts and assumptions to estimate the probability of harm to human health or the environment that may result from particular decisions. An environmental risk assessment process provides environmental decision makers with a clear and consistent way to evaluate whether a risk exists, assess the magnitude of the risk, and predict the possible negative consequences of accepting the risk.

Calculating the risk to humans of a particular activity, chemical, technology, or policy is difficult. There are several tools to help assess the risk. Scientists may use probabilities based on past experience to estimate risks. There are also environmental risks that do not directly affect human health. For example, if human activities cause the extinction of a species, there is a negative environmental impact, but the direct impact on humans may not be obvious. Similarly, government policy decisions that are not sustainable may deplete resources for future generations.

Scientists often use models to estimate the risks associated with new technologies or policies for which there is no established history. For example, animals are often used to model the risk of chemical exposure to humans. However, an animal may not react in the same way as a human. Therefore, animal studies are only indicators of human risk.

In other situations computer simulations are used. For example, complex computer models of climate have been used to assess the effects of energy policies. Most risk assessments can only estimate the probability of negative effects. For example, people may be more or less sensitive to certain chemicals than the laboratory animals studied. Also, some people are more sensitive to certain compounds than others. What may present no risk to one person may be a high risk to others.

Because of all these uncertainties about risk, governments usually make decisions that are more cautious in order to protect public health and safety. For example, government regulations concerning such things as pesticides, air pollutants, and water contaminants set conditions of use and exposure limits that have a large margin of safety. Thus, if animal studies show a negative effect from a certain dose of chemical, the allowable dose for humans is set at a lower level.

Risk assessment is also used to set priorities about which chemicals, technologies, or situations urgently need regulation. Those that have the highest potential to cause damage to health or the environment receive attention first, while those perceived as having minor impacts receive less immediate attention. Medical waste is perceived as high-risk, so laws have been created to minimize the risk, while the risk associated with the use of fertilizer on lawns is considered minimal and is not regulated everywhere.

Many of the most important threats to human health and the environment are highly uncertain. A risk assessment can help governments determine research priorities and make plans that take into account scientific research and public concern for environmental protection.

Managing Environmental Risk

Environmental risk management involves selecting the most appropriate regulations for a specific issue. This decision-making process must consider scientific data and social, economic, and political concerns. The purpose of risk management is to reduce the probability of a negative outcome or reduce the negative outcome's severity if it occurs. Decision makers must understand the probability and consequences of the risk and the factors that could increase or decrease the risk. For example, automobile accidents are a leading cause of accidental death. Once this fact is recognized, actions can be taken to decrease the risks of an automobile accident. Examples include warning signs, speed limits, and laws against drunk driving. Actions can also be taken to decrease the harm to a person who is involved in an automobile accident, for example, using air bags and seat belts.

A risk management plan includes:

1. Evaluating the scientific information regarding various kinds of risks
2. Deciding how much risk is acceptable
3. Deciding which risks should be given the highest priority
4. Deciding where the greatest benefit can be achieved, given the funds available
5. Deciding how the plan will be enforced and monitored

The first step is evaluating the scientific evidence to assess how great the risk is. With environmental issues (such as hazardous waste, climate change, or acid rain) the scientific information can be controversial. For example, issues like these require scientists to project into the future to estimate the severity of future effects. Scientific estimates of these effects can vary widely, and it can be difficult to know which scientific projections to believe.

For each situation, decision makers must examine the scientific evidence and ask what degree of risk is acceptable. They must also integrate economic and political factors into the decision. The next step is to select the issues that are of the highest priority and assign people and resources to solve the problems. It is important to define the problem clearly so that the most suitable regulations, laws, or recommendations can be put in place to deal with the situation. Even after regulations have been put in place, however, there is often still controversy because the stakeholders involved have different views about the issue and how the regulations are enforced.

True and Perceived Risks

Perceptions play a large role in all environmental issues. For example, a few decades ago there was a great deal of fear about asbestos in schools. The public demanded that all asbestos be removed from schools to protect children. Many people had the perception that children were at risk of lung cancer and death due to the asbestos. This simply was not the case. Asbestos can cause lung cancer and other diseases, but only if a person is exposed to regular doses (by breather or ingesting the asbestos) over many years. In fact, asbestos does not become a problem unless it is disturbed or removed during renovation or demolition. Most schools did remove all asbestos, when the best solution would have been to leave it in place and encase it for safety. In the case of asbestos, the perceived risk was much greater than the actual risk. Becoming educated about risks enables us to make better decisions about environmental and human health issues. This saves time and money and allows us to maximize the resources available to reduce actual risks to ourselves and the environment.

Risk estimates by "experts" and by the "public" differ significantly on many environmental problems. Almost every daily activity—driving, walking, or working—involves some element of risk. People often overestimate the frequency and seriousness of dramatic, well-publicized causes of death and underestimate the risks from more familiar causes that claim lives one by one. In general, the public generally perceives risks that are outside their control (such as nuclear power plant accidents) as greater than risks that are in their control (such as drinking alcohol or smoking). In addition, newer technologies (such as genetic engineering) are perceived as having greater risks than more familiar technologies (such as automobiles).

This discrepancy between scientific and public perceptions is one of the most difficult problems that decision makers and public health scientists must face. Numerous studies have shown that in the last 20 years, environmental hazards that truly affect public health are not the ones that receive the highest attention. For example, indoor air pollution receives relatively little attention compared with outdoor pollution, yet it probably accounts for as much, if not more, poor health. Hazardous waste dumps, on the other hand, attract much attention and resources. The same chemicals can be found in common consumer products (such as household cleaners and pesticides), but this raises less concern from the public.

This mismatch between real and perceived risks has significant consequences. If money is used to reduce risks that have little health or environmental impact, there is less to spend on interventions that address more significant risks. This leads to an important question. Should the government focus available resources and technology where they can have the greatest impact on human and ecological well-being, or should it focus them on problems about which the public is most upset? When deciding how to distribute funding, compromise is the only option. Risk experts must learn to look beyond the numbers and probabilities to the broader picture, and the public must be supplied with more data to enable them to make better informed judgments.

It is not economically possible to eliminate all risk. Sometimes risk identification is all that is possible or required. A risk elimination process can be desirable but is not always beneficial. In order to eliminate risk, the cost of the product or service often increases. Many environmental issues are difficult to evaluate from a purely economic point of view, but economics is one of the tools used to analyze any environmental problem.

Environmental Economics

Economics is the study of how people choose to use resources to produce goods and services, and how these goods and services are distributed to the public. In other words, economics is a process that determines how resources should be used. In many respects, environmental problems are largely economic problems. It is not possible to view environmental issues outside normal economic processes that are central to our way of life. As a result, businesses today must pay attention to the economics of environmental issues. Many actions of industry have a negative effect on the environment. Governments can increase regulations to help lessen this environmental damage, but that could negatively impact industry and weaken the economy. It is important to know some basic economic concepts to understand the interaction between environmental issues and economics.

Resources

Economists look at resources as the available supply of something that can be used. Resources are classified as labour, capital, and land. *Labour* refers to human resources. *Capital* is anything that enables the production of goods and services (for example, technology and knowledge). *Land* can be thought of as the natural resources of the planet. *Natural resources* are materials and processes that humans can use for their own purposes but cannot create. Some examples of natural resources are: the agricultural productivity of the soil, rivers, minerals, forests, wildlife, and weather. The landscape is also a natural resource. For example, waterfalls can generate hydroelectric power, and beautiful scenery can encourage tourism.

Natural resources are usually categorized as either renewable or nonrenewable. Renewable resources can be formed or regenerated by natural processes. Soil, vegetation, animals, air, and water are renewable because they can naturally be repaired, regenerated, or cleansed when their quality or quantity is reduced. However, just because a resource is renewable, it does not mean that it is inexhaustible. Renewable resources can be degraded and overused. Nonrenewable resources are not replaced by natural processes, or the time it takes to replace them is much longer than a human lifespan. For example, iron ore and fossil fuels are nonrenewable on human timescales. Once nonrenewable resources are used up, they are gone. We must find substitutes or we must do without.

Supply and Demand

An important concept in economics is *scarcity* (meaning that something is in short supply or very limited in number). Scarcity exists whenever the demand for something exceeds its supply. We live in a world of general scarcity, where resources are limited in terms of the desires of humans to consume them. For each type of good (product) or service there is a price or cost. The price of something reflects its supply (how much or little there is of it) and society's demand for it.

The *supply* is the amount of a good or service people are willing to sell at a given price. *Demand* is the amount of a good or service that consumers are willing and able to buy at a given price. The *price* of a good or service is its monetary value. For any good or service, there is a constantly shifting relationship among supply, demand, and price. When demand exceeds supply, the price rises. The increase in price results in a chain of economic events. When the price of something increases, some people will seek alternatives or to decide not to use that product or service. When this happens, demand for the product or service decreases.

For example, food production depends greatly on petroleum for the energy to plant, harvest, and transport food crops. In addition, petrochemicals are used to make fertilizer and chemical pest-control agents. If the demand for energy exceeds the supply, the price of petroleum increases. As petroleum prices rise, farmers reduce their petroleum use. Perhaps they farm less land or use less fertilizer or pesticide. Because farmers are using less energy, they will produce less food, so the supply of food decreases. Thus, an increase in petroleum prices results in an increase in food prices. As the prices of certain foods rise, consumers seek less costly foods.

When the supply of a product exceeds the demand, producers must lower their prices to get rid of the product. Some of the producers might go out of business. This can happen to farmers when they have a series of good years. Production is high, prices fall, and some farmers go out of business.

Appendix A

Assigning Value to Natural Resources and Ecosystem Services

Nature provides both natural resources and ecosystem services. Both are valuable, but it is easier to assign an economic value to natural resources. We assign value to natural resources based on our perception of their relative scarcity. We are willing to pay for goods or services we value highly and are unwilling to pay for things we think there is plenty of. For example, we will readily pay for a warm, safe place to live but likely are unwilling to pay for the air we breathe.

If a natural resource has always been rare, it is expensive. Pearls and precious metals (like gold) are expensive, because they have always been rare, as well as highly prized. If the supply of a resource is very large and the demand for it is low, the resource may be thought of as free. Sunlight, oceans, and air are often not even thought of as natural resources because their supply is so large. However, modern technologies have allowed us to exploit natural resources to a much greater extent than our ancestors could. Resources that were once considered limitless are now rare. For example, in the past, land and its covering of soil were considered a limitless natural resource, but as the population grew, we began to realize that land is a finite, nonrenewable resource. The economic value of land is highest in metropolitan areas, where open land is unavailable. Unplanned, unwise, or inappropriate use can result in severe damage to the land and its soil.

Even renewable resources can be overexploited. If the overexploitation is severe and prolonged, the resource itself may be destroyed. For example, overharvesting of fish, wildlife, or forests can change the natural ecosystem so much that it cannot recover. As a result, a resource that should have been renewable becomes a depleted nonrenewable resource.

In 2010, a United Nations report stated that the world has vastly underestimated the economic value of nature in developing nations. Ecosystems such as fresh water, coral reefs, and forests can be a major part of the livelihood of people living in developing countries. An example of this is coral reefs. It is estimated that as many as 500 million people worldwide depend on fisheries and tourism for their living. Economists value coral reefs at between $30 billion and $172 billion per year. If reefs collapse because of climate change and other factors, huge numbers of people might be forced to migrate away from coastal areas to make a living, which could lead to further environmental, political, and economic problems.

Ecosystem Services

Ecosystem services are the benefits of nature to households, communities, and economies. This term conveys the important idea that ecosystems are a tangible source of economic wealth. However, it can be difficult to measure that wealth.

Environmental goods and services are not traded in conventional markets; therefore, it is difficult to put a price on them. For example, how can you give a value to a beautiful view? Of course, just because something doesn't have a price doesn't mean it has no value. The challenge is to get people to reveal the value they place on it. One way to figure out what value people place on an ecosystem service is to look at their behaviour. For example, homes built near beautiful scenery sell for more than homes without scenery. When people spend time and money travelling to enjoy natural resources, they reveal the value they place on those resources.

Another difficulty is placing units or quantities on ecosystem services. A grocery store is full of cans, boxes, loaves, and bunches, but environmental goods and services don't come in convenient, quantifiable units. To analyze ecosystem services, we have to understand the link between ecological outcomes and economic consequences. It is important to assign some sort of unit to ecosystem services so that natural scientists and economists can describe ecological changes in the same way.

Environmental Costs

When resources are exploited, there are environmental costs. These include air pollution, water pollution, plant and animal extinctions, depletion of resources, and loss of scenic quality. As mentioned above, it is not easy to convert these environmental costs to a monetary value. This is especially true with the loss of "aesthetics" such as a beautiful scenery, relaxing surroundings, or recreational opportunities. In addition, environmental costs are often *deferred costs*, since they may not be recognized until a later date. For example, when a dam is built to provide hydroelectric power, costs to the environment—such as loss of habitat and changes in water quality—may not be taken into account by planners. Soil erosion is another example of a deferred cost. The damage done by practices that increase soil erosion may not be felt immediately, but the cost becomes obvious to future generations.

Many of the environmental challenges facing the world today arise because modern production techniques and consumption patterns transfer waste disposal, pollution, and health costs to society. These impacts are borne by someone other than the individuals who use a resource. This is referred to as an *external* cost. For example, when a logging company removes so many trees from a hillside that mudslides occur and streams are destroyed, the logging company has transferred a cost to the public. Another example is hazardous waste sites produced by industries that no longer exist. The cleanup of these abandoned sites becomes the responsibility of government and taxpayers.

Usually a variety of environmental costs accompany resource use. The extraction of mineral resources is a good example. All mining operations involve the separation of the valuable mineral from the surrounding rock. The surrounding rock must then be disposed of in some way. This waste rock (mine tailings) is usually piled on Earth's surface. Not only are tailings an eyesore, but also some contain substances that are harmful to humans and other organisms.

Many types of mining operations require large quantities of water for the extraction process. The quality of this water is degraded, so it is unsuitable for drinking, irrigation, or recreation. Since mining disturbs the natural vegetation in an area, water may carry soil particles into streams and cause erosion. Some mining procedures, such as strip mining, rearrange the top layers of the soil, which lessens or eliminates its productivity for a long time.

Most environmental costs have both deferred and external aspects. A good example of a problem that has both a deferred and an external cost is the damage caused by burning high-sulfur coal to produce electricity. The sulfur compounds released into the atmosphere lead to acid rain. The acid rain in turn causes a decline in the growth of forests and damage to buildings and other structures. The cost of acid rain is a deferred cost, because the damage accumulates over time. There is also an external cost, because most damage is not paid for by the utility company. Instead the cost of the damage is transferred to others. The public now has fewer scenic vistas and recreational land, and forestry companies have fewer trees to harvest. Individual property owners must pay repair costs for buildings and other structures.

Environmental costs also may include lost opportunities or values because the resource can no longer be used for another purpose. For example, if homes are built in a forested region, it can no longer be used as a natural area for hiking or hunting. Similarly, when land is converted to roads and parking lots, the opportunity is lost to use the land for farming or other purposes.

Pollution Costs and Prevention

A key environmental cost is pollution. Pollution is any addition of matter or energy that degrades the environment for humans and other organisms. When we think about pollution, however, we usually mean something that people produce in large enough quantities that it interferes with our health or well-being. Two main factors that affect the amount of damage done by pollution are the size of the population and the development of new technology that causes new forms of pollution.

When the human population was small and people lived in a simple manner, the wastes produced were biological and in small enough amounts that they usually did not cause a pollution problem. People used what was naturally available and did not manufacture many products. Humans, like any other animal, fit into their natural ecosystems. Their waste products were biodegradable materials that were broken down into simpler chemicals, such as water and carbon dioxide, by the action of decomposer organisms.

Human-initiated pollution became a problem when human populations became so concentrated that their waste materials could no longer be broken down as fast as they were produced. As the population increased, people began to establish villages, towns, and cities. The release of large amounts of smoke, biological waste, and trash faster than they could be absorbed and dispersed resulted in pollution, which led to unhealthy living conditions.

Throughout history, humans have tried to improve their living conditions. In general, we rely on science and technology to improve our quality of life. However, technological progress can also generate new sources of pollution. For example, the development of the steam engine allowed machines to replace animal power and human labour but required more fuel and increased the amount of smoke and other pollutants in the air. The modern chemical industry has produced many useful synthetic materials (plastics, pesticides, medicines), but it has also produced toxic pollutants. It is important to recognize that it is impossible to eliminate all the negative effects produced by humans and our economic processes. The difficult question is to determine the levels of pollution that are acceptable.

As people recognize the significance of environmental costs, these costs are being converted to economic costs. It takes money to clean up polluted water and air or to reclaim land that has been degraded, and the people who cause the damage should not be allowed to defer the cost or escape paying for the necessary cleanup or remediation. There are really two types of costs associated with pollution: pollution control costs and pollution prevention costs. Pollution control costs include public money used to correct pollution damage once it has occurred. Pollution prevention costs are paid by government or industry to prevent or limit pollution that is a result of some activity. For example, a local government might incur a pollution prevention cost when it installs better equipment to treat its sewage before releasing it back into the environment.

The philosophy of pollution prevention is that pollution should be prevented or reduced at the source whenever feasible. It is increasingly being shown that preventing pollution can cut business costs and thus increase profits.

Common pollution prevention techniques include the following.

- improved processes that use energy and materials more efficiently

- alternative processes (for example, low- or no-chlorine pulping for paper)
- in-process material recovery (for example, water reuse or heavy metals recovery)
- using substitutes for heavy metals and other toxic substances
- using alternative fuels and renewable energy

Cost-Benefit Analysis

Because resources are limited and there are competing demands for most resources, a process must be in place to help people decide the most appropriate use of a scarce resource. Cost-benefit analysis is a quantitative method of assessing the costs and benefits to make decisions about how best to use a resource or what the best solution to a problem might be. In most developed countries, major projects (especially government projects) require a cost-benefit analysis that takes into account environmental impacts and regulations. People also use cost-benefit analysis to determine whether a new policy would generate more social costs than social benefits.

Steps in cost-benefit analysis include the following.

1. identification of the project to be evaluated
2. determination of all impacts, favorable and unfavorable, present and future, on all of society
3. determination of the value of those impacts
4. calculation of the net benefit: (total value of positive impacts) − (total value of negative impacts)

Table 1 gives examples of the kinds of costs and benefits involved in improving air quality. Although it is not a complete list, the table indicates the kinds of considerations that go into this kind of analysis. Some of these are easy to measure in monetary terms. Others are not.

Table 1 Costs and Benefits of Improving Air Quality

Costs	Benefits
Installation and maintenance of new technology:	Reduced deaths and disease
Scrubbers on smokestacks	Fewer respiratory problems
Automobile emissions control	Reduced plant and animal damage
Redesign of industries and machines	Lower cleaning costs for industry and public
Additional energy costs to industry and public	More clear, sunny days; better visibility
Retraining of employees to use new technology	Less eye irritation
Costs associated with monitoring and enforcement	Fewer odour problems

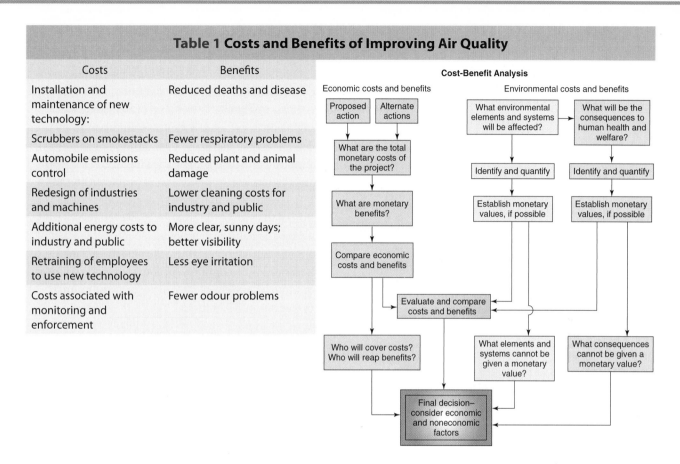

An example of a cost-benefit analysis could be a decision to buy a hybrid car. Hybrid cars can be expensive compared to similar non-hybrid cars. Even with the gas savings you would get, you might not recover the higher purchase price of the hybrid car for several years. However, this quick analysis misses a number of the hybrid's other economic benefits. For example, hybrids hold their value better than non-hybrid cars. This means that you are likely to get more money out of the car when you decide to sell it. In addition, you might be able to benefit from savings such as lower insurance premiums or government tax credits for hybrid car owners. As this hybrid car example shows, a cost-benefit analysis can be very complex and can involve many different variables.

Concerns about the Use of Cost-Benefit Analysis

Critics argue that a cost-benefit analysis is not always appropriate because not everything can be analyzed from an economic point of view. Some people believe that if the only measure of value is economic, many simple non-economic factors (such as beauty or

cleanliness) will be undervalued unless they are given an economic price tag.

Another criticism of cost-benefit analysis is that someone (the analyst) must decide which factors will be considered in the analysis. In theory, cost-benefit analysis should count all benefits and costs associated with a particular issue; however, this is not always done. For example, if a cost is spread thinly over a large population, it may not be recognized as a cost at all. Debates over how to count benefits and costs for future generations, inanimate objects (such as rivers), and nonhumans (such as endangered species) are also common.

Comparing Economic and Ecological Systems

For most natural scientists, current crises such as biodiversity loss, climate change, and many other environmental challenges are symptoms of an imbalance between the socioeconomic system and the natural world. Although humans have always affected the natural world, it is clear that this effect is much greater now than ever before.

Matching economic processes with environmental resources is difficult, because economic systems and ecological systems function in such different ways. One example that illustrates the conflicting views of economics and ecology is the loss of biodiversity. Economic decisions often ignore the environmental context of a species or the interconnections between resource quality and ecosystem functions. For example, from an economic point of view, the value of land used for beef production is measured according to its economic output (beef). Yet the environmental health of the land may be affected by intensive beef production long before the economic output decreases. As long as yields are maintained, these environmental changes go unnoticed by economic measurements and are considered unimportant to land-use decisions.

Another distinction between economics and the environment is the great difference in the *time frame* of markets and ecosystems. Many environmental processes take place over tens of thousands and even millions of years. The time frame for market decisions is short. It may be as short as seconds for stock trades or as long as a few years for the development and construction of a factory.

The concept of *place* is another issue. For ecosystems, place is critical. Ground water is a good example. Local conditions of soil quality, geology, precipitation, and vegetation all contribute to the size and location of ground water reservoirs in that place. These factors cannot simply be transferred from one location to another—they are unique to that place. For economic activities, place is increasingly irrelevant. Local ecological features rarely enter into economic calculations. Production can be transferred to the location where production costs are lowest.

Another difference between economics and the environment is that they are measured in different units. Market economics is measured in terms of money. Natural systems are measured in physical units such as energy, carbon dioxide absorption, centimeters of rainfall, or parts per million of contamination. Focussing only on the economic value of resources while ignoring environmental health can mask serious changes in environmental quality or function.

Problems with Common Property and Resources

When everyone shares ownership of a resource, there is a strong tendency to overexploit and misuse that resource. Often, common public ownership of a resource is like having no owner at all. For example, in pre-industrial England, there were areas of pastureland (called the "commons") provided free by the king to anyone who wished to graze cattle. There are no problems on the commons as long as the number of animals is small in relation to the size of the pasture. However, each herder wants to enlarge his or her herd as much as possible. As the size of each herd grows, the density of animals increases until the commons becomes overgrazed. The result is that everyone eventually loses as the animals die of starvation. Even though the end result should be perfectly clear, no one acts to avert the disaster.

Earth is one big commons stocked with air, water, and irreplaceable mineral resources. This "people's pasture" must be used "in common," but it has very real limits. Each nation attempts to extract as much from the commons as possible without regard to other countries. Furthermore, industrial nations consume far more than their fair share of the total world resources each year— much of them imported from less-developed nations.

A modern example of this problem is the overharvest of marine organisms. Since no one owns the oceans, many countries feel they have the right to exploit the fisheries resources. As in the case of grazing cattle, individuals seek to get as many fish as possible before someone else does. Currently, the UN estimates that nearly all of the marine fisheries of the world are being fished at or above capacity.

Another example of this is the shared fisheries in the Great Lakes region. Commercial fisheries, recreational fishers, Aboriginal fishers, and regulatory agencies in both the United States and Canada have tried for many years to deal with overexploitation of the Great Lakes fishery. The issues involved are complex: invasive and exotic species, pollution, and overfishing are just a few examples. Fishing regulations and zones were not designed from an ecosystem approach. They were established by political decisions, sometimes haphazardly and with complete disregard for biological realities.

Issues with common resources also operate on an individual level. Most people are aware of air pollution, but they continue to drive their automobiles. Many families claim to need a second or third car. It is not that these people are anti-environmental; most would be willing to drive smaller or fewer cars if everyone else did, and they could make do with only one small car if public transport were adequate. But people frequently get "locked into" behaviours that harm the environment, waiting for others to take the first step.

Green Economics

The world has witnessed three economic transformations in the past few centuries. First came the Industrial Revolution, then the technology revolution, then our modern era of globalization. The world now stands at the threshold of another great change: the age of green economics.

The evidence is all around us, often in unexpected places. Brazil, for example has become one of the biggest players in green economics, drawing 44% of its energy needs from renewable fuels. (The world average is 13%.) We often hear how China will soon surpass the United States as the world's largest emitter of greenhouse gases. Less well known, however, are China's more recent efforts to confront major environmental problems. China is on track to invest $10 billion in renewable energy—second only to Germany. It has become a world leader in solar and wind power. China has pledged to reduce energy consumption by 20% by 2014—not far removed from Europe's commitment to a 20% reduction in greenhouse gas emissions by 2020.

Some estimates show that growth in global energy demand could be cut in half over the next 15 years simply by using existing technologies. The Intergovernmental Panel on Climate Change (IPCC) reports such demand cuts could be achieved in very practical ways, from tougher standards for air conditioners and refrigerators and improved efficiency in industry, building, and transport. It estimates that overcoming serious climate change may cost as little 0.1% of global GDP (gross domestic product) a year over the next three decades.

Growth does not need to suffer and, in fact, may accelerate. A German study predicts that more people will be employed in Germany's environmental technology industry than in the auto industry by the end of the next decade. The UN Environment Program estimates that global investment in zero greenhouse energy will reach $1.9 trillion by 2020.

Appendix A

Systems and the Environment

Systems are a central concept in environmental science. A system is a network of interdependent components and processes, with matter and energy flowing from one component of the system to another. For example, the term ecosystem represents a complex assemblage of animals, plants, microscopic organisms, and their environment, through which materials move and energy is transferred.

The idea of systems is useful because it helps to organize thinking about the extremely complex phenomena around us. For example, an ecosystem might consist of countless organisms and their physical surroundings. Keeping track of all the components and their relationships is an impossible task. One way to simplify your approach to thinking about the system is to reorganize its many components into broader categories—for example, as an interacting collection of plants, herbivores, carnivores, and decomposers (**Figure 1**).

Figure 1 A system can be described in very simple terms.

We can use some general terms to describe the components of a system. A simple system consists of components (called state variables or compartments), which store resources such as energy or matter. A simple system also includes flows, or pathways by which those resources move from one state variable to another. In

Figure 1, the plant and animals represent state variables. The plant represents many different plant types, all of which are things that store solar energy and use photosynthesis to create carbohydrates from carbon, water, and sunlight. The rabbit represents many kinds of herbivores, all of which consume plants and then store energy, water, and carbohydrates until they are used, transformed, or consumed by a carnivore. We can describe the flows in terms of herbivory, predation, or photosynthesis—all of which are processes that transfer energy and matter from one state variable to another.

It might seem overly cold and analytical to describe a rabbit or a plant as a state variable, but it is also helpful to do so. By simplifying natural complexity in this way, it becomes easier to analyze and diagnose disturbances or changes in a system. For example, if rabbits become too numerous, herbivory can become too rapid for plants to sustain. Overgrazing can lead to collapse of the system.

Characteristics of Systems

Open system are those that receive inputs from their surroundings and produce outputs that leave the system. Almost all natural systems are open system. In principle, a closed system exchanges no energy or matter with its surroundings, but these are rare. Some environmental scientists talk about pseudo-closed systems—those that exchange only a little energy but no matter with their surroundings.

Throughput is a term used to describe the energy and matter that flow into, through, and out of a system. Larger throughput might expand the size of state variables. For example, you can consider your home economy in terms of throughput. If you get more income, you have the option of increasing your state variables (bank account, new phone, etc.). Usually an increase in income is also associated with an increase in outflow (the money spent on a new phone, for example). In a grassland, inputs of energy (sunlight) and matter (carbon dioxide and water) are stored in biomass. If there is lots of water, the biomass storage might increase (in the form of additional plants). If there's little input, biomass might decrease (grass could become short or sparse). Eventually, stored matter and energy may be exported (by fire, grazing, land clearing). The exported matter and energy can be thought of as throughput.

A grassland is an open system: it exchanges matter and energy with its surroundings (atmosphere and soil, for example). In theory, a closed system would be completely isolated from its surroundings, but in fact, all natural systems are at least partly open. A fish tank is an example of a system that is less open than a grassland, because it can exist with only sunlight and carbon dioxide inputs. (See **Figure 2**.)

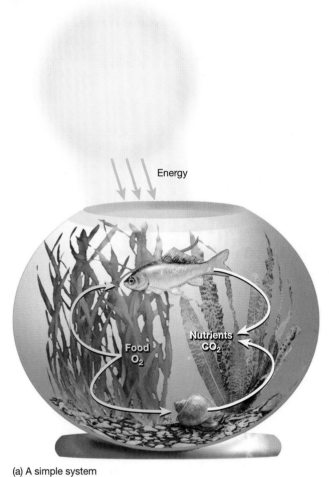

(a) A simple system

Sunlight

Photosynthesis
Plant growth

Plant biomass

Produces waste, CO_2
+

Feeding
O_2

Fish biomass

Blocks sunlight
−

Consumes plants
−

(b) A more complex system

Figure 2 Comparing models of two systems

Systems and Feedback

Systems also experience positive and negative feedback mechanisms. A positive feedback is a self-perpetuating process. In a grassland, a grass plant grows new leaves, and the more leaves it has, the more energy it can capture for producing more biomass. In other words, in a positive feedback mechanism, increases in a state variable (biomass) lead to further increases in that state variable (more biomass). In contrast, a negative feedback is a process that suppresses change. If grass grows very quickly, it may produce more leaves than can be supported by available soil moisture. Without enough moisture, the plant begins to die back.

In climate systems, the concept of positive and negative feedbacks is important. For example, as warm summers melt ice in the Arctic, newly exposed water surfaces absorb heat, which leads to further melting, which leads to further heat absorption, and so on. This is positive feedback. In contrast, some scientists think that clouds have a negative feedback effect. The assumption is that clouds block some solar energy, which reduces evaporation. Thus, slow the warming process.

Systems and Stability

Negative feedbacks tend to maintain stability in a system. Systems are said to exhibit homeostasis—a tendency to stay more or less stable and unchanging. Disturbances are events that destabilize or change a system. There can be many kinds of disturbance in a grassland, for example. Severe drought can set back the community, so it takes more time to recover. Many grasslands also experience occasional fires. Fires stimulate grass growth by clearing accumulated litter and by recycling nutrients, but it also destroys shrubs and trees that might be encroaching on the grassland. Disturbances like this are a normal part of all natural systems. All systems change in nature.

Grassland plots show resilience, which is an ability to recover from disturbance. In fact, studies suggest that species-rich plots show more resilience than species-poor plots. Sometimes severe disturbance can lead to a state shift, in which conditions do not return to "normal." For example, a climate shift that dramatically reduces rainfall could lead to a transition from grassland to desert. Plowing up grassland to plant crops is basically a state shift from a complex system to a single-species (monoculture) system.

Appendix B: Answers to Multiple Choice Questions

Answers are provided for multiple choice questions in chapter and unit reviews.

Unit 1 Environmental Challenges and Solutions

Unit Review
1. b
2. e
3. b
4. d
5. c

Unit 2 Energy Use and Conservation

Chapter 3 Energy Resources
1. d
2. d
3. c
4. b
5. b
6. d
7. a
8. b

Chapter 4 Energy Use for a Sustainable Future
1. e
2. d
3. b
4. b
5. e
6. c
7. e
8. b

Unit Review
1. b
2. a
3. e
4. d
5. c

Unit 3 Sustainable Agriculture and Forestry

Chapter 5 Sustainable Food Production
1. a
2. e
3. c
4. a
5. b
6. c
7. a
8. d

Chapter 6 Sustainable Forestry
1. b
2. c
3. e
4. b
5. a
6. a
7. d
8. a

Unit Review
1. d
2. c
3. e
4. a
5. d

Unit 4 Managing and Reducing Waste

Chapter 7 Solid and Liquid Waste Management
1. a
2. b
3. d
4. a
5. c
6. e
7. b
8. c

Chapter 8 Managing Hazardous Wastes
1. b
2. c
3. b
4. e
5. a
6. d
7. c
8. a

Unit Review
1. b
2. a
3. c
4. e
5. e

Unit 5 Human Health and the Environment

Chapter 9 Air Quality and Environmental Health
1. d
2. b
3. c
4. c
5. b
6. e
7. a
8. d

Chapter 10 Water Quality and Environmental Health
1. b
2. d
3. c
4. e
5. c
6. c
7. d
8. c

Unit Review
1. b
2. e
3. a
4. d
5. e

GLOSSARY

This glossary provides the definitions of the key terms that are shown in **boldface** type in the text. Definitions for terms that are *italicized* within the text are included as well. Each glossary entry also shows the number(s) of the section(s) where you can find the term.

abiotic refers to the non-living components of an environment (Unit 1 Essential Science Background)

agriculture the practice of raising plants and livestock for food or other human needs (5.2)

Air Quality Health Index (AQHI) a system to indicate air quality and its associated health risks (9.2)

algal bloom a rapid increase in the algal population in a body of water when the concentrations of nutrients like phosphorus and nitrogen become too high (5.2, 10.1)

analytical thinking a problem-solving process based on asking "How can I break this problem down into it smaller and more manageable parts?" (2.1)

anthropogenic made (produced) by humans (9.1)

aquaculture the breeding, raising, and harvesting of animals in specially designed aquatic environments (5.5)

aquifer a porous layer of rock saturated with ground water (Unit 5 Essential Science Background)

atmosphere the gaseous part of Earth concentrated within 10 km but extending hundreds of kilometres higher (Unit 1 Essential Science Background)

bioaccumulation the process by which pollutants collect in the cells and tissues of organisms (10.1)

biodiversity the number and variety of organisms in a certain region (Unit 3 Essential Science Background)

biohazard hazardous medical waste such as blood products, body organs, old medicines, and used medical supplies (8.1)

biomagnification the increase in concentration of pollutants in tissues of organisms that are at successively higher levels in a food chain or food web (10.1)

biomass energy chemical energy in non–fossil fuel organic materials (3.3)

bioremediation use of organisms to remove or detoxify harmful waste (8.2)

biosphere all areas in the geosphere, atmosphere, and hydrosphere inhabited by and that support life (Unit 1 Essential Science Background)

biotic refers to the living components of an environment (Unit 1 Essential Science Background)

carrying capacity the largest size of a population that its environment can support (Unit 1 Essential Science Background)

chemical flocculation a step in tertiary sewage treatment in which chemicals are used to bind to nutrients so they settle out of solution (7.2)

chlorinated hydrocarbon hazardous chemical used to make products such as pesticides, solvents, and pipes (8.1)

clearcutting a silviculture method in which most or all the trees from a chosen area are removed (6.1)

coliform bacteria intestinal bacteria whose presence in a water sample indicates fecal contamination (10.1)

companion planting planting two or more plant species close together so that some benefit occurs (5.3)

compost produced when micro-organisms in soil break down organic matter in the presence of oxygen (5.3)

composting a process in which organisms such as worms, insects, and bacteria decompose organic matter (7.1)

conservationist approach a mindset that promotes human well-being but considers a wider range of long-term human good in managing the environment (Chapter 2 Launch)

contaminant a substance that occurs in concentrations higher than would be expected (9.1)

cover crop a crop planted to cover the soil and reduce soil erosion and water run-off; for example, clover, rye, and alfalfa (5.3)

cradle-to-grave analysis assessment of all the environmental impacts that occur during a product's life cycle (Unit 4 Project: An Issue to Analyze)

creative thinking a problem-solving process based on asking "How can I approach this problem in ways that are 'outside the box'?" (2.1)

criteria air contaminant (CAC) substance released directly into the atmosphere in amounts that pose a threat to health; primary air pollutants used by governments to monitor air quality (9.1)

critical thinking skills associated with thinking independently, systematically, and analytically (2.2)

crop rotation the practice of growing different crops at different times on the same land (5.3)

 D

desertification the conversion of arid and semi-arid lands into deserts by inappropriate farming practices or overgrazing (5.4)

development approach a mindset that holds that as much nature as necessary should be converted for human use (Chapter 2 Launch)

drip irrigation a method of irrigation that delivers water directly to the roots of plants (5.2)

 E

ecological footprint a measure of the impact of a population or individual on its environment (Unit 1 Essential Science Background)

ecosystem services benefits that organisms (including humans) receive from the environment and its resources (Unit 1 Essential Science Background)

electrostatic precipitator a device that uses electric charge to remove particulates from industrial emissions (9.2)

energy conservation choices and changes in behaviour that enable people to use less energy without sacrificing the services energy provides (4.2)

energy converter a device that transforms energy from one form to another for a specific purpose (Unit 2 Essential Science Background)

energy recovery technology method of generating energy from waste (also called waste-to-energy technology) (7.1)

energy resource a resource that provides energy to bring about movement or change (3.1)

energy transformation involves input energy entering a system and output energy leaving a system (Unit 2 Essential Science Background)

environment the non-living and living (abiotic and biotic) surroundings that affect an organism's survival (1.1)

environmental science the study of how humans interact with the environment (2.1)

eutrophication process in which bodies of water receive excess nutrients that stimulate growth of algae (10.1)

even-aged forest a forest in which the ages of the trees are within 10–20 years of each other (6.1)

e-waste electronic devices that enter the waste stream (electronic waste) (8.1)

extended producer responsibility (EPR) program where producers are responsible for the end of life management of their products (8.2)

 F

fish kill the death of large numbers of fish and aquatic organisms due to low oxygen levels in the water (5.2)

fission the action of splitting an atom into two or more parts (3.1)

forest an ecosystem in which the dominant plants are trees (6.1)

 G

genetically modified organism (GMO) an organism with a genetic make-up (DNA) that has been altered by scientists (5.2)

geosphere the solid, mainly rocky part of Earth (also called lithosphere) (Unit 1 Essential Science Background)

geothermal energy thermal energy that is captured from Earth's interior (3.3)

global climate change a long-term change in Earth's climate (4.1)

grassroots initiative people working together at a local level to bring about positive change in society (9.2)

greenhouse effect a process that traps outgoing infrared radiation in Earth's atmosphere (4.1)

greenhouse gas gas that traps infrared radiation in Earth's atmosphere (4.1)

green manure a method of fertilizing the soil by growing vegetation on a field and then ploughing it into the topsoil (5.3)

Green Revolution the worldwide increase in agricultural production due to the introduction of new plant species and farming methods (Unit 3 Essential Science Background)

ground water water that fills the spaces in an underground rock layer called an aquifer (Unit 5 Essential Science Background)

 H

hazardous substance a harmful substance that requires special handling (8.1)

hazardous waste a discarded substance that is or contains a flammable, corrosive, reactive, and/or toxic substance (8.1)

heat island effect an effect in which urban areas are higher in temperature than surrounding rural areas due to the heat absorbed by concrete and asphalt (6.2)

heavy metals highly dense metal elements; includes mercury, arsenic, and lead (8.1)

hydraulic fracturing involves pumping fluid into the ground under high pressure to cause layers of shale deep underground to fracture (also called fracking) (3.2)

hydro energy the energy of running or falling water (3.3)

hydrogen fuel cell a technology that converts chemical energy in pressurized hydrogen into electrical energy (4.2)

hydrosphere all the water (liquid and solid) that exists on and within the geosphere (Unit 1 Essential Science Background)

humus organic material resulting from the breakdown of plant and animal remains (5.1)

integrated pest management (IPM) a system that uses biological organisms, chemical substances, and crop rotation to help keep pest populations under control (5.3)

interdisciplinary refers to a discipline that draws on knowledge from other disciplines (2.1)

irrigation adding water to an agricultural field to allow certain crops to grow where the lack of water would normally prevent their cultivation (5.2)

kinetic energy the energy of motion (Unit 2 Essential Science Background)

landfill a disposal site for solid waste where the waste is buried between layers of soil, filling in low-lying ground (7.1)

landfill gas gas produced from decomposition of material in landfills (7.1)

leachate liquid from landfills composed of chemicals from garbage (7.1)

loam a type of soil with large spaces for air and water drainage and clay particles that hold nutrients and water; ideal soil for agriculture (5.1)

logical thinking a problem-solving process based on asking "How can methodical reasoning help me think clearly?" (2.1)

mature tree a tree that has grown to reach its greatest economic value for its size and use (6.1)

methane hydrates natural gas trapped within crystals of frozen water (3.2)

monoculture growth of a single crop, usually on a large area of land (5.2)

mulch general term for protective ground cover (5.3)

native plant a plant that has been growing naturally in an ecosystem without any action, past or present, from humans (6.2)

natural capital a measure of the value of a forest that includes ecosystem services as part of its economic value (6.1)

naturalization an environmentally sustainable technique used to create or re-create natural landscapes (6.2)

nonpoint source pollutant a pollutant that comes from a source that is difficult to identify and control, because it is scattered or diffuse (9.1)

nonrenewable energy resource an energy resource that is non-replaceable once consumed (3.1)

nutrient an element, chemical compound, or ion that organisms need for growth, cellular maintenance, and life processes (Unit 1 Essential Science Background)

nutrient cycles the continuous flows of nutrients in and out of stores (Unit 1 Essential Science Background)

old-growth forest a forest that has developed for at least 120 years without a severe disturbance such as a fire, windstorm, or logging (6.1)

organic product a product that has been produced according to standards defined by government or independent agencies (5.3)

overexploit to harvest so much of a resource that its existence is threatened (5.5)

overgrazing occurs when livestock are allowed to eat so much of the plants on rangelands that the ecological health of the habitat is damaged (5.4)

paradigm a world view that is accepted by most people in a society and that influences their thoughts, behaviour, and actions (1.2)

parent material the material in which soil is formed from (Unit 3 Essential Science Background)

persistent pesticide pesticide that does not break down and remains in the environment for a long time (8.1)

pesticide any chemical used to kill or control populations of unwanted fungi, animals, or plants (5.2)

phantom power energy that is consumed by some devices when they are turned off and in standby mode (4.2)

photochemical smog haze produced over heavily populated areas when pollutants from vehicle exhaust and industrial plants react chemically in the presence of sunlight (9.1)

photovoltaic (PV) cell a technology that converts solar energy directly into electrical energy (3.3)

point source pollutant pollutant that comes from a definite source that can be easily identified (9.1)

pollutant a substance that is or has the potential to be harmful to organisms (9.1)

polychlorinated biphenyl (PCB) a type of chlorinated hydrocarbon (8.1)

polyculture an agricultural practice in which diverse species are raised in the same area (5.3)

potential energy stored energy; energy reserved for future use (Unit 2 Essential Science Background)

preservationist approach a mindset that holds that large portions of nature should be preserved intact as habitat for non-human organisms (Chapter 2 Launch)

primary air pollutant a substance released directly into the atmosphere in amounts that pose a health threat (9.1)

primary sewage treatment a step in municipal sewage treatment that involves physically separating large solids from sewage (7.2)

product life cycle all aspects involved in making, distributing, selling, using, and disposing of a product (Unit 4 Essential Science Background)

radioactive material any material that exhibits radioactivity (3.2)

radioactive waste any waste that exhibits radioactivity; extremely hazardous and must be managed in special ways (8.1)

recycling collecting and reprocessing materials so they can be made into new products (7.1)

reflective thinking a problem-solving process based on asking "How can I use my experiences and learning to help me make judgments about what has happened and what should happen?" (2.1)

renewable energy resource an energy resource that is available on a continuous basis (3.1)

reservoir underground deposit of oil found in sedimentary rock (3.2)

resource recovery the first step in the recycling process in which items suitable for recycling are removed from the waste stream (7.1)

resources everything in the environment that supports and sustains the life of organisms (1.1)

rotational grazing confining animals to a small area of pasture for a short time, before shifting them to a new location (5.4)

scrubber a device that moves gases through a liquid spray that reacts with pollutant chemical compounds to remove them from industrial emissions (9.2)

secondary air pollutant a substance that results from primary air pollutants interacting with one another and with other atmospheric substances (9.1)

secondary sewage treatment a step in municipal sewage treatment that involves the biological breakdown of dissolved organic compounds (7.2)

secondary succession the recolonization of an area where soil has remained intact but an ecological disturbance has occurred, such as a forest fire (Unit 3 Essential Science Background)

selective cutting a silviculture method that involves cutting and removing medium-aged or mature trees individually or in small clusters every 10 to 20 years (6.1)

sewage any materials rinsed down a drain or flushed down a toilet (7.2)

shelterwood system a silviculture method that involves removing trees in a series of cuts over a period of 10 to 30 years (6.1)

sick building syndrome the adverse health effects due to time spent in a building (9.1)

silviculture a branch of forestry related to the development and management of forests (6.1)

soil a mixture of mineral grains, air, water, and organic material that support plant life (5.1)

soil profile the series of horizontal layers of soil (5.1)

soil texture soil property determined by the size of the mineral particles in the soil (5.1)

solid waste any solid or semi-solid material that is discarded (7.1)

source reduction reducing waste or its ability to cause harm through the design, manufacture, purchase, use, and re-use of a product (7.1)

spray irrigation a method of watering crops that uses a sprinkler system to spray water into the air above the plants (5.2)

storm water wastewater that drains from lawns, driveways, roofs, roads, and other urban surfaces (7.2)

strip mining a method of mining that removes overlying soil and rock to access the resource (same as surface mining) (3.2)

surface irrigation a method of supplying water to crops by having the water flow over the field in canals or ditches (also called flood irrigation) (5.2)

surface mining a method of mining that removes overlying soil and rock to access the resource (same as strip mining) (3.2)

sustainability using Earth's services in ways and at levels that can continue forever (Unit 1 Essential Science Background)

sustainable able to continue to exist indefinitely (1.2)

sustainable agriculture producing food to meet the needs of the present without compromising the ability of future generations to meet their needs (5.3)

sustainable energy system a system in which the perception, production, and use of energy ensure that energy is sustainable (4.1)

 T

tailings mining waste left after the mechanical or chemical separation of minerals from crushed ore (3.2)

tertiary sewage treatment a step in municipal sewage treatment that involves the filtering out of nutrients by passing sewage through a natural wetland or an artificial filtering system (7.2)

thermal treatment processing of solid waste at high temperatures (7.1)

tidal energy energy of the regular movement of incoming and outgoing ocean tides (3.3)

till to plough or turn over soil in preparation for planting (Unit 3 Essential Science Background)

timber wood that is used for construction and carpentry (6.1)

top soil an upper layer of soil that contains nutrients and organic material (5.1)

 U

underground mining a method of mining that involves digging deep below Earth's surface to access resources (3.2)

uneven-aged forest a forest made up of trees with vastly different ages, which results in a complex mix of forest layers (6.1)

urban forestry the long-term planning, planting, and maintenance of forests, trees, and green spaces in urban environments (6.2)

 V

vermicomposter small composter that relies on worms to break down organic materials (7.1)

 W

waste stream the movement of waste from its sources to its final destination (7.1)

wastewater any waste that occurs in or can be changed to liquid form (7.2)

watershed the area of land that drains into a body of water (Unit 4 Essential Science Background)

wind energy the energy of moving air (3.3)

INDEX

iv-x 4loops/iStock; pp2-3 Chones/Shutterstock; pp10-11 Kevin Lozaw Photography; p12 Chones/Shutterstock; p13 top left clockwise AP Photo/Mike Groll/The Canadian Press, James Nesterwitz/Alamy, Terrance Klassen/Alamy; p16 AP Photo/South Dakota State University; p18 akg-images/The Image Works; p20 AP/Lise Aserud/ The Canadian Press; p21 AP Photo/Bebeto Matthews/The Canadian Press; p22 Easter Island Statue Project/Jo Anne Van Tilburg; p23 Dark Sky Preserve Program of the Royal Astronomical Society of Canada; p26-27 Kevin Lozaw Photography; pp28-29 Don Johnston/All Canada Photos; p31 Aliki Sapountzi/aliki image library/Alamy, Paul Nicklen/ National Geographic Stock, Nigel Cattlin/Science Source, Sean Kilpatrick/The Canadian Press; p36 Leontura/iStock; p37 Taylor S. Kennedy/National Geographic Stock, fstop123/iStock, Dave Chidley/ The Canadian Press, Patrick Doyle/The Canadian Press; p38 left Photo by Karen Larter. Used by permission of Dunbarton High School, Photo by Dunbarton Environment Club. Used by permission of Dunbarton High School; p39 Photo by David Gordon. Used by permission of Dunbarton High School; p40 PA Archive/Press Association Images/The Canadian Press; p42-43 NASA/Goddard Scientific Visualization Studio; pp44-45 Don Johnston/All Canada Photos; p46 Photo by Creative Services University of Alberta. Used by permission of Dr. David Schindler, Photo by E. Debruyn. Used by permission of Fisheries and Oceans Canada; p47 left kzenon/iStock, Brian McClister/iStock, Leontura/iStock; p48 olada/iStock; pp52-53 NASA Earth Observatory/NOAA/DoD; pp56-57 Bill Brooks/ Masterfile; p58 top left clockwise McGraw-Hill Companies, moodboard/Corbis, Biophoto Associates/Science Source, Don Farrall/ Stockbyte/Getty Images; p59 left clockwise imagewerks/Getty Images, McGraw-Hill Companies, Brand X Pictures/PunchStock, Image Plan/ Corbis, Nature-Elements/Alamy, Design Pics/John Doornkamp; p60 left Jupiterimages, Don Farrall/Getty Images, Santokh Kochar/Getty Images, Bettman/Corbis, Patrick Landmann/Science Source; p61 SSPL/NRM/The Image Works, Mary Evans/Michael Cole Automobilia Collection, AECL–official use only|à usage exclusif - EACL; p62 Yvette Cardozo/Alamy; p64 Biophoto Associates/Science Source; p65 left Jonathan Hayward/The Canadian Press, Yva Momatiuk & John Eastcott/Minden Pictures; p66, 69 Mark Thiessen/National Geographic Stock; p73 DigitalGlobe via Getty Images; p74 US Coast Guard; p76 Sarnia, Ontario, Canada (80MW) First Solar, Inc.; p77 Barrett & Mackay Photo; p78 Michael Durham/Minden Pictures; p81 Andrew Vaughan/The Canadian Press; p82 Canadian Broadcasting Corporation; p86-87 Bill Brooks/Masterfile; p88 worac/iStock; p89 Robert Ellis/iStock; pp90-91 Francis Vachon/The Canadian Press; p93 Photograph by Bruce F. Molnia, U.S. Geological Survey; p94 Adapted from NASA Goddard Space Flight Center Scientific Visualization Studio; p97 Photo by David Crane/Spectral Q; p98 Richard Stouffer/ iStock; p101 left Ingram Publishing, Sharon Meredith/iStock; p103 left Stockbyte/PunchStock, kali9/iStock, Harrison Eastwood/Getty Images; p104 left Alexey Bushtruk/iStock, Alexey Bushtruk/iStock, Jean-Francois Vermette/iStock, janda75/iStock; p105 Natural Resources Canada; p106 top Tracey Tanaka, Hero Images/Corbis; p107 David Tanaka; p108 Islemount Images/Alamy; p110 Canadian Hydrogen and Fuel Cell Association (CHFCA); p111 Samantha Craddock/123rf; p112 Hydro One; p116-117 Francis Vachon/The Canadian Press; p118 left Colin Mcconnell/GetStock, Ermin Gutenberger/iStock; p120 Photo courtesy of Vibha Singh, mustafa deliormanli/iStock; p121 left kristian sekulic/iStock, Pedro Castellano/ iStock, Huntstock, Inc/Alamy; p122 Photo courtesy of Generac Power Systems; p126-127 NASA Goddard Space Flight Center; p128 Pixtal/ age fotostock; p130 Lew Zimmerman/iStock; p132 top 4loops/iStock, SimplyCreativePhotography/iStock; p133 top Phil Degginger/Alamy, Barrett and MacKay Photo; p139 Photo courtesy of Dr. Andrew G. Reynolds, Brock University; p140 DS70/iStock; p141 Heike Kampe/ iStock, Jennifer Grahan/US Geological Survey; p143 Noel Hendrickson/Blend Images/Corbis; p144 top left clockwise Photo by Jeff Vanuga, USDA Natural Resources Conservation Service, kevin miller/iStock, Photo by Lynn Betts, USDA Natural Resources Conservation Service; p146 kevin miller/iStock; p147

constantgardener/iStock; p148 Photo by Tim McCabe, USDA Natural Resources Conservation Service; p149 Courtesy of Dr. Mike Hoffman, Cornell University; p150 Stephan Zabel/iStock; p151 David Tanaka; p152 Svalbard Global Seed Vault/Mari Tefre; p154 David Tanaka, Baris Karadeniz/iStock; p155 dirk ercken/iStock; p156 Centers for Disease Control and Prevention/National Center for Emerging and Zoonotic Infectious Diseases (NCEZID) Division of High-Consequence Pathogens and Pathology (DHCPP); p157 USDA/ Natural Resources Conservation Service; p158 Baris Karadeniz/iStock; p161 Pat & Chuck Blackley/Alamy; p162 PhotoStock-Israel/Alamy; p163 Nicole Fruge/San Antonio Express/ZUMA/Corbis; p163 Nicole Fruge/San Antonio Express/ZUMA/Corbis; p164 Dave Moyer/ McGraw Hill Companies, Scott Speakes/Corbis; p166 Organics image library/Alamy; p168-169 4loops/iStock; p171 top clockwise Lee Foster/Alamy, Andrew Simpson/iStock, Diarmuid Toman/Alamy; p172 Doug Sherman/Geofile; p173 David Tanaka; pp174-175 Photo courtesy of Tremco Roofing and Building Maintenance, photo by Kent Waddington; p176 left Chris Harris/All Canada Photos, Grambo Photography/All Canada Photos; p177 top Barrett and McKay Photo, Oksana Struk/iStock; p179 top james Davis/Alamy, David Tanaka, Bill Brooks/Alamy, Harrison Shull/Aurora Open/Corbis; p180 Calvin Larsen/Science Source; p181 top Gerry Ellis/Minden Pictures, Dave Powell, USDA Forest Service, Bugwood.org; p182 F. Stuart Westmorland/Photo Researchers, Inc.; p183 PEFC Canada, Sustainable Forestry Initiative, Mark Steinmetz; p184 Wayne Lynch/ All Canada Photos; p186 left clockwise Bob Gurr/All Canada Photos, David Tanaka, Public Domain/Wikipedia; p188 top right Bart Coenders/iStock, Jennifer Anderson/USDA-Natural Resources Conservation Service, David Tanaka; p189 left Science Source, Cornelia Schaible/iStock; p191 Jim Belsley/National Park Service/US Dept. of the Interior; p192 Reprinted courtesy of the City of Brampton; p193 left cpaquin/iStock, Filbuk/GetStock, USDS-Agricultural Research Service; p195 top Scott Camazine/ Science Source, Steven Valley, Oregon Department of Agriculture, Bugwood.org, Nigel Cattlin/Photo Researchers, Inc; p197 left top clockwise Corbis, Gregory K. Scott/Science Source, James H. Speer, PhD; p198-199 Photo courtesy of Tremco Roofing and Building Maintenance, photo by Kent Waddington; p200 Robert Harding World Imagery/All Canada Photos; p201 Steve Geer/iStock; p202 Ontario Agricultural College Collection, Archival and Special Collections, University of Guelph Library, Photo courtesy of John Bacher and Ed Borczon; p203 left AP Photo/Rich Pedroncelli/The Canadian Press, Lloyd Sutton/Alamy, Brant Ward/San Francisco Chronicle/Corbis; p204 Charia Jones/The Globe and Mail Inc./The Canadian Press; p205 Bill Bachman/Alamy; p207 MCT/Landov; pp208-209 Comstock/Getty Images; pp212-213 The Great Indoors by Aurora Robson. Photo by Nash Baker. Used by permission of the artist; p214 Cameco Corporation; p215 Paul Vasarhelyi/iStock; p219 AP Photo/Richard Drew/The Canadian Press; p220 Libby Welch/ Alamy; p221 Mikhail Kokhanchikov/iStock; p223 chictype/iStock; p224 Nelson Bennett/Business in Vancouver, Kit Houghton/Corbis; p226 Steve & Dave Maslowski/Science Source; p233 David Tanaka; p238-239 The Great Indoors by Aurora Robson. Photo by Nash Baker. Used by permission of the artist; p241 AP Photo/Jorge Saenz/The Canadian Press; pp242-243 John Lehmann/Globe and Mail/The Canadian Press; p245 David Tanaka; p246 Tetra Images/Alamy; p248 Ariel Skelley/Getty Images; p249 top Andrew Lichtenstein/Corbis, O.Dimier/PhotoAlto; p250 NASA-GFSC Image created by Reto Stockli with the help of Alan Nelson, under the leadership of Fritz Hasler; p251 NASA image by Reto Stockli, based on data from NASA and NOAA; p254 David Tanaka; p255 top Photo courtesy of City of Greater Sudbury, Photo courtesy of District Municipality of Muskoka; p258 USDA-Agricultural Research Service; p259 The Regional Municipality of Durham; p264-265 John Lehmann/Globe and Mail/ The Canadian Press; p268 top Photo by Jennifer & David Stark. Used by permission of Souad Sharabani, Photo by Erinn Oxford. Used by permission of Souad Sharabani; p269 left Andy Reynolds/Image Source/Corbis, Adam Gault/Getty Images, Huntstock/Getty Images;